Terahertz
Biomedical Science
& Technology

Terahertz
Biomedical Science
& Technology

Edited by
Joo-Hiuk Son, PhD

CRC Press
Taylor & Francis Group
Boca Raton London New York

CRC Press is an imprint of the
Taylor & Francis Group, an **informa** business

CRC Press
Taylor & Francis Group
6000 Broken Sound Parkway NW, Suite 300
Boca Raton, FL 33487-2742

First issued in paperback 2020

© 2014 by Taylor & Francis Group, LLC
CRC Press is an imprint of Taylor & Francis Group, an Informa business

No claim to original U.S. Government works

ISBN 13: 978-0-367-57612-7 (pbk)
ISBN 13: 978-1-4665-7044-3 (hbk)

Library of Congress Cataloging-in-Publication Data

Terahertz biomedical science and technology / edited by Joo-Hiuk Son.
   p. ; cm.
   Includes bibliographical references and index.
   ISBN 978-1-4665-7044-3 (hardback : alk. paper)
   I. Son, Joo-Hiuk, editor.
   [DNLM: 1. Terahertz Imaging--methods. 2. Terahertz Spectroscopy--methods. WN 180]

RC78.7.S65
616.07'548--dc 3   2014007643

Visit the Taylor & Francis Web site at
http://www.taylorandfrancis.com

and the CRC Press Web site at
http://www.crcpress.com

Dedicated to the scientists
who push the limits to challenge the uncertain.

Contents

SECTION III   Terahertz Biomedical Applications

Foreword

Microwave and far-infrared molecular spectroscopy has a long and distinguished history with significant involvement in at least nine Nobel Prizes* going back to Bloch and Purcell in 1952, right up to Haroche and Wineland in 2012, who relied on a unique submillimeter-wave vector network analyzer invented by noted terahertz pioneer Philippe Goy[1] to study their Rydberg atoms. The generation and coherent detection of subpicosecond electronic transients pioneered by Auston and Nuss[2,3] at Bell Laboratories in the 1980s and by Grischkowsky[4,5] at IBM and the advent of mode-locked Ti:sapphire femtosecond lasers[6] in the early 1990s greatly expanded the field of terahertz spectroscopy and made it accessible to a wide community of researchers spanning fields as diverse as ultrafast chemistry, quantum physics, biology, and medicine.

Over the last two decades, there has been a very strong interest in applying these technically challenging techniques and instruments toward a better understanding of the low energy (molecular level) interactions that dominate chemical and biological processes at ambient and cryogenic temperatures. Of particular interest is "a universal first principles model for water, the most important molecule in the universe" (p. 265),[7] which can only be completed using the full suite of far-infrared sources, detectors, and measurement techniques that are now being developed and utilized in laboratories around the globe.

This very comprehensive compendium of state-of-the-art research articles covering terahertz devices, measurement techniques, and applications—specific to biological and biomedical interests and uses—represents a wonderful introduction and a lasting reference for those who wish to survey and understand the growing interest in this burgeoning technological field. The list of distinguished contributors spans some of the best research groups and the most notable experts in the world today. Specific topics include terahertz sources and detectors, pulsed and continuous wave circuitry, imaging and tomography techniques, guided wave components, spectroscopic studies of water, protein solvates, and biomolecules, as well as biological interactions with THz electromagnetic fields and recent applications in the biomedical field—drug absorption, cancer detection, and nanoparticle probing.

Although the applications of terahertz spectroscopy and imaging in the biological sciences are still extremely limited, and the efficacy of these techniques when compared with more established modalities is still a very open issue of debate, the reader will not be disappointed with the scope and depth of the studies included in this very early text, which captures the enthusiasm and excitement that comes with the foundations and the opening up of a new field of scientific investigation.

---

* F. Bloch and E.M. Purcell (1952)—nuclear magnetic resonance; C. Townes (1964)—masers and microwave spectroscopy; A. Penzias and R.W. Wilson (1978)—cosmic background; A. Schawlow and N. Bloembergen (1981)—laser spectroscopy; R.R. Ernst (1991) and K. Wuthrich (2002)—NMR; A. Zewail (1999)—femtosecond spectroscopy; G. Smoot and J.C. Mather (2006)—microwave background; S. Haroche and D.J. Wineland (2012)—Rydberg atoms.

References

1. P. H. Siegel, Terahertz pioneer: Philippe Goy—If you agree with the majority you might be wrong, *IEEE Transactions on Terahertz Science and Technology*, 3(4), 247–353, July 2013.
2. D. H. Austen and M. C. Nuss, Electrooptic generation and detection of femtosecond electrical transients, *IEEE Journal of Quantum Electronics*, 24(2), 184–197, February 1988.
3. P. H. Siegel, Terahertz pioneer: David H. Auston—Working collectively to combine complementary knowledge, perspectives and talents, *IEEE Transactions on Terahertz Science and Technology*, 1(1), 5–8, September 2011.
4. Ch. Fattinger and D. Gischkowsky, Terahertz beams, *Applied Physics Letters*, 54, 490–492, 1989.
5. P. H. Siegel, Terahertz pioneer: Daniel R. Grischkowsky—We search for truth and beauty, *IEEE Transactions on Terahertz Science and Technology*, 2(4), 377–382, July 2012.
6. D. E. Spence, P. N. Kean, and W. Sibbett, Sub-100 fs pulse generation from a self-modelocked titanium: sapphire laser, *Conference on Lasers and Electro-optics, CLEO*, Technical Digest Series: Optical Society of America, pp. 619–620, 1990.
7. P. H. Siegel, Terahertz pioneer: Richard J. Saykally—Water, water everywhere…, *IEEE Transactions on Terahertz Science and Technology*, 2(3), 265–270, May 2012.

Peter H. Siegel
California Institute of Technology
Pasadena, California

NASA Jet Propulsion Laboratory
Pasadena, California

Preface

Terahertz science and technology has been one of the most investigated research topics over the past two decades, ever since the generation of terahertz signals, in the form of picosecond electrical pulses driven by femtosecond lasers, was demonstrated. To conquer the vacant electromagnetic spectrum, novel generation and detection techniques, together with essential components, have been developed and utilized in numerous applications such as scientific spectroscopy, security screening, and medical imaging.

Along with the development of terahertz technology, there has been a surge in the publication of research papers on the topic, along with many books that have been written and edited to introduce this state-of-the-art technology because it has huge potential, both scientifically and technically. These books* mainly focus on generation and detection techniques, instrumentation methods such as imaging and tomography, and the spectroscopy of condensed matter such as semiconductors and dielectrics. More recent books, which have been published in the last two years, reflect on the advancement of terahertz technology and its applications but allocate limited space for biomedical studies utilizing terahertz waves, although the biomedical field is one of the fastest growing areas in terahertz applications.

Therefore, there is a huge demand for a book that is solely devoted to terahertz biomedical research, which presents the majority of the results obtained so far. The project was initiated by Luna Han, senior publishing editor at Taylor & Francis Group, who contacted me to edit a book on terahertz biomedical research. I have been fortunate enough to recruit pioneers in terahertz research and distinguished scientists in the terahertz biomedical field as authors for this book. The planned book was to be a compilation of their research results, which until recently were being pursued for a long time.

This book consists of three sections. Section I briefly reviews terahertz technology from sources and detectors to imaging techniques and waveguides for scientists and students entering terahertz research. Section II is a compilation of fundamental biological studies conducted using terahertz waves, covering the characteristics of water and liquids, the spectroscopy of biological molecules, the dynamics of proteins, and the biological effects of terahertz radiation. Section III deals with the recent achievements of terahertz imaging for medical applications, especially in cancer diagnosis. Molecular imaging with terahertz waves is also discussed in terms of principle and application. The section concludes with the prospects of medical applications using terahertz technology.

It would not be possible for this book to reach its readers without its remarkable chapter authors. I am deeply grateful to the authors for contributing their knowledge and expertise to enrich this fascinating new field. I sincerely acknowledge the encouragement and suggestions made by Professor Xi-Cheng

---

* D. Mittleman, ed., *Sensing with Terahertz Radiation* (Springer, Berlin, Germany, 2003), K. Sakai, ed., *Terahertz Optoelectronics* (Springer, Berlin, Germany, 2005), S. L. Dexheimer, ed., *Terahertz Spectroscopy* (CRC Press, Boca Raton, FL, 2007), Y.-S. Lee, *Principles of Terahertz Science and Technology* (Springer, Berlin, Germany, 2008), and X.-C. Zhang and J. Xu, *Introduction to THz Wave Photonics* (Springer, Berlin, Germany, 2009).

Zhang at the University of Rochester from the beginning. I also express appreciation to my students, Jae Yeon Park, Kwang Sung Kim, and Heejun Shin of the Terahertz Biomedical Laboratory at the University of Seoul, for their help with the manuscripts. I thank Luna Han, Robert Sims, and their staff members at Taylor & Francis Group and Remya Divakaran at SPi Global for all their hard work to help the book get published. Lastly, I would like to thank my wife, Miyoung, and my daughters, Eunho and Jeeho, for their love and support.

Editor

Joo-Hiuk Son received his BS and MS in electronics engineering from Seoul National University, Seoul, Republic of Korea, in 1986 and 1988, respectively, and his PhD in electrical engineering from the University of Michigan, Ann Arbor, in 1994. His PhD work was on the experimental and theoretical study of transient velocity overshoot dynamics in semiconductors by monitoring femtosecond-laser-driven terahertz pulses. After leaving Michigan, he worked as a postdoctoral scientist at Lawrence Berkeley National Laboratory and the Department of Electrical Engineering and Computer Science, University of California, Berkeley, where he continued his research on the terahertz spectroscopy of semiconductors from 1994 to 1996. In 1997, he joined the Department of Physics at the University of Seoul, Seoul, Republic of Korea, as an assistant professor and was later promoted to professor. In the early years, his research was focused on the characterization of electrical and optical properties of various nanosized materials, including carbon nanotubes and graphenes utilizing terahertz time-domain spectroscopy. In recent years, his interest has shifted to terahertz electromagnetic interactions with biological materials and their applications in medicine. Combining his expertise on nanomaterials and terahertz medical diagnosis, he invented a highly sensitive terahertz molecular imaging technique using nanoparticle probes.

Contributors

Philip C. Ashworth
Cavendish Laboratory
University of Cambridge
Cambridge, United Kingdom

Benjamin Born
Department of Biological
    Regulation
Weizmann Institute
    of Science
Rehovot, Israel

Nikolay Brandt
Faculty of Physics
and
International Laser Center
Lomonosov Moscow State
    University
Moscow, Russia

Massimiliano Cariati
Division of Cancer Studies
School of Medicine
King's College London
London, United Kingdom

Andrey Chikishev
Faculty of Physics
and
International Laser Center
Lomonosov Moscow State
    University
Moscow, Russia

Hyuk Jae Choi
Asan Medical Center
Seoul, Republic of Korea

Hyunyong Choi
Yonsei University
Seoul, Republic of Korea

Simon Ebbinghaus
Lehrstuhl für Physikalische
    Chemie II
Ruhr-Universität Bochum
Bochum, Germany

Shuting Fan
The Hong Kong University
    of Science and Technology
Kowloon, Hong Kong, People's
    Republic of China

Anthony J. Fitzgerald
School of Physics
University of Western Australia
Perth, Western Australia,
    Australia

Deepu K. George
University at Buffalo
Buffalo, New York

Daniel Grischkowsky
Oklahoma State University
Stillwater, Oklahoma

Maarten R. Grootendorst
Division of Cancer Studies
School of Medicine
King's College London
London, United Kingdom

Seungjoo Haam
Department of Chemical and
    Biomolecular Engineering
Yonsei University
Seoul, Republic of Korea

Joon Koo Han
College of Medicine
Seoul National University
Seoul, Republic of Korea

Martina Havenith
Lehrstuhl für Physikalische
    Chemie II
Ruhr-Universität Bochum
Bochum, Germany

Frank A. Hegmann
University of Alberta
Edmonton, Alberta, Canada

Matthias Heyden
Department of Chemistry
University of California
    at Irvine
Irvine, California

Mohammad P. Hokmabadi
Department of Electrical and
    Computer Engineering
University of Alabama
Tuscaloosa, Alabama

Tae-In Jeon
Korea Maritime and Ocean
    University
Busan, Republic of Korea

Cecil Joseph
University of Massachusetts
    Lowell
Lowell, Massachusetts

Alexey Kargovsky
Faculty of Physics
and
International Laser Center
Lomonosov Moscow State
    University
Moscow, Russia

Gurpreet Kaur
Intel Corporation
Hillsboro, Oregon

Kwang Sung Kim
University of Seoul
Seoul, Republic of Korea

Kyung Won Kim
Dana-Farber Cancer Institute
Boston, Massachusetts

Seongsin Margaret Kim
Department of Electrical and
    Computer Engineering
University of Alabama
Tuscaloosa, Alabama

Olga Kovalchuk
University of Lethbridge
Lethbridge, Alberta, Canada

Yun-Shik Lee
Department of Physics
Oregon State University
Corvallis, Oregon

Emma Macpherson
The Chinese University
    of Hong Kong
Shatin, Hong Kong, People's
    Republic of China

Andrea G. Markelz
University at Buffalo
Buffalo, New York

Seung Jae Oh
YUMS-KRIBB Medical
    Convergence Research
    Institute
Yonsei University
Seoul, Republic of Korea

Jae Yeon Park
Department of Physics
University of Seoul
Seoul, Republic of Korea

Arnie Purushotham
Division of Cancer Studies
School of Medicine
King's College London
London, United Kingdom

Jae-Sung Rieh
School of Electrical Engineering
Korea University
Seoul, Republic of Korea

Alexander Shkurinov
Faculty of Physics
and
International Laser Center
Lomonosov Moscow State
    University
Moscow, Russia

Yookyeong Carolyn Sim
Princeton University
Princeton, New Jersey

Joo-Hiuk Son
Department of Physics
University of Seoul
Seoul, Republic of Korea

Jin-Suck Suh
Department of Radiology
Yonsei University
Seoul, Republic of Korea

Lyubov V. Titova
University of Alberta
Edmonton, Alberta, Canada

Vincent P. Wallace
School of Physics
University of Western Australia
Perth, Western Australia,
    Australia

Daekeun Yoon
School of Electrical Engineering
Korea University
Seoul, Republic of Korea

Jongwon Yun
School of Electrical Engineering
Korea University
Seoul, Republic of Korea

Xi-Cheng Zhang
Huazhong University of Science
    and Technology
Wuhan, Hubei, People's
    Republic of China

1

Introduction to Biomedical Studies Using Terahertz Waves

Joo-Hiuk Son
University of Seoul

Terahertz (THz) waves, also called T-rays, refer to electromagnetic radiation in the frequency range of 0.1–10 THz—one terahertz, abbreviated as THz, representing 10^{12} Hz. In the electromagnetic spectrum, THz waves occupy the region between microwaves and the infrared, as shown in Figure 1.1. Until about two decades ago, when THz time-domain spectroscopy technique was invented by Grischkowsky and his colleagues (Fattinger and Grischkowsky 1988, 1989; van Exter et al. 1989), the THz spectrum had remained more or less unexplored and unutilized. However, the development of ultrafast lasers (Son et al. 1992; Spence et al. 1991; Valdmanis and Fork 1986) and derivative techniques such as photoconductive switching and sampling (Auston et al. 1984; Ketchen et al. 1986; Mourou et al. 1981a,b; Smith et al. 1981) and electro-optic generation and detection (Auston et al. 1984; Valdmanis et al. 1982) in the 1980s made it feasible to study the THz waves. In other words, ultrafast technology helped develop efficient THz sources and detectors, though the fabrication of compact THz system is still a challenge.

The 1990s witnessed the development of various other free-space THz generation and detection techniques, other than Grischkowsky's photoconductive switching method. The most notable ones among these were based on optical rectification for THz generation (Zhang et al. 1992) and on electro-optic effect for THz detection (Wu and Zhang 1995). These techniques facilitated the study of THz radiation over much larger bandwidth than that allowed by photoconductive switching. Moreover, they allowed flexibility of measurement, such as that achieved with live imaging (Mickan et al. 2000). Free-space THz spectroscopy systems have been extensively applied in scientific researches, especially in semiconductor physics (Grischkowsky et al. 1990; Roskos et al. 1992; Son et al. 1993, 1994, 1996; van Exter et al. 1990) and gas sensing (Harde et al. 1991, 1994). One of the advantages of THz spectroscopy is that it allows the samples to be directly inserted in the THz beam path and displays intuitive coherent response in the time domain. A readily achievable high signal-to-noise ratio, over 80 dB, is another advantage.

Imaging with THz waves was first demonstrated using a leaf with varying water content and a packaged integrated circuit as shown in Figure 1.2 (Hu and Nuss 1995). Though these experiments were rather simplistic, they were appreciated by researchers for the spectroscopic information the images contained. The information in this case came from the fact that THz waves are sensitive to water molecules and are absorbed by them, while they are transparent to dielectrics such as plastics. Because of the characteristic energies of molecules' rotational and vibrational motions in THz region, as shown in Figure 1.1, chemicals and biological molecules can be identified by their characteristic resonant peaks. Figure 1.3 shows the

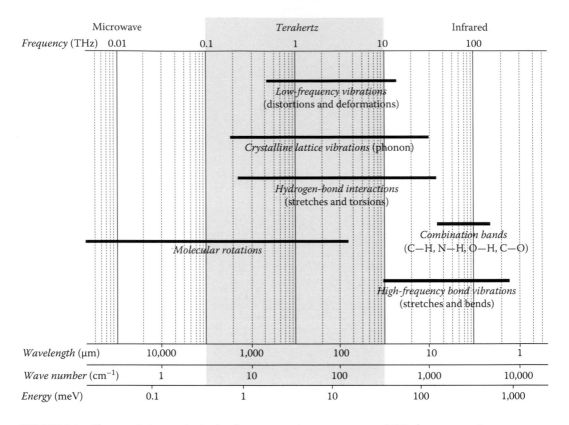

FIGURE 1.1   Characteristic energies in the electromagnetic spectrum around THz frequency region.

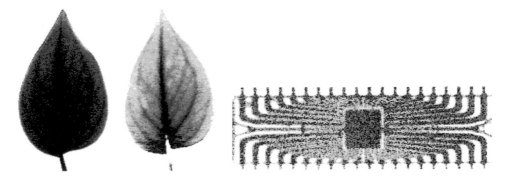

FIGURE 1.2   (See color insert.) Demonstration of imaging with THz waves. (Reprinted from Hu, B. and Nuss, M., Imaging with terahertz waves, *Opt. Lett.*, 20, 1716–1718, 1995. With permission of Optical Society of America.)

THz measurements for some nucleobases—nucleobases are nitrogenous compounds that are constituent components of DNA. The resonant energies coming from the bonds and motions of molecules (Fischer et al. 2002) are clearly displayed in Figure 1.3. The ability for identification of molecules is the most unique and important feature of spectroscopy using THz waves, a feature that electromagnetic waves in other spectral ranges rarely possess. This advantage has been exploited for the detection of explosives (Chen et al. 2004) and illicit drugs (Kawase et al. 2003). The same can be exploited for investigating a whole range of biological molecules.

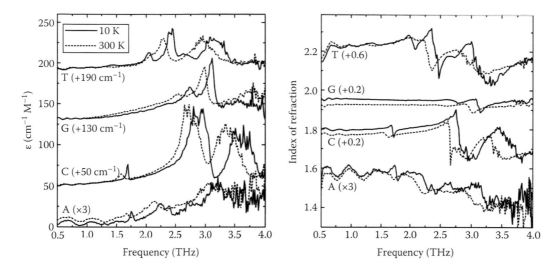

FIGURE 1.3 Resonant features of nucleobases, components of DNAs, in the THz frequency range. (Reprinted from Fischer, B.M., Walther, M., and Jepsen, P.U., Far-infrared vibrational modes of DNA components studied by terahertz time-domain spectroscopy, *Phys. Med. Biol.*, 47, 3807, 2002. With permission of Institute of Physics.)

Another advantage of THz spectroscopy arises from the low-energy photons of THz waves, which range from less than 1 meV to a few tens of millielectronvolts. These energies are well below the ionization energy, and therefore, a molecular system can be studied without being disturbed by a probing tool. In medical imagery, the radiation damage from ionization—as induced, for example, by x-ray—can be avoided using THz waves, due to their low energies. THz waves also give a reasonably good spatial resolution of less than a few hundred microns. This diffraction-limited resolution is maintained even in biological cells and tissues because the dimensions of cells are comparable to or smaller than the wavelength of THz waves. Visible light or infrared light, on the other hand, can hardly have such a resolution due to Rayleigh scattering by cells. These advantages were first exploited in the imaging of human skin cancer (Woodward et al. 2002). Thereafter, many researchers realized the potential of THz radiation in medical applications and investigated it further to assess its success in one of the major medical imaging modalities using human breast and oral cancers (Ashworth et al. 2009; Sim et al. 2013). Further developments in THz imaging for medical diagnosis led to the realization of a THz molecular imaging technique employing nanoparticle probes; this technique dramatically improves the sensitivity of measurements (Oh et al. 2011; Son 2013).

This book, consisting of 19 chapters, is divided into three sections. Section I, comprising five chapters, reviews the state-of-the-art THz technology, right from sources and detectors to endoscopy. Sources include laser-based generation techniques and solid-state devices. These are reviewed in Chapters 2 and 5, respectively. A variety of detectors are also covered in Chapter 2, including femtosecond-laser-based techniques, heterodyne receivers, thermal detectors, and Golay cells. Chapter 3 discusses high-field generation methods, useful in the study of nonlinear dynamics, as well as the realization of large field view imaging systems. Based on these sources and detectors, many THz imaging and tomographic techniques have been developed in pulsed and continuous (CW) modes. These are reviewed in Chapter 4. Chapter 5 covers a variety of solid-state devices and circuits based on diodes and transistors. Chapter 6 summarizes the recent progress made in THz waveguides, which are absolutely essential in the development of THz endoscopes. THz endoscopy is expected to play a key role in the advancement of THz imaging to a novel imaging modality because THz waves cannot reach the internal organs without the help of an endoscope.

Section II, comprising Chapters 7 through 13, compiles the fundamental biological results of studies using THz waves, starting from the characteristics of bulk water and liquids in Chapter 7. Water is not only a major constituent of biological cells but also a scientifically important subject. Biological water, such as that involved in solvation dynamics around a protein molecule, has also been studied by THz waves. The results of these studies are presented in Chapter 8. Chapter 9 describes the THz spectroscopy of various biological molecules, including DNAs and RNAs. The structure–function relation in protein molecules and the protein dielectric response, studied using THz waves, are reviewed in Chapters 10 and 11, respectively. The results on nonlinear interaction of proteins and amino acids are presented in Chapter 12. The last chapter of Section II, Chapter 13, discusses the effects on biological cells and skin tissues due to irradiation by THz pulses. Also discussed are the related safety issues.

Section III, comprising Chapters 14 through 19, is devoted to the applications of THz radiation in medicine. Chapter 14 describes the dynamic imaging of drug absorption in skin exploiting the sensitivity of THz waves to pharmaceutical materials. The current status of the THz imaging technique for diagnosing cancers, which has been investigated for a decade, is reviewed for skin, breast, and oral cancers in Chapters 15 through 17, respectively. Although the results show the promising feasibility, a few issues need to be addressed and improved upon before taking the technique to clinics, especially the issue of the contrast between the benign and malignant tissues. To enhance the measurement contrast or sensitivity, nanoparticle contrast agent probes have been adopted for THz imaging, as other imaging techniques such as magnetic resonance imaging or computed tomography have. The principle and applications of THz molecular imaging technique using such nanoparticle probes make the contents of Chapter 18. This section concludes with Chapter 19 on the potential medical applications of THz radiation.

The editor believes that this book is a good starting point for scientists and students who want to explore new fields of biological sciences and medical applications with THz technology.

References

Ashworth, P. C., E. Pickwell-MacPherson, E. Provenzano et al. 2009. Terahertz pulsed spectroscopy of freshly excised human breast cancer. *Optics Express* 17: 12444–12454.

Auston, D. H., K. P. Cheung, J. A. Valdmanis, and D. A. Kleinman. 1984. Cherenkov radiation from femtosecond optical pulses in electro-optic media. *Physical Review Letters* 53: 1555.

Chen, Y., H. Liu, Y. Deng et al. 2004. Spectroscopic characterization of explosives in the far-infrared region. Presented at *Defense and Security*, pp. 1–8.

Fattinger, C. and D. Grischkowsky. 1988. Point source terahertz optics. *Applied Physics Letters* 53: 1480–1482.

Fattinger, C. and D. Grischkowsky. 1989. Terahertz beams. *Applied Physics Letters* 54: 490–492.

Fischer, B. M., M. Walther, and P. U. Jepsen. 2002. Far-infrared vibrational modes of DNA components studied by terahertz time-domain spectroscopy. *Physics in Medicine and Biology* 47: 3807.

Grischkowsky, D., S. Keiding, M. van Exter, and C. Fattinger. 1990. Far-infrared time-domain spectroscopy with terahertz beams of dielectrics and semiconductors. *Journal of the Optical Society of America B* 7: 2006–2015.

Harde, H., N. Katzenellenbogen, and D. Grischkowsky. 1994. Terahertz coherent transients from methyl chloride vapor. *Journal of the Optical Society of America B* 11: 1018–1030.

Harde, H., S. Keiding, and D. Grischkowsky. 1991. THz commensurate echoes: Periodic rephasing of molecular transitions in free-induction decay. *Physical Review Letters* 66: 1834.

Hu, B. and M. Nuss. 1995. Imaging with terahertz waves. *Optics Letters* 20: 1716–1718.

Kawase, K., Y. Ogawa, Y. Watanabe, and H. Inoue. 2003. Non-destructive terahertz imaging of illicit drugs using spectral fingerprints. *Optics Express* 11: 2549–2554.

Ketchen, M., D. Grischkowsky, T. Chen et al. 1986. Generation of subpicosecond electrical pulses on coplanar transmission lines. *Applied Physics Letters* 48: 751–753.

Mickan, S., D. Abbott, J. Munch, X.-C. Zhang, and T. Van Doorn. 2000. Analysis of system trade-offs for terahertz imaging. *Microelectronics Journal* 31: 503–514.

Mourou, G., C. Stancampiano, A. Antonetti, and A. Orszag. 1981a. Picosecond microwave pulses generated with a subpicosecond laser-driven semiconductor switch. *Applied Physics Letters* 39: 295.

Mourou, G., C. V. Stancampiano, and D. Blumenthal. 1981b. Picosecond microwave pulse generation. *Applied Physics Letters* 38: 470–472.

Oh, S. J., J. Choi, I. Maeng et al. 2011. Molecular imaging with terahertz waves. *Optics Express* 19: 4009–4016.

Roskos, H. G., M. C. Nuss, J. Shah et al. 1992. Coherent submillimeter-wave emission from charge oscillations in a double-well potential. *Physical Review Letters* 68: 2216.

Sim, Y. C., J. Y. Park, K.-M. Ahn, C. Park, and J.-H. Son. 2013. Terahertz imaging of excised oral cancer at frozen temperature. *Biomedical Optics Express* 4: 1413.

Smith, P., D. Auston, A. Johnson, and W. Augustyniak. 1981. Picosecond photoconductivity in radiation-damaged silicon-on-sapphire films. *Applied Physics Letters* 38: 47–50.

Son, J.-H. 2013. Principle and applications of terahertz molecular imaging. *Nanotechnology* 24: 214001.

Son, J.-H., S. Jeong, and J. Bokor. 1996. Noncontact probing of metal-oxide-semiconductor inversion layer mobility. *Applied Physics Letters* 69: 1779–1780.

Son, J.-H., T. B. Norris, and J. F. Whitaker. 1994. Terahertz electromagnetic pulses as probes for transient velocity overshoot in GaAs and Si. *Journal of the Optical Society of America B* 11: 2519–2527.

Son, J.-H., J. V. Rudd, and J. Whitaker. 1992. Noise characterization of a self-mode-locked Ti:sapphire laser. *Optics Letters* 17: 733–735.

Son, J.-H., W. Sha, J. Kim et al. 1993. Transient velocity overshoot dynamics in GaAs for electric fields ≤200 kV/cm. *Applied Physics Letters* 63: 923–925.

Spence, D. E., P. N. Kean, and W. Sibbett. 1991. 60-fsec pulse generation from a self-mode-locked Ti:sapphire laser. *Optics Letters* 16: 42–44.

Valdmanis, J. and R. Fork. 1986. Design considerations for a femtosecond pulse laser balancing self phase modulation, group velocity dispersion, saturable absorption, and saturable gain. *IEEE Journal of Quantum Electronics* 22: 112–118.

Valdmanis, J. A., G. Mourou, and C. W. Gabel. 1982. Picosecond electro-optic sampling system. *Applied Physics Letters* 41: 211.

van Exter, M., C. Fattinger, and D. Grischkowsky. 1989. Terahertz time-domain spectroscopy of water vapor. *Optics Letters* 14: 1128–1130.

van Exter, M. and D. Grischkowsky. 1990. Optical and electronic properties of doped silicon from 0.1 to 2 THz. *Applied Physics Letters* 56: 1694–1696.

Woodward, R. M., B. E. Cole, V. P. Wallace et al. 2002. Terahertz pulse imaging in reflection geometry of human skin cancer and skin tissue. *Physics in Medicine and Biology* 47: 3853.

Wu, Q. and X. C. Zhang. 1995. Free-space electro-optic sampling of terahertz beams. *Applied Physics Letters* 67: 3523–3525.

Zhang, X. C., X. Ma, Y. Jin et al. 1992. Terahertz optical rectification from a nonlinear organic crystal. *Applied Physics Letters* 61: 3080–3082.

I

Terahertz Technology

2

Terahertz Sources and Detectors

Hyunyong Choi
Yonsei University

Joo-Hiuk Son
University of Seoul

2.1 Introduction

The exploration of veiled fields has been the driving force for new scientific discovery and the development of novel techniques. The very far-infrared (FIR) spectral range remains one of the least explored regions of the electromagnetic (EM) spectrum. Until the 1970s, the terahertz (THz) range (100 GHz to 3 THz, 3000–100 μm) was called *THz gap* owing to the lack of efficient sources and detectors (Gallerano and Biedron 2004; Tonouchi 2007; Zhang and Xu 2009; Zhang et al. 2005) for this region. Recently developed powerful sources and sensitive detectors, operating in the THz spectral region, are aimed at rapidly shrinking the *THz gap*. Such developments have led to the introduction of several industrial products spanning diverse areas including biomedical application.

Many techniques have been developed to overcome the generation limits of both photonics and electronics. In the photonic technique, the use of a quantum cascade laser (QCL) for the specific frequency region at a high temperature of around 200 K has been reported. In the electronic technique, solid-state oscillators such as Gunn diodes, impact avalanche and transit time (IMPATT) diodes, and Schottky diode multipliers generate sub-THz radiation. Furthermore, the backward wave oscillators (BWOs) and free-electron lasers (FELs), based on free-electron sources, are being studied. Presently, various commercial equipments based on both photonic and electronic generation are available for THz spectroscopy and imaging applications in the biomedical field.

Though the development of THz continuous-wave (CW) sources has continually progressed, the achievable working frequency and output power remain limited. Conventional Gunn diode and diode multiplier technologies can operate at the initial frequencies in the THz range, but not much beyond it. BWOs have been demonstrated to have limited frequency tunability and low output power ~1 μW. QCLs have generated several milliwatts (mW) of power in the 5 THz range, although their working temperature remains that of a liquid nitrogen atmosphere or lower. Several coherent THz frequency domain spectroscopy systems have been implemented using photomixers and thermal detectors (Demers et al. 2007). They are low cost and time efficient as compared to using the femtosecond laser system. They are considered suitable for very small-size portable systems.

Still the most popular THz generation and detection methods are the photoconductive and optical rectification techniques using femtosecond laser pulses. In the late 1980s, THz time domain spectroscopy (THz-TDS) was introduced as a powerful spectroscopic technique that could be used to explore THz optical properties. The THz-TDS system uses a femtosecond laser along with photoconductive generation and optical rectification. Various photoconduction and optical rectification techniques have been developed (Han et al. 2000; Rice et al. 1994; Smith et al. 1988; Zhang et al. 1992). Recently, it has been demonstrated that broadband THz signals can be obtained by the photoionization of air (Wilke et al. 2002).

The recent developments of various sources and detectors suggest that they have great potential for applications in the field of biomedical studies including imaging and tomography systems of biomaterials. In this chapter, we discuss generation and detection methods on pulsed THz generation using femtosecond lasers and introduce various CW sources and detectors.

2.2 Terahertz Sources

Although there are many methods for the generation of THz signals, THz sources may be broadly classified into pulse and CW sources. Pulsed THz sources generate broadband THz signals using femtosecond lasers. The three main techniques are photoconductive generation, optical rectification, and air photonics generation. These three techniques are frequently used in conjunction with the THz spectroscopy system. The photomixer and difference-frequency generator are optoelectronic devices that generate CW THz signals. The representative compact solid-state sources are GaN-based Gunn diodes, diode-based frequency multipliers, and QCLs. These solid-state sources have been developed recently within a short span of time owing to the commercialization of THz systems. The BWOs and FEL are sources of highly different sizes that use accelerated electrons. We briefly review here the various generation methods and their functional mechanisms. A detailed discussion on high-field THz generation and compact solid-state sources is provided in Chapters 3 and 5.

2.2.1 Femtosecond Laser-Based Pulsed Terahertz Generation Technique

2.2.1.1 Femtosecond Lasers

The femtosecond laser is well known to enable the most effective generation of THz pulses. However, the ultrafast laser pulse itself has opened a new area of laser applications in science and technology (Sibbett et al. 2012). Generally, a laser system consists of a gain medium and a cavity, and the mode of the laser is decided by the optical length of the laser cavity and the resonance frequencies. Mode-locking is a fundamental method that can be used to generate femtosecond pulses. Ultrashort femtosecond pulses can be generated when the phases of the modes are the same and the time distribution of the intensity repeats periodically, thereby giving rise to constructive interference of the modes. The first femtosecond laser having a pulse duration below 100 fs was realized using mode-locking in 1981 (Fork et al. 1981). It was dye laser, however, that had stability and maintenance problems; thus, a femtosecond solid-state laser was required. The picosecond pulse was first generated by mode-locking in the diode pumping Nd:YAG, Nd:YLF solid-state laser in the late 1980s; P. Moulton utilized Ti:sapphire crystal as a gain medium that could generate femtosecond pulses over a very wide spectrum. This provided the opportunity to develop femtosecond solid-state lasers that would replace dye lasers. Eventually, a pulse generation of 60 fs was achieved by a self-mode-locked Ti:sapphire laser by D. E. Spence (Spence et al. 1991).

High-power femtosecond laser pulses have been utilized for research in nonlinear optics including second-harmonic generation, sum-frequency generation, parametric oscillation, and amplification. A transient polarization caused by a laser pulse is a source of radiation in the term in the Maxwell equations. The fast polarization changes caused by high-speed photoconductive antennas and optical rectification in electrooptic (EO) crystals are utilized to generate broadband coherent radiation in the veiled EM spectrum. The transient polarization of 100 fs laser beam radiates a pulsed EM wave at a frequency of about 3 THz. Until recently, femtosecond laser-based THz sources have been nearly ideal candidates for use in coherent light sources. In the following section, we review the various techniques used for the generation of THz pulses from this ultrafast pulse.

2.2.1.2 Photoconductive Antenna

Photoconductive switching techniques are the most common methods for THz pulse generation. Photoconductive switching, which consists of focused femtosecond laser beam pulses and dc-biased photoconductive antennas, generates transient polarization. When the femtosecond laser pulses focus on the photoconductive antenna, it is electrically shortened and the applied voltage is reduced. A transient polarization process is followed by the ultrafast creation of electron–hole pairs in the semiconductor. A schematic diagram is shown in Figure 2.1a. The photo-generated electrons and holes are accelerated by the bias field, and the current density is expressed as (Zhang and Xu 2009)

$$J(t) = N(t)e\mu E_{dc}, \tag{2.1}$$

where

  $N(t)$ is the density of photocarriers
  e is the electronic charge
  μ is the carrier mobility
  E_{dc} is the bias of the electric field

A fast current pulse $J(t)$ is set by the laser pulse duration and resistance–capacitor (RC) time constant of the antenna structure (Abdullaev et al. 1972; Smith et al. 1988). The generated THz is proportional to the differential of the photocurrent in accordance with the following equation:

$$E_{THz} \propto \frac{\partial}{\partial t} J(t). \tag{2.2}$$

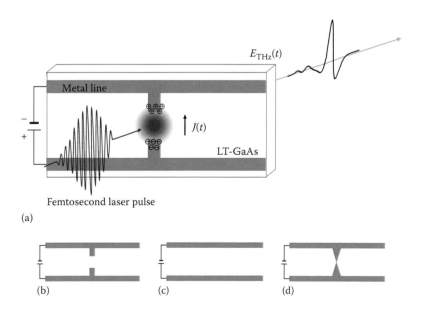

FIGURE 2.1 (a) The schematic diagram of photoconductive antenna generation and various antenna structures: (b) dipole antenna, (c) coplanar line antenna, and (d) bow-tie antenna.

The spectral range of the generated THz EM waves depends on various conditions, such as the antenna structure, the shape and power of the input laser pulse, the bias voltage, and the characteristics of the antenna substrate. For obtaining a broad spectrum, in particular, the full width at half maximum (FWHM) of the input laser is a very important key parameter. Since the input laser generates the electric current on the substrate, a short laser pulse width can be used to induce short THz EM pulses. This is also applied when using the detection method.

Auston was the pioneer of this generation method, and Grischkowsky developed it into a time domain spectroscopic system (Auston et al. 1984; Fattinger and Grischkowsky 1988, 1989). Generally, the spectral range of the photoconductive technique lies between sub-THz frequencies and 3 THz, and it has a good signal-to-noise ratio of over 10,000:1. Three types of antennas are widely used—the dipole antenna, the coplanar line antenna, and bow-tie antenna—which are shown in Figure 2.1b–d. The generated THz pulses and its Fourier-transformed amplitude spectrums with dipole, coplanar line, and bow-tie antenna detected by the dipole antenna are shown in Figure 2.2. The difference between dipole antenna and coplanar line antenna is the presence of the dipole gap. The small width of the dipole gap and the distance between the coplanar lines can enable the generation of a wide spectral range, but the generation power is reduced when compared with the wide dipole gap antenna and the corresponding distance between the coplanar lines. The width of the dipole gap and the distance between the coplanar lines are affected as there is a trade-off between their generation power and the spectral widths. The bow-tie antenna emits higher power by several factors than the dipole antenna especially at lower frequency range (Tani et al. 1997).

The substrates used in the photoconductive antenna are GaAs, low-temperature-grown GaAs (LTG-GaAs), and silicon on sapphire (SOS) (Hamster et al. 1993, 1994; van Exter et al. 1989). These materials have high resistivity, reasonable carrier mobility, and short lifetime. High resistivity leads to sufficient breakdown voltage across the antenna and high carrier mobility increases the efficiency. LTG-GaAs is the most widely used substrate because of its short lifetime ($\tau_e = 1\,\mathrm{ps}$, $\tau_h = 4\,\mathrm{ps}$).

2.2.1.3 Optical Rectification

This approach uses the ultrafast change in polarization in nonlinear optical crystals (Carrig et al. 1995; Huber et al. 2000; Kawase et al. 2001; Lee et al. 2001; Verghese et al. 1998). The high-peak femtosecond laser pulse generates a coherent carrier distribution using the photoconductive effect and induces time-dependent coherent polarizations in EO crystals. However, the carrier distributions and polarizations maintain their coherent states only for a short time (approximately from a sub-picosecond range to a few picoseconds) due to carrier–phonon and carrier–carrier scattering.

Optical rectification is the second-order polarization of EO crystals. An incident laser pulse is used as a source to produce the second-order nonlinear effect. The applied electric field E affects the polarization of nonlinear crystals. The polarization P induced by the electric field E is expressed as (Boyd 2003)

$$P = \varepsilon_0\left(\chi^{(1)}E + \chi^{(2)}EE + \chi^{(3)}EEE + \cdots\right), \tag{2.3}$$

where $\chi^{(n)}$ is the nth-order nonlinear susceptibility tensor. Optical rectification uses the nonlinear crystal, which has a nonzero second-order susceptibility tensor $\chi^{(2)}$. Suppose that if an incoming electric field has only two frequency components ω_1 and ω_2, then E can be expressed by $E = E_1 e^{-i\omega_1 t} + E_2 e^{-i\omega_2 t} + c.c.$ where c.c. stands for complex conjugate. The second-order polarization is given by

$$P = \varepsilon_0\,\chi^{(2)}EE = \varepsilon_0\,\chi^{(2)}\left[2\left(|E_1| + |E_2|\right)e^0 + \left(|E_1|^2 e^{-i2\omega_1 t} + |E_2|^2 e^{-i2\omega_2 t}\right)\right.$$
$$\left. + 2E_1 E_2 e^{-i(\omega_1 + \omega_2)t} + 2E_1 E_2^* e^{-i(\omega_2 - \omega_1)t}\right]. \tag{2.4}$$

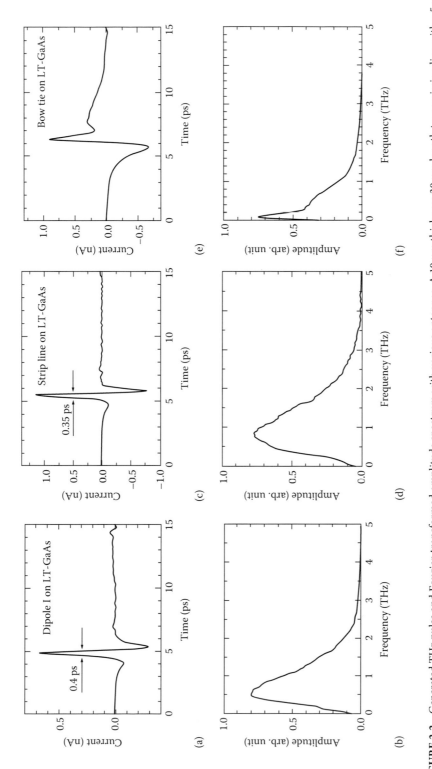

FIGURE 2.2 Generated THz pulse and Fourier-transformed amplitude spectrum with various antennas. A 10 μm thickness 30 μm length transmission line with a 5 μm photoconductive gap dipole antenna (a) and (b), a 10 μm wide line with an 80 μm separation coplanar line antenna (c) and (d), and bow-tie antenna, which is 1 mm long with a photoconductive gap of 10× μm and 90° bow angle (e) and (f). A 10 μm thickness 10 μm length transmission line with a 5 μm photoconductive gap dipole antenna was used for detection. (Reprinted from Tani, K. et al., *Appl. Opt.*, 36, 7853, 1997.)

The polarization has the frequency components of $2\omega_1$ and $2\omega_2$, $\omega_1 + \omega_2$, and $\omega_2 - \omega_1$, which are known as second-harmonic generation, sum-frequency generation, and difference-frequency generation (DFG), respectively. Among these nonlinear optical processes, DFG is the main mechanism of THz pulse generation. In other words, the spectrum of the generated THz pulse corresponds to the difference frequency $\omega_2 - \omega_1$ of incident pulses. Thus, the bandwidth of the generated THz is determined from that of the incident laser pulse. The THz field is proportional to the second derivative of the polarization with respect to time t, that is, $E_{THz} \propto (\partial^2/\partial t^2)P$.

The generated THz pulse is strongly affected by the orientation of the nonlinear susceptibility tensor associated with the incident pulse. Considering the orientation of the crystal, the second-order nonlinear polarization can be written as follows (Boyd 2003):

$$P_x^{(2)} = \sum_{y,z} \varepsilon_0 \chi_{xyz}^{(2)} E_y(\omega) E_z^*(\omega). \tag{2.5}$$

The notations x, y, and z indicate the Cartesian coordinate directions of the incident pulse E. Because most of the crystal structure is highly symmetric, only a few of the 18 susceptibility $d_{xl} = (1/2)\chi_{xyz}^{(2)}$ tensors can survive, that is, many of them vanish. In case of ZnTe, it has a symmetry lattice with zinc blende structure with the nonzero d-matrix components $d_{14} = d_{25} = d_{36}$. The directional interaction between the optical pulse and the crystal structure is a crucial factor. For normal optical incident field, a ZnTe cut (110) orientation can only generate a THz field. The THz field amplitude also depends on the angle ϕ between the optical field and the crystal axes. Figure 2.3 describes the incident field of an optical pulse on ZnTe. According to the reports studied by Chen et al. (2001), a maximum THz electric field is generated with an angle of ϕ 54.7.

The other important point is phase matching between the generated THz pulse and the optical pulse in a nonlinear crystal. If the phase-matching condition is satisfied, the optical pulse coherently generates a THz field in the crystal, and the THz field can be enhanced by propagating within the crystal thickness. This phase-matching condition has the form of $\Delta k = (k_2 - k_1) - k_{THz} \approx 0$ where k_1 and k_2 are wave vectors of the optical pulse and k_{THz} is a wave vector of the generated THz pulse. By using the relation $k = (2\pi/\lambda) = (\omega n(\omega)/c_0)$, we can rewrite the phase-matching condition as (Nahata et al. 1996)

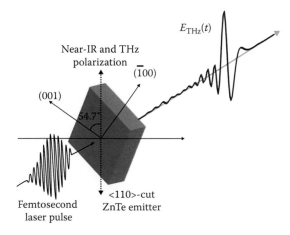

FIGURE 2.3 Schematic diagram of phase-matched THz generation with ZnTe (110).

$$n(\omega_2)\omega_2 - n(\omega_1)\omega_1 = n(\omega_{THz})\omega_{THz}, \tag{2.6}$$

where the interaction reaches π, that is, $\Delta k L_c = \pi$, and the phase-matching condition is not satisfied, thereby reducing the radiation efficiency of THz. From this, it can be inferred that the thickness of the nonlinear crystal should be shorter than the coherent length L_c. When the THz frequency is much lower than the optical laser pulse, the group velocity $v_g(\omega) = (\partial\omega/\partial k)$ of the femtosecond laser pulse and the phase velocity $v_p(\omega) = (\omega/k)$ of the THz pulse have a relation similar to that for phase matching. Control of the beam polarizations and nonlinear crystal angle enables phase matching.

The most commonly used nonlinear crystal is ZnTe, which offers a bandwidth in the 0.1 ~ 3 THz range. In addition to high chemical stability, ZnTe provides a good phase-matching condition with an FIR incident laser pulse. Besides ZnTe, some nonlinear crystals are used as alternatives to generate ultrabroadband THz pulse. For instance, GaSe, which has a very broad velocity matching range originating from a large birefringence, can be used to generate ultrabroadband THz pulses (~30 THz). Ideally, a 10 fs laser pulse can generate ~100 THz bandwidth in a lossless and dispersionless nonlinear medium. However, such a wide THz spectral range generated from a nonlinear medium has not yet been reported.

2.2.1.4 Photoionization of Air

When high-power pulse laser beams are focused on air, the atoms in air are ionized and attain a plasma condition. Excited electrons of the ionized atoms produce ~100 kV/cm 0.1 ~ 30 THz high-power broadband THz EM waves by a ponderomotive force as shown in Figure 2.4. The generation of THz signals is given by (Xie et al. 2006)

$$E_{THz}(t) \propto \chi^3 E_{2\omega}(t)E_\omega(t)E_\omega(t)\cos(\varphi), \tag{2.7}$$

where
   $E_\omega(t)$ is the electric field of the fundamental laser
   $E_{2\omega}(t)$ is the second-harmonic generation of $E_\omega(t)$ by the BBO crystal
   φ is the phase shift

There are many issues that call for an improvement of the previous pulsed THz generation methods. Generally, patterns and substrates of photoelectric antennas limit their bandwidth. The method of optical rectification loses bandwidth by dispersion and phonon phenomena in an EO crystal, and it has limitations on generation power because there is a damage threshold of the crystal when high-power laser beams are radiated onto the crystal. The method of photoionization of air, however, is relatively free from the bandwidth loss or limitation on generation power, and there are many advantages such as control of the bandwidth of the THz signal by varying a pulse width of the pump laser. Moreover, this method enables measurement of the THz signal without a loss of bandwidth when it uses air breakdown

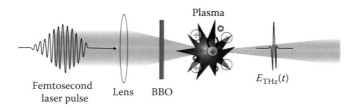

FIGURE 2.4 Schematic diagram of air photonics generation.

coherent detection (ABCD) as a detection method, which is similar to the generation method (Dai et al. 2011; Lu and Zhang 2011). This method is represented by an equation of four-wave mixing with the generating THz signal and probe beam as given in the following:

$$E_{2\omega}(t) \propto \chi^3 E_{THz}(t) E_\omega(t) E_\omega(t) \cos(\varphi).$$ (2.8)

Air photonics has constraints on the control of the THz signal, and these problems cause instability of the signal since air, which is the major medium of this method, has an inhomogeneous flow density. Therefore, studies to solve these problems are being carried out using various types of gases, particularly high-pressure gases.

2.2.2 Optoelectronic Continuous-Wave Sources

2.2.2.1 Photomixer

Photomixing is a common application in photoconductors. It entails two frequency-offset lasers and integrated circuits to generate CW THz signals. Two laser beams, which have the same polarization but different frequencies $\omega_\pm = \omega_0 \pm \omega_{THz}/2$, are spatially overlapped and focused onto a photomixer device (Verghese et al. 1997). The photomixer is made of ultrafast photoconductive material, which exhibits photonic absorption and has a short charge carrier lifetime. The photomixer modulates its conductivity at the frequency difference, and an applied electric field allows the conductivity variation to be converted into a current, which is radiated by a pair of antenna. The modulation optical amplitude of the photomixer is given by (Preu et al. 2011; Sakai 2005)

$$P = P_1 + P_2 + 2\sqrt{mP_1P_2}\cos(\omega_{THz}t),$$ (2.9)

where

P_1 and P_2 are the powers of the two incident lasers
m is the mixing efficiency value between 0 and 1

A typical photoconductive photomixer is made of low-temperature GaAs with a patterned metalized layer, which is used to form an electrode array and radiating antenna. Recently, the log-spiral antenna is the most widely used because of its broad tenability (Verghese et al. 1997).

The advantages of this technique are that it is continuously tunable over the frequency range from 300 GHz to 3 THz and spectral resolutions of the order of 1 MHz can be achieved. As the frequency tunability and spectral resolution of the output THz are determined on the incident laser quality, a 1 THz signal is gained at 780 and 1550 nm when the difference in wavelength at the frequency ω_0 is 2.05 and 8.05 nm, respectively (Brown 2003). The reported maximum power operating at the wavelength of 1.55 μm of the laser beam is 20 mW and 10 μW at 100 GHz and 1 THz, respectively. However, the usual output power is below 1 μW.

2.2.2.2 Difference-Frequency Generation

DFG is a commonly used second-order nonlinear optical process. It generates THz waves using highly nonlinear materials such as GaSe, GaP, and organic ionic salt crystal 4-dimethylamino-*N*-methyl-4-stilbazolium-tosylate (DAST) (Geng et al. 2010). Similar to the photomixing method, the two optical beams at frequencies $\omega_\pm = \omega_0 \pm \omega_{THz}/2$ are incident on a nonlinear crystal, and the output frequency is converted into the difference between the two input frequencies. The second-order nonlinear phenomenon has already been mentioned in Section 2.2.1.3. When the frequencies of the two optical beams are similar, the electric field is expressed as

$$E(t) = E_0 \sin\left(\left(\omega_0 + \frac{\omega_{THz}}{2}\right)t\right) + E_0 \sin\left(\left(\omega_0 - \frac{\omega_{THz}}{2}\right)t\right) = 2E_0 \cos\left(\omega_{THz}t\right)\sin\left(2\omega_0 t\right).$$ (2.10)

The second oscillating frequency, $2\omega_0$, does not modulate the polarization significantly. Thus, the second-order polarization is given by

$$P = \varepsilon_0\,\chi^{(2)}EE = \varepsilon_0\,\chi^{(2)}E^2\cos\left(\omega_{\mathrm{THz}}t\right)^2 = \frac{1}{2}\varepsilon_0\,\chi^{(2)}E^2\left(1+\cos\left(\omega_{\mathrm{THz}}t\right)\right). \tag{2.11}$$

Similar to optical rectification, the generated THz pulse is related to the second derivative of polarization with respect to time t; THz waves are thus generated. The second-order nonlinear susceptibility $\chi^{(2)}$ determines the THz gain within the tunable wavelength range. The output power and conversion efficiency are mainly affected by the quality of the nonlinear crystals and the two frequency-offset laser beams. The tuning range is enhanced by using dual-wavelength oscillations from an electronically tuned Ti:sapphire laser with an acousto-optic device (Kawase et al. 1999).

2.2.3 Compact Solid-State THz Sources

2.2.3.1 GaN-Based Gunn Diodes

A Gunn diode, also known as a transferred electron device (TED), is used in high-frequency electronics. Its internal construction is unlike other diodes in that it consists only of n-doped semiconductor material, whereas most diodes consist of both p- and n-doped regions. In the Gunn diode, three regions exist: two of them are heavily n-doped on each terminal, with a thin layer of lightly doped material in between (Panda et al. 2009). When a voltage is applied to the device, the electrical gradient is largest across the thin middle layer. Conduction takes place as in any conductive material with current being proportional to the applied voltage. Eventually, at higher field values, the conductive properties of the middle layer are altered, increasing its resistivity and preventing further conduction whereby current starts to fall. This means a Gunn diode has a region of negative differential resistance. Its largest use is in electronic oscillators to generate microwaves in applications such as radar speed guns and microwave relay transmitters.

The frequency can be tuned mechanically, by adjusting the size of the cavity or, in case of yttrium iron garnet (YIG) spheres, by changing the magnetic field. Gunn diodes are used to build oscillators in the 10 GHz to THz frequency range.

2.2.3.2 Diode-Based Frequency Multipliers

The Schottky diodes have gone through the huge progress in the field of multiplied THz source. They multiply the incoming microwave and generate the around 0.1 THz signal. By scheme of first balanced doubler circuit, which was developed by Erickson, a whisker diode is the only key component and there is no filter that divides the incoming and the generated waves (Erickson 1990). Because it facilitates power handling, the planar diode was used as a successful frequency doubler. The maximum power obtained was nearly 95 mW with 45% efficiency (Porterfield et al. 1999).

Another type of multiplication of microwaves relies on third-harmonic generation. The Schottky diode that employs GaAs is an efficient component of a frequency tripler circuit. Due to the high switching rates of up to 10 GHz, artificial tuners are not necessary. In the basic tripler block, a wave of frequency ω_0 is incident on the Schottky diode array after passing through a waveguide and receiving antenna. Like nonlinear materials, a Schottky diode array converts incoming waves into harmonic waves. As a result, the output waves of frequency $3\omega_0$ are emitted through a waveguide. Recently, Porterfield applied the tripler at 220 and 440 GHz (Porterfield 2007). It is reported that the 440 GHz frequency tripler radiated wave that had 9 mW peak power with 12% efficiency. A higher output power was measured for the 220 GHz tripler, whose peak power was 23 mW with 16% efficiency.

2.2.3.3 Quantum Cascade Lasers

QCLs have been recently introduced and developed as a THz source. Recent advances in nanotechnology have made possible the realization of this semiconductor-based THz source. The concept is based

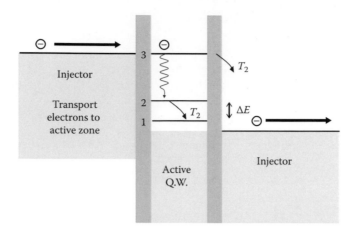

FIGURE 2.5   Schematic representation of three-level quantum cascade structure.

on a superlattice suggested way back in 1970. This idea of a device was realized at Bell Laboratory in 1994 (Faist et al. 1994; Kazarinov and Suris 1971). The device, fabricated by molecular beam epitaxy, has been developed as a source of THz; the first THz QCL was demonstrated in 2002 (Ajili et al. 2002; Köhler et al. 2002).

It can emit relatively high-power CW and pulsed waves with the output power reaching up to hundreds of mWs (Williams 2007). In general, QCLs are operated at low temperatures. The emission efficiency decreases with increasing temperatures (Indjin et al. 2003; Mátyás et al. 2010). This disadvantage has been addressed and resolved recently (Belkin et al. 2007).

QCLs are made up of heterojunctions, which are junctions of lattice-matched semiconductors having different band gaps. The structure is depicted in Figure 2.5 and consists of an injector, an injection barrier, and an active region. This structure is repeated to form the device; the potential structure is aligned with the external bias. As in an ordinary optical laser, one period of the active region has a 3-level subband structure. The injected electron gives rise to radiative emission with intersubband transition in the active region. This electron propagates to the barrier region, and the transmitted electron is injected into the next structure. This process is repeated like a cascade of light emissions.

Recent research aims at decreasing the threshold current and lasing frequency while increasing the operating temperature to room temperature and expanding the frequency range. In order to achieve these goals, intensive research on new designs for grids, structures, and waveguides is required.

2.2.4 Free-Electron-Based Sources

2.2.4.1 Backward Wave Oscillator

A BWO is an electric vacuum device that is used to generate tunable CW THz waves. It consists of an electron gun, a slow wave structure in a vacuum tube, and an external device for applying the magnetic field as shown in Figure 2.6. The electron beam, which is generated in a heated cathode, is focused by a magnetic field and moved to the anode through a decelerating structure such as a grating. As a result, a traveling EM wave is produced in the opposite direction. The waveguide is coupled to the traveling EM wave and transfers it out into free space. BWOs offer high output power, good wave-front quality, wavelength tunability, and a high signal-to-noise ratio. Followed by these features, BWOs are used as THz sources for imaging systems (Chen et al. 2012; Dobroiu et al. 2004).

Recently, it has been reported that the operating frequency range varies from the sub-THz frequency range (Xu et al. 2011) to 1 THz (Xu et al. 2012). Yet, the spectral range covered by one BWO is typically a hundred GHz, and it needs many BWOs and frequency multipliers to cover the THz frequency region

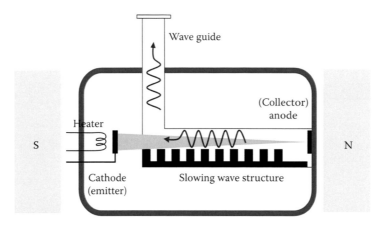

FIGURE 2.6 Schematic diagram of BWO.

(Grüner and Dahl 1998). The output frequency depends on the electron speed, which is regulated by the voltage applied between the electrodes. However, the frequency is also determined by the degree of conformity between the phase velocity of the surface wave on the grating and the electron speed. In order to increase the operational frequency, a highly homogeneous magnetic field and the new slow wave structure, which reduces the circuit ohmic loss and reflection, are required.

2.2.4.2 Free-Electron Lasers

In principle, a FEL is not an actual laser. It uses a relativistic electron beam and accelerator instead of the stimulated emission of photons. This is true even for compact THz FELs, which are being studied in several countries. A FEL is not a small system as it has an electron beam accelerator and a wiggler array. The electron beam moves freely through a wiggler, an array of magnets, which are arranged to supply a periodic, transverse magnetic field, to generate coherent EM radiation. The periodically alternating magnetic field forces a sinusoidal oscillation of the electrons as shown in Figure 2.7. In this part of the oscillator, a beam of electrons is accelerated to almost the speed of light, and the acceleration of electrons along this path results in the release of photons. The wavelength of the light emitted can be tuned by changing the magnetic field strength of the magnetic array or by adjusting the energy of the electron beam. A FEL can produce high-power radiation with an average brightness of more than six orders of magnitude higher than from typical PC antenna emitters. So, it has significant potential in applications

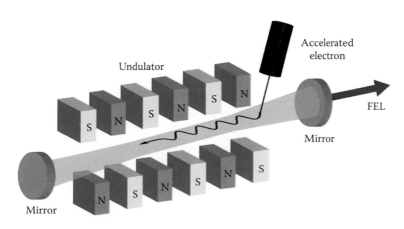

FIGURE 2.7 Schematic diagram of FELs.

where high-power sources are essential or in the investigation of nonlinear THz spectroscopy such as in biological and medical studies (Doria et al. 2002) and for the exploration of vibrational and conformational molecular transitions (Grosse 2002; Xie et al. 2001).

2.3 Terahertz Detectors

Similar to THz sources, THz detectors also use many different mechanisms. Since THz pulses have very fast electrical signals, they cannot easily detect normal electronic devices. To obtain fast electronic signals, the optoelectronic method is mainly used. The most widely used detecting methods are photoconductive detection, which uses photoconductive switching, and EO detecting, which uses the characteristics of EO crystals. A heterodyne receiver detects the CW THz signal by beating with a reference THz signal. It obtains the same amplitude and phase for coherent detection. Unlike in these coherent measurement techniques, thermal detectors only measure the amplitude. Commonly used thermal detectors are bolometers, pyroelectric detectors, and Golay cells. Both THz pulse detection and CW detection methods are briefly reviewed herein.

2.3.1 Femtosecond Laser-Based THz Detection Technique

2.3.1.1 Photoconductive Detectors

The photoconductive antennas induce the dc photocurrent depending on cross correlation between the THz EM waves and the photocarriers, which are generated by the incident femtosecond laser beam. The THz field induced average photocurrent $J(\tau)$s expressed (Hamster et al. 1993; Zhang and Xu 2009) as

$$\bar{J}(\tau) = \bar{N}e\mu E(\tau), \tag{2.12}$$

where
  $\bar{N}$ is the average electron density
  τ is the time delay between the THz EM pulses and the probe laser pulses

The current amplifier reads the photocurrent gating by detecting the laser pulse. The schematic diagram is presented in Figure 2.8.

The photoconductive detection of broadband THz EM pulse is based on the substrate material and antenna structures similar to those used for generation. The short carrier lifetime of the substrate is also important for high sensitivity and low noise detection of the THz EM pulses.

Photoconductive sampling has high signal-to-noise ratio (over 1:100,000) and a low background noise caused by low, dark current. It is also very sensitive. However, photoconductive detection has a limit of about 4.5 THz. Frequency spectrum detection also has constraints such as the RC time delay in the photoconductive dipole antenna and the phonon absorption of the antenna substrate material.

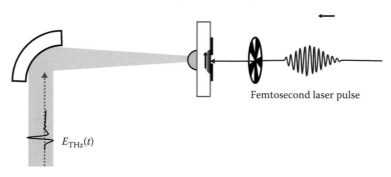

Femtosecond laser pulse

$E_{\mathrm{THz}}(t)$

FIGURE 2.8   The schematic diagram of photoconductive detector.

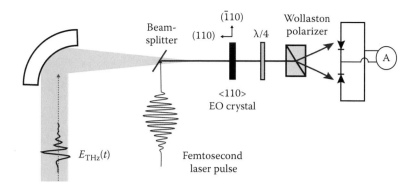

FIGURE 2.9 The schematic diagram of EO sampling.

2.3.1.2 Electrooptic Sampling

In general, EO sampling technique offers a wider spectrum range signal than the photoconductive sampling technique. Depending on the used EO crystal, this technique detects different range spectra with the bandwidth reaching up to 100 THz.

The EO sampling technique requires an EO crystal, a quarter-wave ($\lambda/4$) plate, a Wollaston polarizer, and a balanced photodetector as shown in Figure 2.9. EO sampling uses the Pockels effect of the EO crystals, which is related to the second-order nonlinear polarization (Boyd 2003). When the static electric field is applied to the EO crystal, it induces birefringence. In this case, the THz EM waves act as a static electric field, and the optical laser beam is used for the detection of their birefringence.

The optical laser beam and the THz EM waves are induced in the EO crystal at once. The polarization of the optical beam changes to become elliptical as it propagates through the EO crystal. The quarter-wave plate causes a phase shift of $\lambda/4$ in the incident optical beam so on passing through it, the linear polarized beam changes to be circularly polarized. The Wollaston prism divides the circular polarized beam to p and s linear polarized beams. These two beams are incident on the balanced detector, and we can measure the electric field of THz.

The combination of several nonlinear crystals such as ZnTe, GaP, and GaSe enables the full range from below 1 to 30 THz to be covered. Using ZnTe and GaSe, the frequency regions covered are 0.1–3 THz and 10–30 THz with high nonlinearity. GaP partially covers this gap, but its power is lower than ZnTe due to its lower nonlinearity. The generated THz pulse and Fourier-transformed amplitude spectrums with a GaSe crystals at various phase-matching angles are presented in Figure 2.10.

2.3.2 Heterodyne Receivers

Heterodyne receivers measure the amplitude and phase information of CW THz signals by frequency beat and down-conversion. Frequency beat is the interference between two signals of slightly different frequencies. This device mainly consists of a local oscillator, a mixer, and a spectrometer as shown in Figure 2.11. The local oscillator generates the reference signal, which has a frequency similar to that of the THz signal. Recently, various CW THz sources such as BWOs, QCLs, optically pumped THz gas lasers, and diode-based frequency multipliers are used as local oscillators. The mixer, which receives the reference and THz signal, delivers a down-conversion signal at the difference frequency of the two signals. This down-conversion signal in the microwave range is called the intermediate frequency (IF). The IF signal is amplified and then detected by the spectrometer.

There are three kinds of mixers: Schottky diode mixers, tunnel junction (superconductor–insulator–superconductor [SIS]) mixers, and hot electron bolometer mixers. These mixers have complicated characteristics depending on their material and structure (Hübers 2008). At millimeter wavelengths,

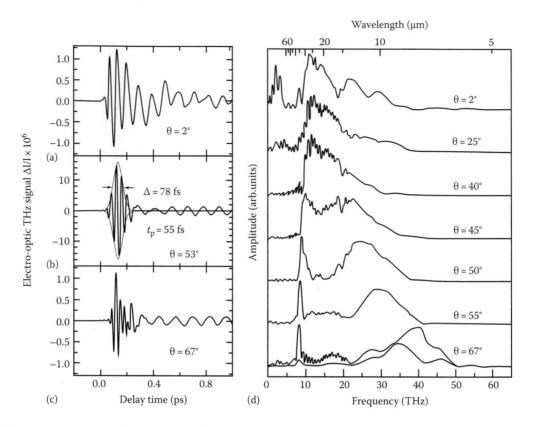

FIGURE 2.10 Generated THz pulse and Fourier-transformed amplitude spectrum with a 90 μm thick GaSe crystal by optical rectification of 10 fs laser pulses at various phase-matching angles as detected with a 10.3 μm thick ZnTe EO sensor. The phase-matching angles are (a) 2°, (b) 53°, and (c) 67°. The thin line in (b) represents a Gaussian fit to the field envelope with a FWHM of Δ = 78 fs, corresponding to pulses with a FWHM of the intensity as short as $t_p = 55$ fs. (d) Normalized amplitude spectra of the measured EO THz signal for the 90 μm thick GaSe emitter at different phase-matching angles. (Reprinted from Huber, R. et al., *Appl. Phys. Lett.*, 76, 3191, 2000.)

SIS mixers offer the best sensitivities, closely approaching the quantum limit. A heterodyne detector has a complex structure than a direct detector, but it enables the measurement of the amplitude and phase information with higher sensitivity. This coherent detection method has 100 dB times higher signal-to-noise ratio than direct detection.

High-resolution heterodyne detection is an important technique in astronomy and atmospheric science (Kulesa 2011). This technology is involved in large astronomy projects such as the ESA's Herschel Space Observation and the Atacama Large Millimeter Array to measure the astronomical emissions from various states of stars and molecular clouds. The fine line frequencies are characterized by the molecular species and their rotational transition. Due to the same reason, heterodyne detection enables the investigation of the Earth's atmosphere.

2.3.3 Thermal Detectors

2.3.3.1 Bolometers

A bolometer measures the power of EM radiation using highly sensitive temperature-dependent electrical resistance. It is applied to observe the broad spectral range from THz to x-rays; therefore, it has been widely used for long for detection in the areas of astronomy, particle physics, and mm waves. In order to have good sensitivity from the mm wave to the THz frequency region, liquid He atmosphere is required.

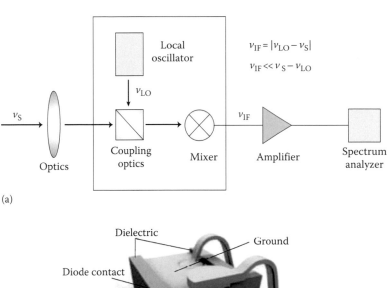

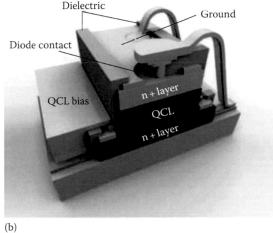

FIGURE 2.11 (a) Scheme of a heterodyne receiver and an integrated transceiver. The principle of operation of a heterodyne receiver. ν_{LO} is the frequency of the local oscillator, ν_S is the frequency of the signal, and ν_{IF} is the beat frequency generated by the mixer. ν_{IF}, the difference between ν_{LO} and ν_S, is often easier to amplify and process than ν_S because it is at a much lower frequency. (b) Structure of the integrated transceiver. The Schottky diode is on top of the QCL waveguide ridge. The integrated transceiver replaces the discrete local oscillator and mixer units as well as the coupling optics shown in (a). (Reprinted from Hübers, H. W., *Nat. Photon.*, 4, 503, 2010.)

A bolometer consists of an absorber with a thermometer, a thermal reservoir (heat reservoir), and a thermal link as shown in Figure 2.12 (Chasmar et al. 1956; Jones 1953). The absorber acts as a detector of thermal energy. It generally uses various materials such as semiconductors, superconductors, and metals, which are dependent on the measured spectral range. When the EM radiation energy is incident on the absorber with the thermometer, the absorber raises its temperature and transfers its changed thermal energy to a thermal reservoir by using a thermal link (Low 1961). The detection time of a bolometer, which is dependent on the thermal time constant between the absorber and the thermal reservoir, is faster than those of other thermal detectors.

2.3.3.2 Pyroelectric Detectors

Pyroelectricity is the ability of certain materials to change thermal energy to electrical energy (Sebald et al. 2008). It can be observed that pyroelectric materials are spontaneously polarized because their unit cells have electrical dipole moments with a certain axis in the materials (Porter 1981).

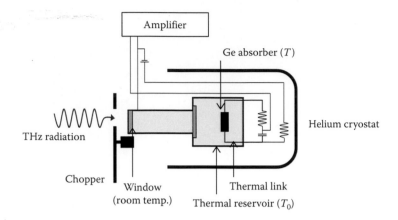

FIGURE 2.12   Schematic of a bolometer.

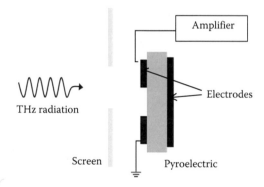

FIGURE 2.13   Schematic of a pyroelectric detector.

A pyroelectric detector is made from a pyroelectric material with metal electrodes on opposite sides, as shown in Figure 2.13. Typically, triglycine sulfate (TGS), deuterated triglycine sulfate (DTGS), lithium tantalate (LiTaO₃), and barium titanate (BaTiO₃) are used as pyroelectric materials in pyroelectric detectors. The top of the electrode has an absorber for measuring the spectral region. If the electrodes are transparent in the spectral range, then pyroelectric material can be used as the absorber in the spectral range. Poling of the pyroelectric materials to be used in pyroelectric detectors is done to make their domains parallel to each other. However, no observable potential difference is found between the two opposite electrodes, because the internal polarization in the poled pyroelectric materials is always balanced by a surface charge that accumulates via various leakage paths between the two faces before application of the external electric field. If the external field (or radiation) is applied to the pyroelectric detectors, which means the heat is applied to the detectors, the capacitance in the pyroelectric detector is directly changed because the polarization in the poled pyroelectric materials changes by an amount that is determined by the temperature change and the pyroelectric coefficient.

2.3.3.3 Golay Cells

A Golay cell is an optoacoustic detector that consists of a gas-filled cavity covered with infrared absorbing material and a flexible membrane. When infrared radiation is absorbed, it heats the gas, causing it to expand. The resulting increase in pressure deforms the membrane. A schematic of Golay cell is

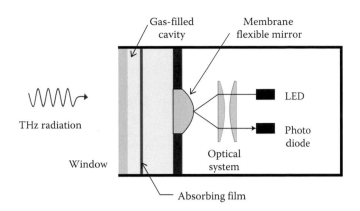

FIGURE 2.14 Schematic of a Golay cell.

shown in Figure 2.14. Light reflected off the membrane is detected by a photodiode, and the motion of the membrane produces a change in the signal on the photodiode (Golay 1947). A Golay cell has slow response time compared to a bolometer, but it is still used for observing THz EM waves because it has better sensitivity and it does not need cryogenic conditions.

References

Abdullaev, G. B., L. A. Kulevskii, A. M. Prokhorov et al. 1972. GaSe, a new effective material for nonlinear optics. *JETP Letters* 16: 90–95.

Ajili, L., G. Scalari, D. Hofstetter et al. 2002. Continuous-wave operation of far-infrared quantum cascade lasers. *Electronics Letters* 38: 1675–1676.

Auston, D., K. Cheung, and P. Smith. 1984. Picosecond photoconducting Hertzian dipoles. *Applied Physics Letters* 45: 284–286.

Belkin, M. A., F. Capasso, A. Belyanin et al. 2007. Terahertz quantum-cascade-laser source based on intracavity difference-frequency generation. *Nature Photonics* 1: 288–292.

Boyd, R. W. 2003. *Nonlinear Optics*. San Diego, CA: Elsevier Science.

Brown, E. 2003. THz generation by photomixing in ultrafast photoconductors. *International Journal of High Speed Electronics and Systems* 13: 497–545.

Carrig, T. J., G. Rodriguez, T. S. Clement, A. Taylor, and K. R. Stewart. 1995. Scaling of terahertz radiation via optical rectification in electro-optic crystals. *Applied Physics Letters* 66: 121.

Chasmar, R. P., W. H. Mitchell, and A. Rennie. 1956. Theory and performance of metal bolometers. *Journal of the Optical Society of America A* 46: 469–477.

Chen, G., J. Pei, F. Yang et al. 2012. Terahertz-wave imaging system based on backward wave oscillator. *IEEE Transactions on Terahertz Science and Technology* 2: 504–512.

Chen, Q., M. Tani, Z. Jiang, and X. C. Zhang. 2001. Electro-optic transceivers for terahertz-wave applications. *Journal of the Optical Society of America B* 18: 823–831.

Dai, J., J. Liu, and X. C. Zhang. 2011. Terahertz wave air photonics: Terahertz wave generation and detection with laser-induced gas plasma. *IEEE Journal of Selected Topics in Quantum Electronics* 17: 183–190.

Demers, J. R., R. T. Logan, and E. R. Brown. 2007. An optically integrated coherent frequency-domain THz spectrometer with signal-to-noise ratio up to 80 dB. Presented at *IEEE International Topical Meeting on Microwave Photonics*, pp. 92–95.

Dobroiu, A., M. Yamashita, Y. N. Ohshima et al. 2004. Terahertz imaging system based on a backward-wave oscillator. *Applied Optics* 43: 5637–5646.

Doria, A., G. P. Gallerano, and E. Giovenale. 2002. Free electron broad-band THz radiator. *Nuclear Instruments and Methods in Physics Research Section A: Accelerators, Spectrometers, Detectors and Associated Equipment* 483: 461–465.

Erickson, N. 1990. High efficiency submillimeter frequency multipliers. Presented at *IEEE MTT-S International Microwave Symposium Digest*, pp. 1301–1304.

Faist, J., F. Capasso, D. Sivco et al. 1994. Quantum cascade laser: An intersub-band semiconductor laser operating above liquid nitrogen temperature. *Electronics Letters* 30: 865–866.

Fattinger, C. and D. Grischkowsky. 1988. Point source terahertz optics. *Applied Physics Letters* 53: 1480–2.

Fattinger, C. and D. Grischkowsky. 1989. Terahertz beams. *Applied Physics Letters* 54: 490–492.

Fork, R., B. Greene, and C. Shank. 1981. Generation of optical pulses shorter than 0.1 psec by colliding pulse mode locking. *Applied Physics Letters* 38: 671.

Gallerano, G. and S. Biedron 2004. Overview of terahertz radiation sources. *Proceedings of the 2004 FEL Conference*, pp. 216–221.

Geng, Y., X. Tan, X. Li, and J. Yao. 2010. Compact and widely tunable terahertz source based on a dual-wavelength intracavity optical parametric oscillation. *Applied Physics B: Lasers and Optics* 99: 181–185.

Golay, M. J. E. 1947. Theoretical consideration in heat and infra-red detection, with particular reference to the pneumatic detector. *Review of Scientific Instruments* 18: 347–56.

Grosse, E. 2002. THz radiation from free electron lasers and its potential for cell and tissue studies. *Physics in Medicine and Biology* 47: 3755.

Grüner, G. and C. Dahl. 1998. *Millimeter and Submillimeter Wave Spectroscopy of Solids.* Heidelberg, Germany: Springer.

Hamster, H., A. Sullivan, S. Gordon, and R. Falcone. 1994. Short-pulse terahertz radiation from high-intensity-laser-produced plasmas. *Physical Review E* 49: 671.

Hamster, H., A. Sullivan, S. Gordon, W. White, and R. W. Falcone. 1993. Subpicosecond, electromagnetic pulses from intense laser-plasma interaction. *Physical Review Letters* 71: 2725–2728.

Han, P. Y., M. Tani, F. Pan, and X. C. Zhang. 2000. Use of the organic crystal DAST for terahertz beam applications. *Optics Letters* 25: 675–677.

Huber, R., A. Brodschelm, F. Tauser, and A. Leitenstorfer. 2000. Generation and field-resolved detection of femtosecond electromagnetic pulses tunable up to 41 THz. *Applied Physics Letters* 76: 3191.

Hübers, H. W. 2008. Terahertz heterodyne receivers. *IEEE Journal of Selected Topics in Quantum Electronics* 14: 378–391.

Hübers, H. W. 2010. Terahertz technology: Towards THz integrated photonics. *Nature Photonics* 4: 503–504.

Indjin, D., P. Harrison, R. Kelsall, and Z. Ikonic. 2003. Mechanisms of temperature performance degradation in terahertz quantum-cascade lasers. *Applied Physics Letters* 82: 1347–1349.

Jones, R. C. 1953. The general theory of bolometer performance. *Journal of the Optical Society of America A* 43: 1–10.

Kawase, K., M. Mizuno, S. Sohma et al. 1999. Difference-frequency terahertz-wave generation from 4-dimethylamino-N-methyl-4-stilbazolium-tosylate by use of an electronically tuned Ti:sapphire laser. *Optics Letters* 24: 1065–1067.

Kawase, K., J. Shikata, K. Imai, and H. Ito. 2001. Transform-limited, narrow-linewidth, terahertz-wave parametric generator. *Applied Physics Letters* 78: 2819.

Kazarinov, R. F. and R. A. Suris. 1971. Possibility of the amplification of electromagnetic waves in a semiconductor with a superlattice. *Soviet physics: Semiconductors* 5: 707–709.

Köhler, R., A. Tredicucci, F. Beltram et al. 2002. Terahertz semiconductor-heterostructure laser. *Nature* 417: 156–159.

Kulesa, C. 2011. Terahertz spectroscopy for astronomy: From comets to cosmology. *IEEE Transactions on Terahertz Science and Technology* 1: 232–240.

Lee, Y. S., T. Meade, T. B. Norris, and A. Galvanauskas. 2001. Tunable narrow-band terahertz generation from periodically poled lithium niobate. *Applied Physics Letters* 78: 3583–3585.

Low, F. J. 1961. Low-temperature germanium bolometer. *Journal of the Optical Society of America A* 51: 1300–1304.

Lu, X. and X. C. Zhang. 2011. Balanced terahertz wave air-biased-coherent-detection. *Applied Physics Letters* 98: 151111.

Mátyás, A., M. A. Belkin, P. Lugli, and C. Jirauschek. 2010. Temperature performance analysis of terahertz quantum cascade lasers: Vertical versus diagonal designs. *Applied Physics Letters* 96: 201110.

Nahata, A., A. S. Weling, and T. F. Heinz. 1996. A wideband coherent terahertz spectroscopy system using optical rectification and electro-optic sampling. *Applied Physics Letters* 69: 2321.

Panda, A. K., G. N. Dash, N. C. Agrawal, and R. K. Parida. 2009. Studies on the characteristics of GaN-based Gunn diode for THz signal generation. Presented at *APMC 2009 Microwave Conference*, pp. 1565–1568.

Porter, S. 1981. A brief guide to pyroelectric detectors. *Ferroelectrics* 33: 193–206.

Porterfield, D. W. 2007. High-efficiency terahertz frequency triplers. Presented at *IEEE/MTT-S International Microwave Symposium*, pp. 337–340.

Porterfield, D. W., T. W. Crowe, R. F. Bradley, and N. R. Erickson. 1999. A high-power fixed-tuned millimeter-wave balanced frequency doubler. *IEEE Transactions on Microwave Theory and Techniques* 47: 419–425.

Preu, S., G. Döhler, S. Malzer, L. Wang, and A. Gossard. 2011. Tunable, continuous-wave Terahertz photomixer sources and applications. *Journal of Applied Physics* 109: 061301.

Rice, A., Y. Jin, X. Ma et al. 1994. Terahertz optical rectification from⟨ 110⟩ zinc-blende crystals. *Applied Physics Letters* 64: 1324–1326.

Sakai, K. 2005. *Terahertz Optoelectronics*. Berlin, Germany: Springer.

Sebald, G., E. Lefeuvre, and D. Guyomar. 2008. Pyroelectric energy conversion: Optimization principles. *IEEE Transactions on Ultrasonics, Ferroelectrics and Frequency Control* 55: 538–551.

Sibbett, W., A. A. Lagatsky, and C. T. A. Brown. 2012. The development and application of femtosecond laser systems. *Optics Express* 20: 6989–7001.

Smith, P. R., D. H. Auston, and M. C. Nuss. 1988. Subpicosecond photoconducting dipole antennas. *IEEE Journal of Quantum Electronics* 24: 255–260.

Spence, D. E., P. N. Kean, and W. Sibbett. 1991. 60-fsec pulse generation from a self-mode-locked Ti:sapphire laser. *Optics Letters* 16: 42–44.

Tani, M., S. Matsuura, K. Sakai, and S. Nakashima. 1997. Emission characteristics of photoconductive antennas based on low-temperature-grown GaAs and semi-insulating GaAs. *Applied Optics* 36: 7853–7859.

Tonouchi, M. 2007. Cutting-edge terahertz technology. *Nature Photonics* 1: 97–105.

van Exter, M., C. Fattinger, and D. Grischkowsky. 1989. High-brightness terahertz beams characterized with an ultrafast detector. *Applied Physics Letters* 55: 337–339.

Verghese, S., K. McIntosh, and E. Brown. 1997. Highly tunable fiber-coupled photomixers with coherent terahertz output power. *IEEE Transactions on Microwave Theory and Techniques* 45: 1301–1309.

Verghese, S., K. McIntosh, S. Calawa et al. 1998. Generation and detection of coherent terahertz waves using two photomixers. *Applied Physics Letters* 73: 3824.

Wilke, I., A. M. MacLeod, W. Gillespie et al. 2002. Single-shot electron-beam bunch length measurements. *Physical Review Letters* 88: 124801.

Williams, B. S. 2007. Terahertz quantum-cascade lasers. *Nature Photonics* 1: 517–525.

Xie, A., A. F. G. van der Meer, and R. H. Austin. 2001. Excited-state lifetimes of far-infrared collective modes in proteins. *Physical Review Letters* 88: 18102.

Xie, X., J. Dai, and X. C. Zhang. 2006. Coherent control of THz wave generation in ambient air. *Physical Review Letters* 96: 75005.

Xu, X., Y. Wei, F. Shen et al. 2011. Sine waveguide for 0.22-THz traveling-wave tube. *IEEE Electron Device Letters* 32: 1152–1154.

Xu, X., Y. Wei, F. Shen et al. 2012. A watt-class 1-THz backward-wave oscillator based on sine waveguide. *Physics of Plasmas* 19: 013113.

Zhang, X. C. and J. Xu. 2009. *Introduction to THz Wave Photonics*. New York: Springer.

Zhang, X. C., X. Ma, Y. Jin et al. 1992. Terahertz optical rectification from a nonlinear organic crystal. *Applied Physics Letters* 61: 3080–3082.

Zhang, Y., Y. W. Tan, H. L. Stormer, and P. Kim. 2005. Experimental observation of the quantum Hall effect and Berry's phase in graphene. *Nature* 438: 201–204.

3

Tabletop High-Power Terahertz Pulse Generation Techniques

Yun-Shik Lee
Oregon State University

3.1 Introduction

Two methods are commonly used for broadband terahertz (THz) pulse generation: (1) transient photocurrent excitations in a biased photoconductive (PC) antenna and (2) optical rectification (OR) in a nonlinear crystal, both utilizing femtosecond lasers. While technological advances of the relatively unexplored spectral band are still in progress, the THz radiation sources have been used for a wide range of applications such as cancer diagnosis, identification of chemical and biological agents, security imaging, and noncontact inspection of packaged goods among many others (Lee 2009; Zhang and Xu 2009). Employing broadband THz pulses and time-resolved THz detection techniques, THz time-domain spectroscopy has become a standard method to obtain complex optical constants of matter in the THz region without imposing on Kramers–Kronig analysis.

Broader applications of the THz technology have been, however, stymied by low output power of the conventional THz sources. Low power of THz sources limits THz spectroscopy to be applied only to probing linear optical properties of material systems. For example, the linear optical spectra of biological macromolecules do not exhibit pronounced features in the THz region; thus, it is difficult to identify and characterize them in the linear spectroscopy regime. Nonlinear THz spectroscopy using high-field THz sources has a potential to shed light on macromolecular dynamics from a new perspective exploiting nonlinear spectroscopy techniques such as saturated absorption and photon echo spectroscopy (Tanaka et al. 2011). Low power of THz sources is also one of the main reasons that technical development of THz imaging is still in its infancy. At present, most of the THz imaging systems employ raster scanning with a single THz detector and mechanical translational stages. For these systems, it usually takes a few minutes to acquire a single 2D image. High-power THz sources are essential components for a real-time THz imaging system, which will greatly improve image-taking speed.

Previously, large-area optical excitations of a biased PC antenna (Darrow et al. 1992) or a ZnTe crystal (Blanchard et al. 2007; Löffler et al. 2005) have been used for generating high-power THz pulses. The radiation source of a large-aperture PC antenna is the surge current of photoexcited carriers induced by a bias field. The output THz fields exhibit a universal behavior of saturation at high optical excitations (Darrow et al. 1992; Taylor et al. 1993). The maximum THz pulse energy is limited by the capacitance and the breakdown voltage of the PC antenna, that is, the THz pulse energy cannot exceed the electrical energy stored in the closed gap of electrodes. The output THz fields are about 10 kV/cm (Darrow et al. 1992; Taylor et al. 1993). OR in a large-area ZnTe crystal also has been used for high-power THz generation. Pulse energies up to 1.5 μJ were obtained, but the energy conversion efficiency was only 3×10^{-5}. The relatively low conversion efficiency is mainly due to strong two-photon absorption in ZnTe (Vidal et al. 2011).

In this chapter, we discuss two recently developed techniques for efficient generation of high-field THz pulses: (1) THz generation by OR of tilted optical pulses in lithium niobate and (2) THz generation in laser-induced gas plasma. Output THz pulse energy from these sources exceed 1 μJ, and optical-to-THz energy conversion efficiency reaches 10^{-3} corresponding to ~30% of photon conversion efficiency.

3.2 Terahertz Generation by Optical Rectification of Tilted Optical Pulses in Lithium Niobate

3.2.1 Material Properties of Lithium Niobate

Lithium niobate ($LiNbO_3$) is a widely used nonlinear optical crystal for a variety of applications in photonics and optoelectronics. The versatile material has favorable properties for such applications: it is transparent over a broad spectral range from mid-IR to UV (350–5200 nm) and has large optical nonlinearity, ferroelectricity, and piezoelectricity. The large electro-optic coefficient, $d_{eff} = 168$ pm/V (Wu and Zhang 1996), indicates that $LiNbO_3$ has potential for efficient THz generation via OR. An undesirable property of intrinsic $LiNbO_3$ is that the material is highly photorefractive. In order to avoid the problem, MgO-doped $LiNbO_3$ is commonly used for OR. MgO doping increases optical damage threshold and reduces THz absorption in $LiNbO_3$ by suppressing the parasitic nonlinear optical effects. The necessary doping level is less than 2 mol% for stoichiometric $LiNbO_3$, which causes little changes in other linear and nonlinear optical properties of the material (Pálfalvi et al. 2004).

3.2.2 Phase Matching and Optical Rectification in $LiNbO_3$

3.2.2.1 Velocity Mismatch between Optical and THz Pulses

The conventional method of THz generation, in which THz waves are generated by OR under collinear phase-matching condition, cannot be applied for $LiNbO_3$ because THz phase velocity is much different from optical group velocity in $LiNbO_3$ (the extraordinary optical group and THz refractive indices are $n_O = 2.25$ at 800 nm [Nakamura et al. 2002] and $n_T = 4.96$ [Pálfalvi et al. 2005], respectively). The THz radiation source of OR is a second-order nonlinear polarization $P_T^{(2)}(t)$ in a nonlinear crystal. An optically induced local THz field $E_T(t)$ is proportional to the second-order time derivative of $P_T^{(2)}(t)$:

$$E_T(t) \propto \frac{\partial^2 P_T^{(2)}(t)}{\partial t^2} = \chi^{(2)} \frac{\partial^2 |E_O(t)|^2}{\partial t^2}, \tag{3.1}$$

where
  $\chi^{(2)}$ is the second-order nonlinear susceptibility of the crystal
  $E_O(t)$ is the optical field at the location

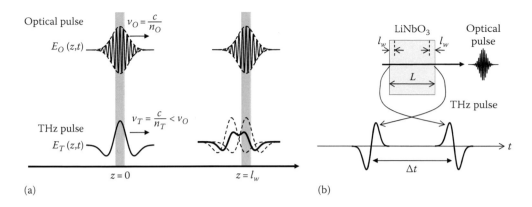

FIGURE 3.1 (a) Velocity mismatch between optical and THz pulses gives rise to destructive interference between the THz fields optically induced at two positions with a walk-off length separation in a LiNbO$_3$ crystal. (b) Only surviving THz radiation from a LiNbO$_3$ slab of a finite thickness is the contributions from the entrance and exit surfaces within a depth of l_w.

The THz waveform resembles the optical pump pulse envelop. When an optical pulse and a THz pulse collinearly propagate in LiNbO$_3$, the optical pulse will precede the THz pulse by the optical pulse duration τ_p after propagating a walk-off length, $l_w = c\tau_p / (n_T - n_O)$. Therefore, THz fields generated at two positions separated by a walk-off length destructively interfere each other as illustrated in Figure 3.1a. In a uniform medium, THz radiation field completely vanishes via continuous destructive interference. When an optical pulse transmits trough a LiNbO$_3$ crystal of a finite thickness L, THz radiation only from the entrance and exit surfaces within a depth of l_w can survive as shown in Figure 3.1b (Xu et al. 1992). The pulses from the two surfaces are temporally separated due to the velocity mismatch between the optical and THz pulses, where the time delay is $\Delta t = (n_T - n_O)L/c$.

3.2.2.2 Tilted-Pulse-Front Phase Matching

A noble method to circumvent the optical/THz velocity mismatch in LiNbO$_3$ is to tilt the optical pulse front toward the direction normal to the Cherenkov cone (Hebling et al. 2002). In a LiNbO$_3$ crystal, a focused femtosecond laser pulse, whose size (~10 μm in all direction) is substantially smaller than the wavelength of THz radiation, acts like a point source moving faster than generated THz waves. The collapsed THz waves of Cherenkov radiation form a shock front in the shape of a cone, as illustrated in Figure 3.2a. The Cherenkov radiation is emitted under a constant angle θ_c with the optical pulse trajectory, given by

$$\theta_c = \cos^{-1}\left(\frac{v_T}{v_O}\right) = \cos^{-1}\left(\frac{n_O}{n_T}\right) \cong 63.0°, \tag{3.2}$$

where v_T and v_O are THz phase and optical group velocities, respectively. When the pulse front of an optical beam is tilted to be perpendicular to the Cherenkov radiation and the beam size is substantially larger than the wavelength of THz radiation, the optical pulse front and THz waves move at the same speed, $v_T = v_O \cos\theta_c$, in the direction of Cherenkov radiation as shown in Figure 3.2b. Under this velocity matching condition, OR continuously supplies THz radiation in phase, and hence the THz radiation is coherently amplified as the optical pulse propagates in the LiNbO$_3$ crystal.

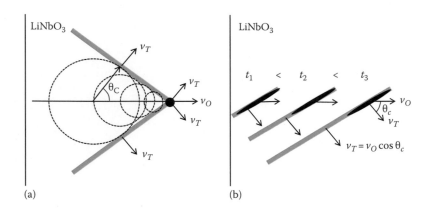

FIGURE 3.2 (a) A focused femtosecond laser pulse generates THz radiation forming a Cherenkov cone with an emission angle θ_c. (b) An optical pulse of tilted pulse front aligned to the Cherenkov cone copropagates with THz radiation with the same speed, $v_T = v_O \cos \theta_c$. (From Stepanov, A.G. et al., *Appl. Phys. Lett.,* 83, 3000, 2003.)

A diffraction grating is utilized to tilt an optical pulse front. Figure 3.3a shows a schematic diagram of ray optics: how an optical pulse front perpendicular to its propagation direction is tilted after diffraction off a grating. The tilt angle θ is determined by the formula

$$\sin \theta = \gamma[1 - \sin^2 \alpha + 2\gamma \sin \alpha]^{-1/2}, \tag{3.3}$$

which is obtained by calculating the path length difference with the grating equation for $m = +1$, first positive order,

$$\sin \alpha + \sin \beta = \frac{\lambda}{d} \equiv \gamma, \tag{3.4}$$

where
   α is an incidence angle
   β is a diffraction angle
   λ is an optical wavelength
   d is the grating period

When an optical pulse enters into a LiNbO$_3$ crystal, the tilt angle decreases due to refraction as shown in Figure 3.3b. Matching the tilt angle in LiNbO$_3$ to the Cherenkov angle θ_c, we obtain the relation between θ and θ_c:

$$\tan \theta = n_O \tan \theta_c. \tag{3.5}$$

In an experimental configuration to acquire $\theta_c = 63.0°$, that is, $\theta = 77.2°$ at $\lambda = 800$ nm and $n_O = 2.25$, the incidence and diffraction angles are $\alpha = 44.8°$ and $\beta = 61.8°$ for a 2000 line/mm grating. It should be noted that the tilt angle in LiNbO$_3$ also depends on the magnification factor of the imaging optics between the grating and the crystal. The tangent of the tilt angle is inversely proportional to the magnification ratio M, that is, Equation 3.5 should be modified as $\tan \theta = n_O/M \tan \theta_c$.

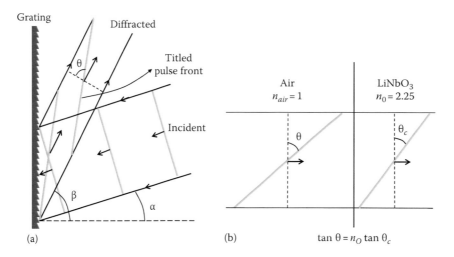

FIGURE 3.3 (a) An optical pulse front is tilted by a diffraction grating due to path length differences at different positions. (b) When an optical pulse enters into a LiNbO₃ crystal, the tilt angle reduces because of refraction.

3.2.2.3 Theory of Optical Rectification in LiNbO₃

OR is a second-order nonlinear optical process. The second-order nonlinear polarization induced by a monochromatic optical plane wave of frequency ω is expressed as

$$P_i^{(2)}(0) = \sum_{j,k} \varepsilon_0 \chi_{ijk}^{(2)}(0,\omega,-\omega)E_j(\omega)E_k^*(\omega), \tag{3.6}$$

where $\chi_{ijk}^{(2)}$ is the second-order nonlinear susceptibility tensor element for LiNbO₃. The tensor has eight nonzero and three independent elements. In the contracted notation $d_{il} = \chi_{ijk}^{(2)}/2$, the d-matrix of LiNbO₃ is written as

$$\begin{pmatrix} 0 & 0 & 0 & 0 & d_{31} & -d_{22} \\ -d_{22} & d_{22} & 0 & d_{31} & 0 & 0 \\ d_{31} & d_{31} & d_{33} & 0 & 0 & 0 \end{pmatrix}. \tag{3.7}$$

LiNbO₃ is a strongly anisotropic nonlinear crystal that d_{33} is much larger than d_{31} and d_{22}. A simple polarization configuration for efficient THz generation in LiNbO₃ is that optical fields are aligned along the crystal z-axis generating a nonlinear polarization in the same direction so that the nonlinear polarization is simply expressed as $P^{(2)} = 2d_{eff}E_O(\omega)^2$, where the effective electro-optic coefficient $d_{eff} = d_{33}$ in this polarization geometry.

Under phase-matching condition in the absence of pump absorption or depletion, the THz conversion efficiency (defined as the ratio of THz power to optical power) including THz absorption in LiNbO₃ (absorption coefficient, α_T) is expressed as the following formula (Hoffmann and Fülöp 2011):

$$\eta_{\text{THz}} = \frac{2\omega_{\text{THz}}^2 d_{eff}^2 L^2 I_O}{\varepsilon_0 n_O^2 n_T c^3} \frac{\sinh^2(\alpha_T L/4)}{(\alpha_T L/4)^2} e^{-\alpha_T L/2}, \tag{3.8}$$

where
  ω_{THz} is the THz frequency
  L is the crystal thickness
  I_O is the optical pump intensity
  n_T and n_O are refractive indices at THz and optical frequencies
  c is the speed of light

TABLE 3.1   Optical Constants for LiNbO$_3$

n_O (800 nm)	n_T (1 THz)	α_T (cm^{-1})	d_{eff} (pm/V)
2.25	4.96	17	168

Table 3.1 lists the optical constants for LiNbO$_3$ for extraordinary polarized (//crystal z-axis) optical and THz waves (Hoffmann and Fülöp 2011). The THz absorption coefficient ($\alpha_T = 17$ cm^{-1} at room temperature) indicates that THz generation in a LiNbO$_3$ crystal is effective only within the 0.5 mm thick layer including the exit surface of THz radiation.

3.2.3 Experimental Results

3.2.3.1 Experimental Setup

Figure 3.4 shows an experimental setup for tilted-pulse-front OR. The setup is composed of a grating, a half-wave plate, a lens, and a LiNbO$_3$ crystal. One surface of the LiNbO$_3$ crystal is cut at the Cherenkov angle θ_c to get THz waves coupled into free space at normal incidence. The half-wave plate rotates the optical pulse polarization from horizontal to vertical direction to have it parallel to the crystal z-axis. The lens transfers the image of a tilted pulse front on the grating into the LiNbO$_3$ crystal. The optical pulse image should be formed within a short distance (<0.5 mm) from the exit surface of the LiNbO$_3$ crystal to curtail THz absorption.

THz generation efficiency of the pulse-front-tilting method is susceptible to its imaging system. In particular, it is crucial to acquire in-depth images of high quality at the focal plane. Theoretical studies show that a telescope imaging system is superior to a single lens system, producing less distortion in THz beam profile and phase fronts (Fülöp et al. 2010; Pálfalvi et al. 2008; Figure 3.5).

3.2.3.2 THz Output from LiNbO$_3$ THz Emitters

THz generation by OR in LiNbO$_3$ needs high-power femtosecond laser pulses for optical pumping because tilted-pulse-front phase matching requires large area of optical excitation. The most commonly used laser systems are 1 kHz Ti–sapphire amplifiers. Figure 3.6 shows typical THz output from a LiNbO$_3$ THz emitter pumped by a 1 kHz Ti–sapphire regenerative amplifier. Broadband THz pulses (central frequency, 1 THz; bandwidth, 1 THz as shown in Figure 3.6b) are generated by OR of femtosecond laser pulses with tilted pulse fronts (central wavelength, 800 nm; pulse energy, 1 mJ; pulse duration, 90 fs; repetition rate, 1 kHz) in a LiNbO$_3$ crystal. Figure 3.6a shows a THz waveform

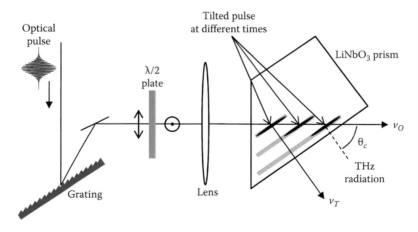

FIGURE 3.4   Experimental setup for THz generation by tilted-pulse-front OR in LiNbO$_3$.

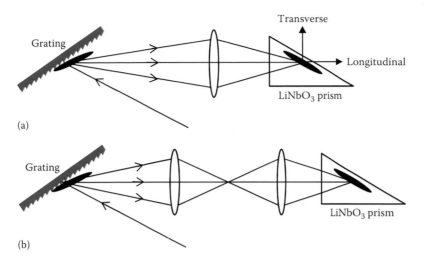

(a)

(b)

FIGURE 3.5 (a) Single lens and (b) telescope imaging systems.

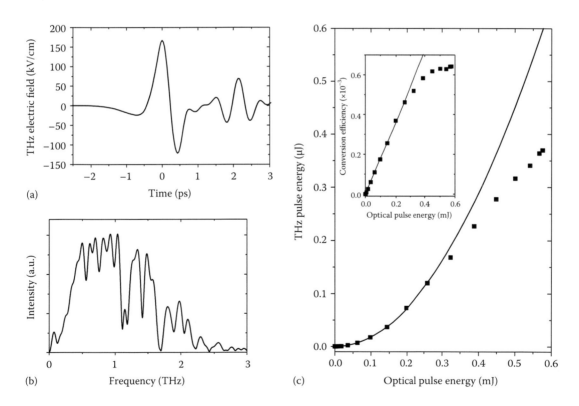

FIGURE 3.6 THz emission from a LiNbO$_3$ THz generator pumped by a 1 kHz Ti–sapphire regenerative amplifier (central wavelength, 800 nm; pulse energy, 1 mJ; pulse duration, 90 fs; repetition rate, 1 kHz). (a) THz waveform measured by electro-optic sampling in a 1 mm ZnTe and (b) corresponding power spectrum obtained by Fourier transform of the THz waveform. (c) Output THz pulse energy versus optical pump pulse energy. The filled squares are experimental results. The solid line indicates an ideal quadratic dependence under the assumption of no pump depletion and nonlinear absorption. The inset shows the optical-to-THz energy conversion efficiency.

TABLE 3.2 Tilted-Pulse-Front THz Generators with Various Ti–Sapphire Amplifiers

	Optical Pump Pulse		Grating Density (mm^{-1})	Imaging System	THz Pulse			
Rep rate	Duration (fs)	Energy			Energy	Frequency (THz)	ηTHz	References
200 kHz	150	2.3 μJ	2000	Telescope	30 pJ	1.8	1.3×10^{-5}	Stepanov et al. (2003)
1 kHz	150	0.5 mJ	2000	Telescope	0.24 μJ		5×10^{-4}	Stepanov et al. (2005)
1 kHz	150	0.65 mJ	1800	Telescope	0.57 μJ	0.9	8×10^{-4}	Jewariya et al. (2009)
1 kHz	85	2 mJ	1800	Telescope	2 μJ	1	1×10^{-3}	Hirori et al. (2011)
10 Hz	400	20 mJ	2000	Single lens	10 μJ	0.5	6×10^{-4}	Yeh et al. (2007)

in the time domain measured by electro-optic sampling in a 150 μm ZnTe crystal. THz field amplitude reaches 160 kV/cm when the optical pump pulse energy is 0.6 mJ. Figure 3.6b shows the power spectrum of the THz pulse obtained by Fourier transform of the waveform. The broadband spectrum is centered around 1 THz and ranges up to 3 THz.

Figure 3.6c shows THz output pulse energy as a function of optical pump pulse energy. The solid square indicates experimental measurements, while the thin solid line represents the ideal quadratic dependence (thin solid line) of OR without pump depletion and nonlinear absorption. The measured THz pulse energy departs from the quadratic dependence around 0.25 mJ of the optical pump pulse energy and exhibits linear dependence above 0.4 mJ. The saturation effects are mainly due to multiphonon absorption enhanced by strong THz fields (Stepanov et al. 2005). Optical-to-THz energy conversion efficiency is shown in the inset of Figure 3.6c. The conversion efficiency reaches 0.64×10^{-3} at 0.57 mJ and flattens above 0.4 mJ due to the saturation effects. The THz generation efficiency is limited by relatively strong THz absorption in LiNbO$_3$ (Lee 2009). The dominant THz absorption mechanism in LiNbO$_3$ is anharmonic decay of the optical phonon into two acoustic phonons, which can be substantially suppressed by cooling down the crystal. It has been reported that optical-to-THz conversion efficiency at 77 K is more than three times higher than that at 300 K (Stepanov et al. 2003).

Various types of Ti–sapphire amplifiers have been used for THz generation by tilted-pulse-front OR in LiNbO$_3$. Table 3.2 summarizes performance and characteristics of several tilted-pulse-front THz generators. The most commonly used laser system is 1 kHz Ti–sapphire amplifiers, with which THz pulse energy is ~1 μJ and energy conversion efficiency reaches ~1×10^{-3}. It is notable that a compact LiNbO$_3$ THz emitter can be built using a fiber laser system. Tilted-pulse-front OR in LiNbO$_3$ has been demonstrated using Yb-doped amplified fiber laser systems (Hoffmann et al. 2007, 2008). A 1 kHz fiber amplifier (wavelength, 1.03 μm; pulse energy, 0.5 mJ; pulse duration, 300 fs) produced 0.1 μJ THz pulses with the energy conversion efficiency of 2.5×10^{-4}.

3.3 Broadband THz Generation by Two-Color Optical Excitation in Gas Plasmas

3.3.1 Overview

Interaction of strong laser pulses with photogenerated gas plasmas can produce electromagnetic radiation covering an extraordinarily broad spectrum ranging from THz waves to x-rays. THz pulses have been observed from laser-induced air plasma, where the THz generation mechanism is the ponderomotive force acting on electrons and ions, which separates the two kinds of charges and creates a charge density gradient (Hamster et al. 1993). When a strong bias field is applied, THz generation efficiency can be relatively high, comparable to that of THz radiation from semiconductor surfaces (Loffler et al. 2000).

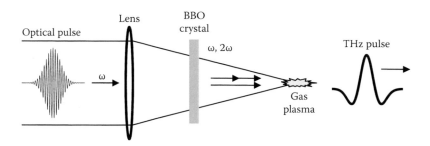

FIGURE 3.7 Broadband THz generation by two-color optical excitation in gas plasma. Asymmetric superimposed laser fields of the fundamental (ω) and SH (2ω) pulses induce transient currents in the plasma-generating broadband THz pulses.

THz generation efficiency of laser–plasma interaction is greatly improved by two-color optical excitation, which is obtained by mixing a fundamental optical wave of frequency ω with a second-harmonic (SH) wave of frequency 2ω (Bartel et al. 2005; Cook and Hochstrasser 2000; Kress et al. 2004). The experimental scheme is sketched in Figure 3.7. SH pulses are generated in a nonlinear optical crystal such as a β-barium borate (BBO) crystal. Asymmetric electric fields of the superposition of the fundamental and the SH waves drive the ionizing electrons generating transient photocurrent (Kim et al. 2007). The laser-induced photocurrent is the source of the THz radiation from the laser–plasma interaction. The THz radiation intensity is susceptible to optical polarizations. In order to optimize THz generation efficiency, the fundamental and SH polarizations must be collinearly polarized (Kress et al. 2004; Xie et al. 2006).

It should be noted that laser-induced gas plasma is used not only for THz emission but also for THz sensing (Dai et al. 2006; Karpowicz et al. 2008; Liu et al. 2010). The detection bandwidth of gas-plasma THz sensing covers the entire THz gap ranging from 0.1 to 30 THz, which is only limited by the optical probe pulse duration. Furthermore, remote sensing beyond tens of meters is available in atmospheric environment, which is nearly impossible for conventional THz sensing technologies. The THz wave air photonics involving THz emission and detection with laser-induced gas plasma is an exciting subfield of THz science and technology. Comprehensive reviews of the topic are found in the references (Dai et al. 2011a,b). In this section, we will focus on THz generation by two-color optical excitation in gas plasma.

3.3.2 THz Generation Mechanism

3.3.2.1 Four-Wave Mixing Approximation

A simple phenomenological model to describe THz generation by two-color optical excitation in gas plasma is based on four-wave mixing (FWM), where two fundamental photons (ω) and one SH photon (2ω) are coupled in the medium producing one THz wave photon (ω_{THz}). The third-order nonlinear polarization in the THz region is expressed by the equation

$$P^{(3)}(\omega_{THz}) = \varepsilon_0 \chi^{(3)}(\omega_{THz}, \omega, \omega, -2\omega + \omega_{THz})E(\omega)E(\omega)E^\star(2\omega - \omega_{THz}), \tag{3.9}$$

where $\chi^{(3)}$ is the third-order nonlinear susceptibility tensor of the gas plasma. In the FWM process, THz frequency is in the range of the optical bandwidth, while the carrier frequencies of the three optical fields are canceled out. Gas plasma is treated as an isotropic nonlinear medium in the FWM model, where $\chi^{(3)}$ has three independent components, $\chi^{(3)}_{xxxx}$, $\chi^{(3)}_{xyxy}$, and $\chi^{(3)}_{xxyy}$. Maximum THz field is obtained for *xxxx*-polarization configuration in which the optical and THz waves are collinearly polarized (Xie et al. 2006).

The THz field in the far field zone is related to the nonlinear polarization as

$$E_{THz}(t) \propto \frac{\partial^2 P^{(3)}(t)}{\partial t^2}. \tag{3.10}$$

The temporal profile of an instantaneously created polarization $P^{(3)}(t)$ is determined by optical pulse envelopes. With the superposed optical field expressed as

$$E(t) = E_\omega(t)\cos\omega t + E_{2\omega}(t)\cos(\omega t + \phi), \tag{3.11}$$

where ϕ is the relative phase between the fundamental and SH fields, the emitted THz field has the relation (Cook and Hochstrasser 2000; Xie et al. 2006)

$$E_{THz}(t) \propto \chi^{(3)}E_\omega(t)^2 E_{2\omega}(t)\cos\phi \propto \chi^{(3)}I_\omega\sqrt{I_{2\omega}}\cos\phi. \tag{3.12}$$

Figure 3.8 shows how the THz field amplitude varies with the relative ϕ. The interference pattern shown in Figure 3.8a fits well with the relation $\cos\phi = \cos(\omega_{400\,nm}\tau)$, where τ is the relative time delay between fundamental and SH pulses. The optical intensity dependence of FWM, $E_{THz}(t) \propto \chi^{(3)}I_\omega\sqrt{I_{2\omega}}$, is demonstrated in Figure 3.9. The THz field amplitude is proportional to the fundamental pulse intensity (Figure 3.9a) and to the square root of the SH pulse intensity (Figure 3.9b).

The FWM model is valid when optical excitation intensity is relatively low ($<10^{14}$ W/cm^2), as shown in Figures 3.8 and 3.9. The perturbative approach, however, starts to break down when the gas medium becomes highly nonlinear with an onset of plasma formation. It is pronounced that THz emission power exhibits a threshold behavior, abruptly increasing above a critical optical intensity. The breakdown of the FWM model is also evident that the output THz field amplitude becomes proportional to ~sin ϕ at high optical excitation intensities. It is notable that the phenomenological FWM model does not clarify the microscopic origin of the nonlinear optical process whether the source of the THz polarization is the free electrons in the plasma or the bound electrons of molecules.

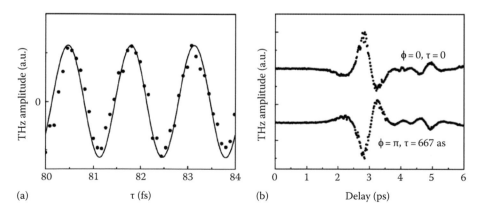

FIGURE 3.8   (a) Interference pattern depending on the relative phase ϕ. The solid line is a fit by $\cos\phi = \cos(\omega_{400\,nm}\tau)$, where τ is the relative time delay between fundamental and SH pulses. (b) The THz waveforms at $\phi = 0$ and π show opposite polarity. (Reprinted from Xie, X., Dai, J., and Zhang, X.-C., *Phys. Rev. Lett.*, 96, 075005, 2006. ©2006 by the American Physical Society.)

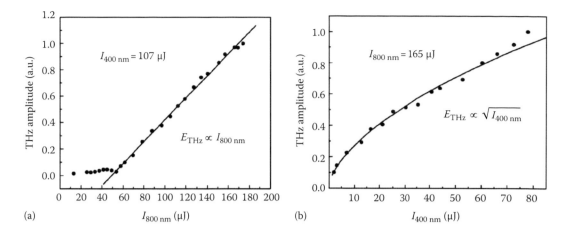

FIGURE 3.9 THz amplitude versus (a) I_ω and (b) $I_{2\omega}$. The solid lines indicate (a) linear and (b) square-root fit. (Reprinted from Xie, X., Dai, J., and Zhang, X.-C., *Phys. Rev. Lett.*, 96, 075005, 2006. ©2006, by the American Physical Society.)

3.3.2.2 Asymmetric Transient Photocurrent Model

At high optical excitation intensity ($>10^{15}$ W/cm^2), interaction of intense laser fields with gas molecules enters into a nonperturbative regime where tunneling ionization is the dominant mechanism to produce highly ionized gas plasma (Augst et al. 1989). The tunneling ionization process is instantaneous and highly nonlinear. When gas molecules is in the presence of strong two-color laser fields, free electrons escaped from the molecules via rapid tunneling ionization from a transient current driven by the asymmetric laser field. The time scale of the photoionization is the laser pulse duration (typically less than 100 fs); therefore, the electromagnetic radiation spectrum ranges from 1 to 10 THz.

Figure 3.10a shows superposed optical fields of fundamental ($\lambda = 800$ nm, $I_\omega = 10^{15}$ W/cm^2) and SH ($\lambda/2 = 400$ nm, $I_{2\omega} = 2 \times 10^{14}$ W/cm^2) when the relative phase ϕ is 0 and $\pi/2$ (Kim et al. 2007). Trajectories of free electrons generated at various laser phases are shown in Figure 3.10b, indicating that optical fields at $\phi = \pi/2$ induce asymmetric drift motion of electrons. The drift electron velocity at a laser phase θ and a relative phase ϕ is expressed as

$$v_d = \frac{eE_\omega}{m_e\omega}\sin\theta + \frac{eE_\omega}{2m_e\omega}\sin(2\theta + \phi). \tag{3.13}$$

Figure 3.10c compares the drift electron velocities at $\phi = 0$ and $\pi/2$ over one period of the laser field from $\phi = -\pi$ to π. Drift velocities near the peak of the laser fields are of interest, where tunneling ionization predominantly takes place. The drift velocity is evenly distributed near the peak of the laser field at $\phi = 0$, while the velocity distribution is lopsided at $\phi = \pi/2$. This result indicates that the transient current of photogenerated electrons vanishes at $\phi = 0$ and is maximum at $\phi = \pi/2$. THz fields emitted from the transient photocurrent therefore has the relation

$$E_{\text{THz}}(t) \propto \frac{dJ_d}{dt} \sim ev_d\frac{dN_e}{dt} \propto \sin\phi, \tag{3.14}$$

where
J_d is the transient current density
N_e is the electron density

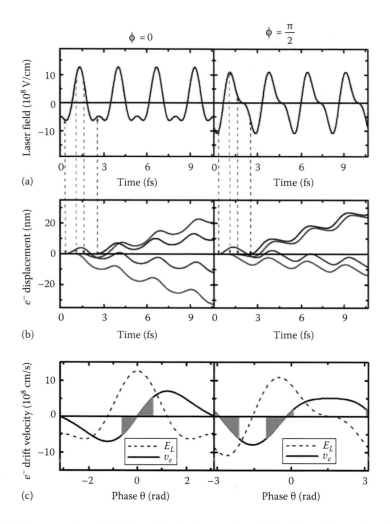

FIGURE 3.10 (a) Superposed optical fields of the fundamental and SH with a relative phase $\phi = 0$ and $\pi/2$. (b) Electron trajectories born at various phases of $\theta = -9\pi/10$, $-\pi/10$, $\pi/10$, and $9\pi/10$. (c) Drift electron velocity versus θ (solid line), overlaid with the laser field (dashed line). (Reprinted from Kim, K.-Y. et al., *Opt. Express*, 15, 4577, 2007.)

Figure 3.11a demonstrates experimental observation confirming the prediction of the asymmetric transient photocurrent model about the relative phase dependence of THz generation. The THz yield is sinusoidal as a function of the BBO-to-plasma distance d. Extrapolation of the data indicates that THz yield vanishes at $d = 0$. Since the relative phase ϕ at the plasma is linearly proportional to d,

$$\phi = \frac{\omega}{c}(n_\omega - n_{2\omega})d, \tag{3.15}$$

where n_ω and $n_{2\omega}$ are refractive indices of air at ω and 2ω, and the relative phase dependence of THz yield is consistent with the transient asymmetric photocurrent model, $E_{THz}(t) \propto \sin\phi$.

3.3.2.3 Quantum Mechanical Model

The classical models based on FWM and transient photocurrent have limited applications valid at different optical excitation intensity regimes. Since tunneling ionization, the essential physical process

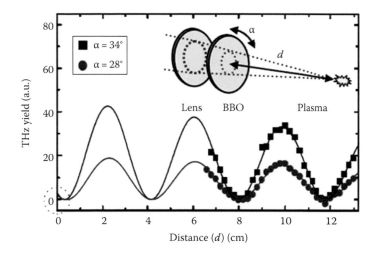

FIGURE 3.11 THz yield versus BBO-to-plasma distance d with two different BBO angles $\alpha = 28°$ and 34°. (Reprinted from Kim, K.-Y., Glownia, J. H., Taylor, A. J., and Rodriguez, G., *Opt. Express*, 15, 4577, 2007.)

of plasma formation, is intrinsically quantum process, a quantum mechanical model is desirable for a comprehensive description of plasma formation via tunneling ionization and light–plasma interaction to generate THz waves. A full-quantum mechanical model has been developed to describe the formation and dynamics of electron wave packets when strong laser pulses interact with gas molecules (Karpowicz and Zhang 2009).

Numerical simulations of the quantum theory reveal that the THz emission process has two steps as illustrated in Figure 3.12a. Initially, electron wave packets created by tunneling ionization form a transient current near the atom that emits THz radiation. Later, propagating away from the atom, electron wave packets are decelerated by inelastic collisions with surrounding atoms and ions generating THz radiation that is coherently added to the initial THz radiation. The quantum mechanical model provides precise numerical analyses for the relative phase dependence of THz emission. The relative phase dependence varies with optical excitation intensity. Figure 3.12b and c demonstrates electron density distributions of argon plasma in the presence of two-color optical fields (peak amplitude is 2×10^{10} V/m) at $\phi = 5\pi/12$ and $11\pi/12$, respectively. At the moderate optical excitation intensity, the asymmetry of electron density is minimum at $\phi = 5\pi/12$ (close to $\pi/2$) and maximum at $\phi = 11\pi/12$ (close to π).

3.3.3 Performance of Gas-Plasma THz Emitters

A clear advantage of THz generation in gas plasmas is that the media are not limited by optical damage. Optical pulses can be focused into a small volume way beyond the optical damage threshold of solid materials. It, however, does not mean that THz generation efficiency is ever enhancing as optical excitation intensity increases. Figure 3.13 demonstrates peak THz field amplitude as a function of incident optical pump pulse energy (Kim et al. 2007). The THz output exhibits a threshold behavior at low excitation intensities below 5 mJ, steep increases in the intermediate region from 5 to 15 mJ, and becomes saturated at high intensities above 15 mJ. The saturation behavior of THz generation is accounted for by strong enhancement of THz absorption in high-density plasma.

The THz power scaling behavior has been examined with various gas species and pressure (Rodriguez and Dakovski 2010). Figure 3.14a shows THz pulse energy as a function of incident optical pulse energy at a gas pressure of 590 torr for air, neon, argon, krypton, and xenon. Argon exhibits maximum optical-to-energy energy conversion efficiency ~1.5 × 10⁻⁴, producing 0.9 µJ THz pulses at an optical pulse energy of 6 mJ.

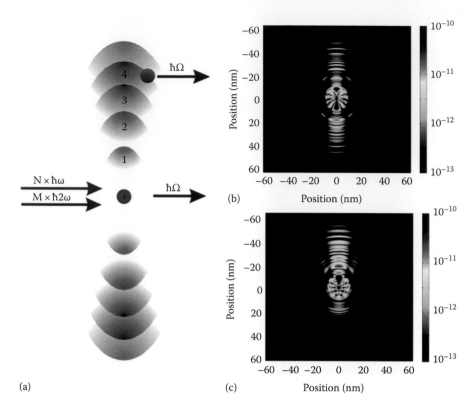

FIGURE 3.12 (a) Schematic diagram of THz emission from photogenerated electron wave packets. (b), (c) Electron density distributions of argon plasma in the presence of strong two-color optical fields at $\phi = 5\pi/12$ and $11\pi/12$, respectively. (Reprinted with permission from Karpowicz, N. and Zhang, X.-C., *Phys. Rev. Lett.*, 102, 093001, 2009. ©2009, by the American Physical Society.)

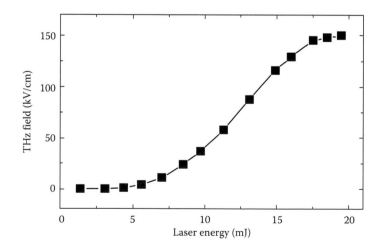

FIGURE 3.13 Peak THz field amplitude versus optical pump pulse energy before frequency doubling. (Reprinted from Kim, K.-Y., et al., *Opt. Express*, 15, 4577, 2007.)

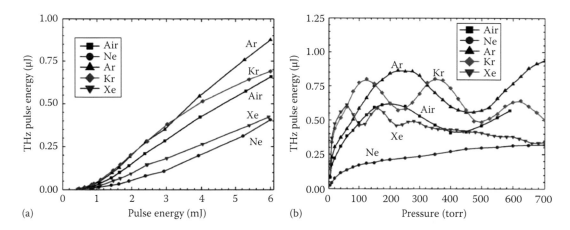

(a) Pulse energy (mJ)

(b) Pressure (torr)

FIGURE 3.14 THz output pulse energy versus (a) optical pulse energy at a gas pressure of 590 torr and (b) pressure at an optical pulse energy of 5.4 mJ for the gases of air, neon, argon, krypton, and xenon. (Reprinted from Rodriguez, G. and Dakovski, G. L., *Opt. Express*, 18, 15130, 2010.)

Figure 3.14b shows the pressure dependence of the THz output at an optical pulse energy of 5.4 mJ as the pressure varies from 5 to 700 torr for the same gas species. The THz output exhibits a periodic fluctuation whose frequency increases as mass increases. The pressure-dependent fluctuations are mainly due to phase slippage of ω and 2ω optical fields induced by index changes during the propagation in the neutral gas before plasma formation. It is also notable that the THz output of the heavy gases, krypton and xenon, rapidly increases in the low-pressure region below 100 torr yet gradually decreases above the saturation pressure. The saturation and falloff of the THz output may be due to phase slippage in the plasma or from plasma defocusing effects.

The tunneling ionization process is virtually instantaneous so that the bandwidth of THz radiation from photoexcited gas plasma is roughly same with that of the optical pump pulse. The THz spectrum can also be significantly broadened by self-phase modulation and ionization-induced blue shifts, if the optical field is sufficiently strong (Kim et al. 2008). It has been demonstrated that ultrabroadband femtosecond pulses (<20 fs) can generate THz pulses whose spectrum reaches 100 THz covering far-IR and mid-IR regions (Matsubara et al. 2012; Thomson et al. 2010). The Fourier-transformed spectrum of THz pulses generated in an air plasma by 10 fs optical pulses has the central frequency and the bandwidth of ~20 THz, while the high-frequency tail reaches 200 THz (Matsubara et al. 2012).

3.4 Conclusion

We have reviewed two powerful methods to generate high-field THz pulses: (1) OR of tilted optical pulses in LiNbO$_3$ and (2) two-color optical excitation in gas plasma. Both methods indeed produce strong THz pulses: pulse energy exceeds 1 μJ and field amplitude reaches 1 MV/cm. The two methods, however, have somewhat different characteristics. Optical-to-THz energy conversion efficiency of LiNbO$_3$ THz emitters is as high as 1×10^{-3} at an optical pulse energy of ~1 mJ, while optimal conversion efficiency of gas-plasma THz emitters is ~1×10^{-4}. Due to the absence of optical damage, gas-plasma THz emitter can endure very high optical excitation intensities above 10 mJ. The central frequency and the bandwidth of THz pulses from LiNbO$_3$ THz emitters are about 1 THz. On the other hand, air-plasma THz emitters can produce ultrabroadband THz pulses whose spectrum extends up to ~100 THz.

References

Augst, S., D. Strickland, D. D. Meyerhofer, S.-L. Chin, and J. H. Eberly. 1989. Tunneling ionization of noble gases in a high-intensity laser field. *Physical Review Letters* 63: 2212.

Bartel, T., P. Gaal, K. Reimann, M. Woerner, and T. Elsaesser. 2005. Generation of single-cycle THz transients with high electric-field amplitudes. *Optics Letters* 30: 2805–2807.

Blanchard, F., L. Razzari, H. C. Bandulet et al. 2007. Generation of 1.5 μJ single-cycle terahertz pulses by optical rectification from a large aperture ZnTe crystal. *Optics Express* 15: 13212–13220.

Cook, D. J. and R. M. Hochstrasser. 2000. Intense terahertz pulses by four-wave rectification in air. *Optics Letters* 25: 1210–1212.

Dai, J., B. Clough, I.-C. Ho et al. 2011a. Recent progresses in terahertz wave air photonics. *IEEE Transactions on Terahertz Science and Technology* 1: 274–281.

Dai, J., J. Liu, and X.-C. Zhang. 2011b. Terahertz wave air photonics: Terahertz wave generation and detection with laser-induced gas plasma. *IEEE Journal of Selected Topics in Quantum Electronics* 17: 183–190.

Dai, J., X. Xie, and X.-C. Zhang. 2006. Detection of broadband terahertz waves with a laser-induced plasma in gases. *Physical Review Letters* 97: 103903.

Darrow, J. T., X.-C. Zhang, D. H. Auston, and J. D. Morse. 1992. Saturation properties of large-aperture photoconducting antennas. *IEEE Journal of Quantum Electronics* 28: 1607–1616.

Fülöp, J. A., L. Pálfalvi, G. Almási, and J. Hebling. 2010. Design of high-energy terahertz sources based on optical rectification. *Optics Express* 18: 12311–12327.

Hamster, H., A. Sullivan, S. Gordon, W. White, and R. W. Falcone. 1993. Subpicosecond, electromagnetic pulses from intense laser-plasma interaction. *Physical Review Letters* 71: 2725.

Hebling, J., G. Almasi, I. Z. Kozma, and J. Kuhl. 2002. Velocity matching by pulse front tilting for large area THz-pulse generation. *Optics Express* 10: 1161–1166.

Hirori, H., F. Blanchard, and K. Tanaka. 2011. Single-cycle THz pulses with amplitudes exceeding 1 MV/cm generated by optical rectification in LiNbO$_3$. *Applied Physics Letters* 98: 091106.

Hoffmann, M. C. and J. A. Fülöp. 2011. Intense ultrashort terahertz pulses: Generation and applications. *Journal of Physics D: Applied Physics* 44: 083001.

Hoffmann, M. C., K.-L. Yeh, J. Hebling, and K. A. Nelson. 2007. Efficient terahertz generation by optical rectification at 1035 nm. *Optics Express* 15: 11706–11713.

Hoffmann, M. C., K.-L. Yeh, H. Y. Hwang et al. 2008. Fiber laser pumped high average power single-cycle terahertz pulse source. *Applied Physics Letters* 93: 141107.

Jewariya, M., M. Nagai, and K. Tanaka. 2009. Enhancement of terahertz wave generation by cascaded χ^2 processes in LiNbO$_3$. *Journal of the Optical Society of America B* 26: A101-A6.

Karpowicz, N., J. Dai, X. Lu et al. 2008. Coherent heterodyne time-domain spectrometry covering the entire "terahertz gap". *Applied Physics Letters* 92: 011131.

Karpowicz, N. and X.-C. Zhang. 2009. Coherent terahertz echo of tunnel ionization in gases. *Physical Review Letters* 102: 093001.

Kim, K.-Y., J. H. Glownia, A. J. Taylor, and G. Rodriguez. 2007. Terahertz emission from ultrafast ionizing air in symmetry-broken laser fields. *Optics Express* 15: 4577–4584.

Kim, K. Y., A. J. Taylor, J. H. Glownia, and G. Rodriguez. 2008. Coherent control of terahertz supercontinuum generation in ultrafast laser–gas interactions. *Nature Photonics* 2: 605–609.

Kress, M., T. Löffler, S. Eden, M. Thomson, and H. G. Roskos. 2004. Terahertz-pulse generation by photoionization of air with laser pulses composed of both fundamental and second-harmonic waves. *Optics Letters* 29: 1120–1122.

Löffler, T., T. Hahn, M. Thomson, F. Jacob, and H. Roskos. 2005. Large-area electro-optic ZnTe terahertz emitters. *Optics Express* 13: 5353–5362.

Lee, Y. S. 2009. *Principles of Terahertz Science and Technology*. New York: Springer.

Liu, J., J. Dai, S. L. Chin, and X.-C. Zhang. 2010. Broadband terahertz wave remote sensing using coherent manipulation of fluorescence from asymmetrically ionized gases. *Nature Photonics* 4: 627–631.

Loffler, T., F. Jacob, and H. Roskos. 2000. Generation of terahertz pulses by photoionization of electrically biased air. *Applied Physics Letters* 77: 453–455.

Matsubara, E., M. Nagai, and M. Ashida. 2012. Ultrabroadband coherent electric field from far infrared to 200 THz using air plasma induced by 10 fs pulses. *Applied Physics Letters* 101: 011105.

Nakamura, M., S. Higuchi, S. Takekawa et al. 2002. Optical damage resistance and refractive indices in near-stoichiometric MgO-doped $LiNbO_3$. *Japanese Journal of Applied Physics* 41: L49–L51.

Pálfalvi, L., J. A. Fulop, G. Almási, and J. Hebling. 2008. Novel setups for extremely high power single-cycle terahertz pulse generation by optical rectification. *Applied Physics Letters* 92: 171107.

Pálfalvi, L., J. Hebling, J. Kuhl, A. Peter, and K. Polgár. 2005. Temperature dependence of the absorption and refraction of Mg-doped congruent and stoichiometric LiNbO3 in the THz range. *Journal of Applied Physics* 97: 123505.

Pálfalvi, L., J. Hebling, G. Almási et al. 2004. Nonlinear refraction and absorption of Mg doped stoichiometric and congruent LiNbO3. *Journal of Applied Physics* 95: 902–908.

Rodriguez, G. and G. L. Dakovski. 2010. Scaling behavior of ultrafast two-color terahertz generation in plasma gas targets: Energy and pressure dependence. *Optics Express* 18: 15130–15143.

Stepanov, A. G., J. Hebling, and J. Kuhl. 2003. Efficient generation of subpicosecond terahertz radiation by phase-matched optical rectification using ultrashort laser pulses with tilted pulse fronts. *Applied Physics Letters* 83: 3000–3002.

Stepanov, A., J. Kuhl, I. Kozma et al. 2005. Scaling up the energy of THz pulses created by optical rectification. *Optics Express* 13: 5762–5768.

Tanaka, M., H. Hirori, and M. Nagai. 2011. THz nonlinear spectroscopy of solids. *IEEE Transactions on Terahertz Science and Technology* 1: 301–312.

Taylor, A. J., P. K. Benicewicz, and S. M. Young. 1993. Modeling of femtosecond electromagnetic pulses from large-aperture photoconductors. *Optics Letters* 18: 1340–1342.

Thomson, M. D., V. Blank, and H. G. Roskos. 2010. Terahertz white-light pulses from an air plasma photo-induced by incommensurate two-color optical fields. *Optics Express* 18: 23173–23182.

Vidal, S., J. Degert, M. Tondusson, J. Oberlé, and E. Freysz. 2011. Impact of dispersion, free carriers, and two-photon absorption on the generation of intense terahertz pulses in ZnTe crystals. *Applied Physics Letters* 98: 191103.

Wu, Q. and X. C. Zhang. 1996. Ultrafast electro-optic field sensors. *Applied Physics Letters* 68: 1604–1606.

Xie, X., J. Dai, and X.-C. Zhang. 2006. Coherent control of THz wave generation in ambient air. *Physical Review Letters* 96: 075005.

Xu, L., X. C. Zhang, and D. H. Auston. 1992. Terahertz beam generation by femtosecond optical pulses in electro-optic materials. *Applied Physics Letters* 61: 1784–1786.

Yeh, K.-L., M. Hoffmann, J. Hebling, and K. A. Nelson. 2007. Generation of 10 μJ ultrashort terahertz pulses by optical rectification. *Applied Physics Letters* 90: 171121.

Zhang, X. C. and J. Xu. 2009. *Introduction to THz Wave Photonics*. New York: Springer.

4

Terahertz Imaging and Tomography Techniques

Hyunyong Choi
Yonsei University

Joo-Hiuk Son
University of Seoul

4.1 Introduction

Terahertz (THz) imaging technique has great potential in the field of nondestructive, noncontact measurements. Followed by the ongoing advances in the development of THz sources and detectors, THz imaging is applied in various fields and indicated in large-scale market products. Since the first illustration of THz imaging by Hu and Nuss (1995), many target materials related to the biomedical material, security, and semiconductor industry have been examined. In addition, there has been a considerable progress in the development of THz-based imaging techniques.

THz imaging has inherent advantages because the fingerprint spectra of most semiconductors and biological materials are in the range of 0.1–10 THz. THz radiation strongly interacts with materials because the order of their wavelength corresponds to that of a molecular-level structure. The THz frequency range accommodates some semiconductor transport parameters, the rotational energy of gas molecules, and that of the collective vibrations mode. In addition, energies associated with hydrogen bonding modes of molecular structures can also be found in the THz frequency range. The THz frequency range is an important spectral area in which the characteristics of materials can be identified (Abbott and Zhang 2007; Ho et al. 2008; Mantsch and Naumann 2010). THz radiation has unique characteristics that are not present in conventional x-ray, ultrasound, infrared, and millimeter range radiations. Because the energy level of a THz wave is about 4.14 meV (much less than the energy ~0.12–120 keV of x-rays), it does not pose an ionization hazard as does x-ray radiation. A THz wave is characterized by its high transmittance for nonmetallic and nonpolar materials and high reactivity for the high-molecular component. Thus, the penetration achieved by THz waves is somewhat similar to that of x-rays, although THz wave penetration does not cause harmful effects such as ionizing radiation.

Based on the nature of their THz source, THz imaging systems can be classified either as a pulsed THz radiation system or as a continuous-wave (CW) THz radiation system. The pulsed THz is generated and detected using the femtosecond laser-based technique. This wave is characterized by a broadband signal in the frequency domain. Pulsed THz imaging has been developed to time-of-flight imaging, which can give the tomographic information and one-shot imaging for video rate real-time imaging. Unlike pulsed THz imaging, CW THz imaging is characterized by a fixed or limited tunable frequency range and does not provide information regarding the phase. Despite these shortcomings, CW THz imaging has potential for application, because it provides fast detection and simple analysis and because it has a compact size and a relatively low cost.

The tomography technique is a noninvasive measurement that is used to obtain information in three dimensions using sequences of transmitting 2D images. As THz waves easily pass through nonmetallic and nonpolar materials, tomography images of objects composed of these materials simply show the inner regions of such objects. There are two types of tomography techniques, those that utilize THz diffraction tomography (DT) and those that are based on THz computed tomography (CT). The scattered or transmitted THz images are measured by rotating the sample to a different projection angle. Following this procedure, the images are reconstructed as 3D images. THz Fresnel lens tomography and THz holography are introduced later in the text as novel techniques. The focal length of the Fresnel lens varies depending on the frequency. After applying Fresnel lens in the diffraction topographic system, it is possible to obtain the images on the same image plane for a sample but from a different position. The THz holography technique differs from conventional holography. Because THz holography can record the amplitude and phase information without splitting the source beam into reference and sample beams, a 3D image can be acquired with it by using a simple system. In addition, digital holography methods have to be used because THz waves are in the invisible range.

In this present chapter, we briefly review the THz imaging technique and discuss recent THz imaging techniques and their applications. The computed and diffracted tomography methods and their 3D image reconstruction algorithms are presented. We then discuss the quality and limitations of each tomography method and introduce the novel tomography technique.

4.2 Review of THz Imaging Techniques

4.2.1 Femtosecond Laser-Based THz Pulse Imaging

4.2.1.1 Raster Scan Imaging

THz time-domain spectroscopy (THz-TDS) is a powerful spectroscopic technique based on the generation and detection of femtosecond THz radiation. It provides high signal-to-noise ratio (SNR), submillimeter resolution measurements of the amplitude, and phase of coherent THz pulses that are transmitted through or reflected from the sample.

A schematic presentation of THz-TDS is shown in Figure 4.1. Conventional broadband THz generation methods are typically based on the use of a femtosecond laser whose pulse duration is less than a hundred of femtoseconds. Femtosecond laser pulses are used to generate and detect a coherent few-cycle THz pulse. THz-TDS measures the amplitude and phase of the THz pulse. The most common methods used to generate and detect THz pulse are photoconductive switching and electro-optic (EO) sampling. The femtosecond laser beam is divided into two paths with a beam splitter, and each beam is focused on the generator and the detector, respectively. The path lengths of the two rays have to be the same because the pulse has to gate simultaneously at the detector. The time-domain THz pulse signal is obtained by varying the time delay between the generator and the detector. These techniques can easily make it possible to achieve a high spectral resolution and a high SNR over 60 dB. Because the THz spectrum is calculated by using the fast Fourier transform (FFT), the spectral resolution is determined by the inverse of the temporal scan range. For the 100 GHz spectral resolution, the time-domain spectrum range of 10 ps (about 1.5 mm of travel in mechanical delay line) is required (Chan et al. 2007). The reference signal $E_{reference}(\omega)$ and sample signal $E_{sample}(\omega)$ provide the complex-valued optical properties that depend

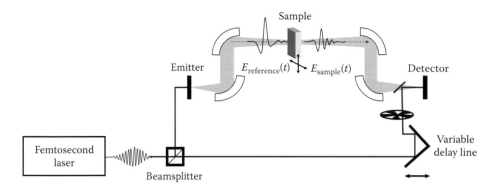

FIGURE 4.1 Conventional THz-TDS system based on femtosecond laser.

on the frequency ω. The complex refractive index of the sample can be obtained from the relation given by (Duvillaret et al. 1996)

$$\frac{E_{sample}(\omega)}{E_{reference}(\omega)} = Te^{-\frac{d\alpha(\omega)}{2}}e^{-i\frac{2\pi}{\lambda}(n(\omega)-1)d} \equiv |R|e^{i\theta},\quad (4.1)$$

where

 T is given by the Fresnel equation $T_{1\to2} = \left(2n_1/(n_1 + n_2)\right)$
 α is the absorption coefficient
 λ is the wavelength
 d is the thickness of the sample

The frequency-dependent complex index values of refraction, $n+ik$, are

$$n(\omega) = 1 + \frac{c}{\omega d}\theta \quad (4.2)$$

$$k(\omega) = -\frac{c}{\omega d}\ln\left(\frac{R}{T}\right),\quad (4.3)$$

where c is the speed of light. Using this analysis, the frequency-dependent complex properties of a sample can be obtained. As mentioned already, a THz spectrum has unique fingerprints for most semiconductors and biological materials. These fingerprints can be used to both identify the unknown sample and to quantize the mixed sample for various application fields.

In the raster scan imaging technique, an image is formed pixel by pixel by controlling the position of the sample with moving stages and by measuring each pixel of the image using the conventional THz-TDS system. The sample is located in the focal plane of an off-axis parabolic mirror. The THz-TDS system determines the THz image quality. The frequency-dependent complex refractive index can be extracted at each pixel. With these properties, THz raster scan imaging can be used to determine useful characteristics of the samples as a noncontact and noninvasive method. The obtained images can be depicted in terms of the amplitude in the time domain, amplitude in the frequency domain, and phase difference. As shown in Figure 4.2, the spatial resolution of such imaging is affected by the frequency. A higher frequency produces better resolution owing to the corresponding shorter wavelength. Selecting a suitable frequency that corresponds to the fingerprint allows chemical mapping to be performed with a component spatial

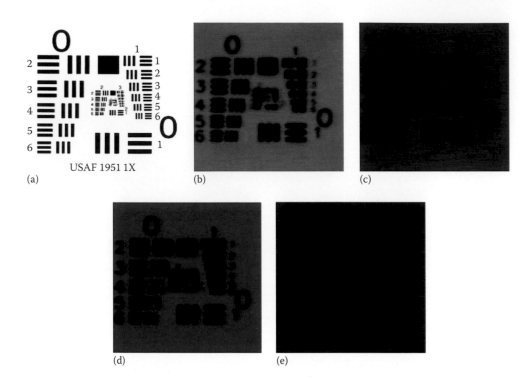

FIGURE 4.2 (a) Optical image of 2″ × 2″ USAF1951 resolution targets. THz images for (b) peak amplitude, (c) 0.75 THz, (d) 1.5 THz, and (e) 3 THz.

pattern analysis (Shen et al. 2005a). This is particularly useful in investigations of strongly inhomogeneous systems, as is the case for many biological samples, including plants and biological cells.

In the first imaging system, THz waveforms were measured at the rate of 12 waveforms per second (Hu and Nuss 1995). This rate is mainly related to the speed of the optical delay line. There are various types of optical delay lines used for high-speed measurements, including linear scanning retroreflectors, piezoelectric fiber stretchers, and helicoid mirrors. And specifically designed rotation mirror arrays are used with the rotational delay line (Xu and Zhang 2004). A rotary optical delay line increases the scan repetition rates to up to 400 waveforms per second with a 2.1 cm scan length (Kim et al. 2008).

4.2.1.2 Time-of-Flight Imaging

The time-domain THz signal, which can only be acquired by a femtosecond laser-based system, allows extraction of the depth information of a sample. When EM waves are incident on an optical medium, some of the beams are reflected from the surface of the medium and the rest propagate into the medium (Born et al. 1999). In this sense, while it is assumed that the target consists of several layers of different media, the reflected THz signal includes information on the internal structural organization of the target. The signal that is acquired by the reflection mode THz system is composed of the signals that are reflected from each of the interfaces in the target (Mittleman et al. 1997b, 1999). A schematic presentation is shown in Figure 4.3.

The resulting signal is a linear sum of the components reflected from each layer:

$$E^{\text{out}} = \sum_{i=1} E_i^{\text{out}} \tag{4.4}$$

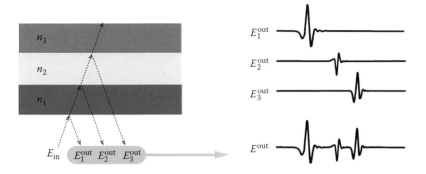

FIGURE 4.3   Reflection of a THz wave from the interfaces of a layered sample.

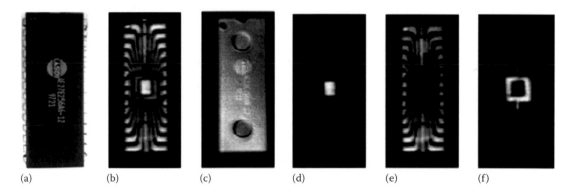

(a)   (b)   (c)   (d)   (e)   (f)

FIGURE 4.4   (a) The visible image, (b) an image of the maximal value of signals, (c–f) images taken at depths of 0.14, 2.06, 2.44, and 3.19 mm. (Reprinted from Cho, S. H. et al., *Opt. Express*, 19, 16401, 2011.)

$$E_i^{\text{out}} = E^{\text{in}} \left(\prod_{j=0}^{i-1} t_{j,j+1}\, p_{j+1}^2\, t_{j+1,j} \right) r_{i-1,i}, \tag{4.5}$$

where

 t, r, and p are the transmission coefficient, reflection coefficient, and propagation factor, respectively
 the subscript notation j, $j+1$ denotes the value at the interface between the jth and $j+1$th layers, respectively

The different coefficients are given as $t_{j,j+1} = \left(2n_j/(n_j + n_{j+1})\right)$, $r_{j,j+1} = \left((n_j - n_{j+1})/(n_j + n_{j+1})\right)$ $p_j = \exp\left[-i\omega \dfrac{n_j}{c_0} d\right]$, where n is the refractive index and d is the thickness of the layer (Born et al. 1999; Duvillaret et al. 1996).

Sectional images can be acquired by using the time delay that results from the wave propagation. Each such image represents an image (at time t) of the 3D data. One example is shown in Figure 4.4 (Cho et al. 2011).

In addition to simple imaging, it is also possible to extract the depth-dependent optical properties by using the deconvolution relation. If the reflected wave can be assumed to be a convolution of the incident wave and optical structural information, we obtain the so-called impulse response:

$$y(t) = x(t) * h(t), \tag{4.6}$$

where $y(t)$, $x(t)$, and $h(t)$ are the measured reflected wave, incident wave, and impulse response function, respectively. The corresponding relation in the frequency domain is

$$y(\omega) = x(\omega)h(\omega) \tag{4.7}$$

$$h(t) = FFT^{-1}\left[\frac{FFT[y(t)]}{FFT[x(t)]}\right]. \tag{4.8}$$

Using the simple process that is described earlier, it is hard to acquire a well-defined signal because of the low SNR due to the noise-amplifying effect; thus, suitable signal conditioning is required. This process consists of noise suppression by properly selected filters:

$$h(t) = FFT^{-1}\left[[\text{Filter}]\frac{FFT[y(t)]}{FFT[x(t)]}\right]. \tag{4.9}$$

Because the signal that is measured by the computer is in the form of a discrete array of length N, practical calculations proceed using the index in the array, rather than the time t. Each of the indices has the time resolution t/N. By using $h(t)$, we can estimate the reflection coefficients and refractive index values for all array indices:

$$r_k = h_k \prod_{j=1}^{k-1}\left(1 - r_j^2\right)^{-1}, \, r_1 = h_1 \tag{4.10}$$

$$n_n = \prod_{j=1}^{n}\frac{1 - r_j}{1 + r_j}. \tag{4.11}$$

The thickness is calculated from the refractive index and the time resolution:

$$d_i = \frac{c_0}{n_i}\Delta t. \tag{4.12}$$

4.2.1.3 One-Shot Imaging

Conventional THz imaging system uses a single-pixel measurement and raster scanning. High-quality images can be obtained using this technique, but the acquisition time is very long due to the serial pixel acquisition. The EO sampling detection enables the increasing of the number of measured pixels using a CCD camera (Jiang and Zhang 1999; Wu et al. 1996). This technique reduces the acquisition time remarkably and results in real-time imaging. Nikon Corporation developed a real-time THz imaging system with large aperture photoconductive antenna, large-area ZnTe crystal, and CCD camera; this system is schematically shown in Figure 4.5a (Usami et al. 2002). The EO sampling technique measures the polarization change on the EO crystal. THz EM waves induce birefringence, which can be detected by using the optical probe beam. The large-area EO crystal and the large probe beam are used to detect the change in the polarization of the expanded THz beam, and the changes are recorded by the crossed polarizers and CCD camera. The EO crystal is placed between two crossed polarizers in order to enhance the ratio between the THz polarization's change and the probe beam. However, the power values of the THz beam and the probe beam limit the detection area. Amplified femtosecond laser systems are necessary in order to generate higher THz field amplitudes and higher optical modulation depths. Figure 4.5b and c shows the snapshot images of Venus flytrap, obtained from the video rate acquisition of 30 frames per second. The THz pulse in the temporal domain can be measured by varying the time delay between the THz wave

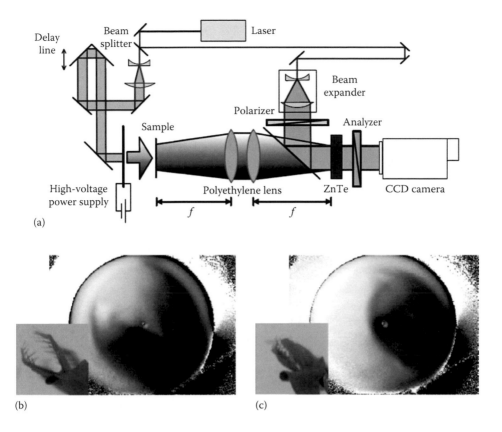

FIGURE 4.5 (a) Schematic presentation of the real-time THz imaging system. Focal length of the polyethylene lens (f) is 10 cm. (b,c) Snapshots from a THz movie. Initial (b) and final (c) states of the motion of a Venus flytrap. The corresponding visible light images are also shown in insets on the lower left side. (Reprinted from Usami, M. et al., *Phys. Med. Biol.*, 47, 3749, 2002.)

and optical probe beam. The frequency-dependent measurement of each pixel is also possible. The resolution improves as the frequency is increased and is about twice the Rayleigh resolution limit (defined by the blurring effects such as diffraction and by the phase mismatch) (Usami et al. 2002).

4.2.2 CW THz Imaging

With the advance of pulsed THz imaging techniques, the use of CW THz imaging system has rapidly grown over the past several decades. Constructing the THz image by pixel-by-pixel raster scanning of the sample is time consuming, whereas fast or real-time operation is required for the application of THz imaging in fields such as medical diagnosis, chemical analysis, and industrial inspection. Compared with THz pulse imaging, CW THz imaging techniques allow fast scanning. Pulsed imaging is more time consuming than CW imaging, because the former is performed with a linear stage for detection adopting a photoconductive detector or EO sampling and a tightly focused beam. The most famous application of CW imaging was demonstrated by Kawase et al. (2003). These authors developed the CW imaging system for detection and identification of drugs hidden in mail envelopes. The sample that they used and the resulting images are presented in Figure 4.6. Using the parametric oscillator and pyroelectric detector, signals in the 1.3–2.0 THz range were measured, and the absorption spectrum enabled detection of the drug at a concentration of ~20 mg/cm². For significantly faster acquisition times, one can use a continuous THz source with a thermal detector, which makes it possible to cover

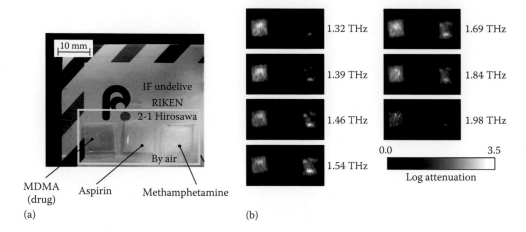

FIGURE 4.6   (a) An optical image of the sample. The small polyethylene bags contain about 20 mg of powder. From left to right: MDMA (drug), aspirin, and methamphetamine. During the imaging, the bags were placed inside the envelope. (b) Seven multispectral images. (Reprinted from Kawase, K. Y. et al., *Opt. Express*, 11, 2549, 2003.)

a large area with single-shot imaging. Array detectors, such as microbolometer arrays and germanium detector arrays, can also be used in THz CW imaging. This technique can also be used to acquire several pixels of an image in a single measurement. However, detectors differ in their temperature sensitivity and temperature working conditions. The commercial microbolometer, which has been developed for thermal imaging, can generate real-time images, such as a visible video camera, whose sampling rate is 30–60 frames per second (Bolduc et al. 2010).

Furthermore, with tunable CW sources that generate a specific frequency in the 0.1–2 THz range, the CW THz imaging system can yield higher-resolution images than the THz pulse imaging technology at that same specific frequency (Tonouchi 2007). Recent advances in semiconductor techniques make it possible to design CW THz imaging systems that are both compact and have simple hardware. A CW THz imaging system that is based on quantum cascade laser (QCL) is a very attractive possibility for the CW THz imaging of samples, because QCLs are very compact (Lee et al. 2006).

4.2.3 Applications of THz Imaging

The applications of THz imaging techniques are very diverse. Several representative fields for THZ imaging applications are as follows:

1. Biomedical: diagnosis of breast cancer and skin cancer, monitoring of a tablet film coating (Ho et al. 2009), drug testing, and THz endoscope medical diagnostics (Ji et al. 2009)

    The fact that the THz radiation does not damage the material makes THz imaging a useful tool in the biomedical field. The resolution of a THz image is comparable to that obtained through magnetic resonance imaging (MRI); however, the equipment that is required is relatively simple. The first THz image had reported the water absorption changes in a leaf (Hu and Nuss 1995). The water absorption change is critical for distinguishing between the healthy tissue and the diseased one. Breast and skin cancer diagnostics utilize this effect in THz imaging (Fitzgerald et al. 2006; Woodward et al. 2002). Disease is detected by analyzing the change in the THz optical constants, as well as the change in cell structure. Frozen tissue from the human brain (Png et al. 2009), porcine tissue (Hoshina et al. 2009), and rabbit liver tumor tissue (Park et al. 2011) all showed different THz amplitudes for different cell types.

    The pharmaceutical applications of THz imaging have been developed to characterize crystalline performance of pharmaceutical medications and solid drugs (Zeitler et al. 2010). The evaluation of coating thickness and uniformity in coated pharmaceutical tablets is included to measure and image the THz chemical properties (Ho et al. 2009). A mixture of multiple drugs was verified by the chemical components of drug using spectral characteristics in the THz range. Moreover, a THz minimized device has been developed that allows imaging the organs within human body. This THz endoscope system is based on the optical fiber technology and it measures THz absorption and refractive index of the side wall of the digestive organ (Ji et al. 2009).

2. Security and military: illicit drug detection in mail, passenger, and baggage screening

Kawase et al. used THz imaging with a tunable THz wave source to develop nondestructive inspection techniques for detection and identification of illegal drugs hidden in mail envelopes (Kawase et al. 2003). Explosives and narcotics have their fingerprints in the THz range; thus, the ability of THZ imaging to identify these chemicals is valuable for applications in security-related fields (Choi et al. 2004; Shen et al. 2005a). Nondestructive 3D chemical mapping and identification of different chemicals buried in the 3D matrix can be performed using THz pulsed imaging (Shen et al. 2005b).

THz imaging has been applied to detect the materials that are hidden within visible-opaque materials such as paper and plastic bags and clothes. This can help the Food and Drug Administration (FDA) discover the harmful drugs or help discover the dangerous weapons in an airport. In airport security applications, THz techniques are already used at the frequency of 500 GHz by utilizing arrays of Schottky barrier diodes with silicon photonic bandgap crystal and heterodyne detection (Tonouchi 2007).

3. Food inspection: monitoring the quality of food and detection of pesticides and antibiotics

Considerable research had been conducted to determine the feasibility of using THz techniques for food inspection. The water content is especially used for the detection because THz absorption is high for polar liquids (Federici 2012). The moisture content in dried food such as pecans (Li et al. 2010), crushed wheat grain (Chua et al. 2005), and oil (Gorenflo et al. 2006) has been explored. By using the different absorption fingerprints, pesticides (Hua and Zhang 2010) and antibiotics (Redo-Sanchez et al. 2011) have been detected, and metallic and nonmetallic foreign bodies (Jördens and Koch 2008; Lee et al. 2012) have been detected in the food matrix. The strong analytic ability offered by THz imaging has been applied to detect materials hidden in visible–opaque materials such as carbohydrate foods and paper and plastic bags (Gowen et al. 2010).

4. Semiconductor industry: noncontact electrical probe

THz imaging systems can achieve fast scanning in seconds through real-time imaging and can also be used to perform the imaging of arrays devised on large areas that are on a scale that is approximately in the millimeter to centimeter range. Electrical properties of semiconductor devices, such as mobility, conductivity, carrier density, and plasma oscillations, can be precisely characterized using the transmission or reflection THz imaging configuration without the need for any contact or preprocesses. THz imaging can be successfully used to measure the electron density without the need to contact the samples (Hu and Nuss 1995; Mittleman et al. 1997a; van Exter and Grischkowsky 1990). Thus, this technology is expected to help improve the yield rates of semiconductor devices through their inspection, such as probing of resistivity or disconnected metal lines within the circuits. THz imaging can also be utilized to improve the efficiency or local overload of units or arrays of devices by mapping the heat distributions (Kiwa et al. 2003). Additionally, the THz near-field microscopy on semiconductor transistors allows accessing the carrier concentration and mobility with a resolution of 40 nm (Huber et al. 2008).

5. Other applications: imaging in astronomy

Sensitive heterodyne spectroscopic techniques implemented in the THz region are important for the mapping of remote objects. These methods are becoming more important for the detection of thermal radiations in planetary and astronomical sciences, investigations of the dynamics of astronomical objects and of the planetary atmosphere (Boreiko and Betz 2009; Roeser et al. 1986), the distribution of gas molecules in the stratosphere (Englert et al. 2000; Mees et al. 1995; Pickett 2006), etc.

Both passive imaging, in which thermal radiation is detected without additional illumination using thermal detector, and the sensing of reflected/scattered radiation with illumination by a narrowband sources are very dependent on the sensitivity of thermal detectors. Therefore, the development of low-noise thermal receivers that effectively operate in the THz region is a top priority in applications that use THz imaging to detect thermal radiation from targets.

An important point in THz imaging is that the resolution of the image is fundamentally limited by the diffraction of the wavelength, similar to what occurs in other conventional imaging systems. Higher-frequency components are better resolved, but because in conventional systems the wavelength of these components corresponds to the range of few hundred microns, the resolution is still insufficiently high for medical applications. To overcome these diffraction limits, a near-field imaging technique was proposed, which improves the resolution limit up to a value of λ/4. A near-field microscopy technique was recently used to achieve a resolution of λ/3000 at 2.54 THz (Huber et al. 2008) for a single frequency and the order of λ/600 or a broadband system (Wachter et al. 2009). In the near future, these advanced techniques will be used in various fields, ranging from theoretical studies to biomedical applications.

4.3 THz Tomography

4.3.1 THz CT

4.3.1.1 Acquisition and Reconstruction of Tomography Signal

In the field of 3D imaging, x-ray CT is an excellent nondestructive 3D imaging technique that can be used in diverse applications such as medical diagnostics, biomedical materials research, and semiconductor manufacturing. An x-ray CT constructs the 3D image using 2D projection images that are acquired under varied incident angles. However, the low x-ray absorption by soft materials is a hurdle for the plastic, paper, and paintings (Recur et al. 2012). THz CT is introduced as a new 3D image modality that can easily be applied to soft materials (Ferguson et al. 2002). The essential idea, to acquire the cross-sectional images of a sample by using THz radiation, is identical to that of the conventional x-ray CT. A conceptual schematic of the THz CT is illustrated in Figure 4.7.

The key algorithm in the reconstruction is Radon transform, which was suggested by an Austrian mathematician Johann Radon. The function $R_\theta(r)$, which measures the projection data along the propagation path r, is a Radon-transformed signal of the projected shadow of the actual sectional image $f(x, y)$:

$$R_\theta(r) = \int\limits_{-\infty}^{\infty}\int\limits_{-\infty}^{\infty} f(x,y)\delta(r - x\cos\theta - y\sin\theta)dx\,dy, \tag{4.13}$$

where
  r is the offset distance from the axis of rotation
  θ is the projection angle
  δ is the Dirac impulse

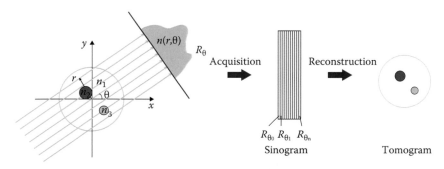

FIGURE 4.7   Principles of CT. The angle-dependent projection R_θ is measured along the 360°. The set of projected 2D images is called the sinogram. A 3D tomogram is reconstructed by the inverse Radon transform using the sinogram.

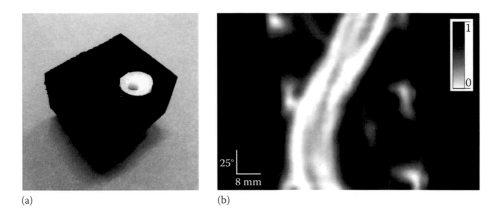

(a) (b)

FIGURE 4.8 (a) Photograph of a sample that was used for the demonstration of THz CT: parallelepiped-like black foam form (41 × 49 mm² in area) with two holes, each of 15 mm diameter (one hole was empty, and another hole contained a Teflon cylinder with a 6 mm diameter cylindrical air hole inside). (b) The sinogram with Nq = 72 projections (lines) and Nr = 128 samples per projection (columns). (Reprinted from Recur, B. et al., *Opt. Express*, 19, 5105, 2011.)

To acquire the tomography image, the projected data are measured along 360° by rotating the sample or emitter–detector pair. The collection of 2D projection images from different directions is called the sinogram. As an example, the mixed material sample and its sinogram with 72 projections are presented in Figure 4.8 (Recur et al. 2011). The sample is a parallelepiped-like black form with two holes: one of the holes is hollow, while another hole is filled by a Teflon cylinder whose refractive index is higher than that of the parallelepiped-like form. It is possible to acquire the 3D image from the sectional images on different longitudinal positions by using the reconstruction algorithm, the inverse Radon transform. The inverse Radon transform is given by

$$f(x,y) = \int\limits_{0}^{\pi}\int\limits_{-\infty}^{\infty} R_\theta(r)\delta(r - x\cos\theta - y\sin\theta)\mathrm{d}r\,\mathrm{d}\theta. \tag{4.14}$$

Sectional images along different axial positions can be used to reconstruct a 3D image that includes the internal structure of the sample.

4.3.1.2 THz CT System

THz CT uses focused collimated beams in conventional THz-TDS systems. The target is mounted on a rotation stage on the beam line to acquire the revolved projection data. As opposed to the conventional x-ray CT, which measures the intensities of CWs, a THz CT uses not only CWs but also pulsed waves. Owing to the fact that a continuous THz source has higher intensity than a pulsed source, the benefits of a high-power imaging device are conferred on the continuous mode (Zhang 2002). A pulsed mode that uses a broadband signal provides more information than a continuous mode. According to the well-known optical relation, signal's amplitude contributes to the absorption, while phase delay is associated with the refractive index (Born et al. 1999; Hecht 1998). This information, obtained by a variety of processing techniques, can be mapped onto 3D space.

4.3.1.3 Quality and Limitations of THz CT

Being the simplest CT reconstruction method, the Radon transform is also called a simple back-projection algorithm (BFP). This approach is characterized by a relatively low accuracy, because an image that is reconstructed using this method necessarily includes blurring. The filtered BFP (back projection of filtered projections), the most widely used algorithm, enhances image quality by proper signal processing and filtering before the reconstruction stage (Nguyen et al. 2005). In addition to adding filters to a simple Radon transform, other iterative reconstruction methods, such as the simultaneous algebraic reconstruction technique (SART) or the ordered subsets expectation maximization (OSEM) algorithm, are used (Recur et al. 2011). The reconstructed images that are obtained using these methods are presented in Figure 4.9 for different numbers of projection images. The results, obtained by using the iterative reconstruction algorithms, are better than those obtained using the filtered BFP. The resolution depends on the number of projection data values, and this dependence is especially prominent at the edges of a target. Depending on the reconstruction method that is used, strong streaks can appear in the case of nonpermeable materials (e.g., when the inside of the sample is metal). These artifacts are removed by selecting suitable signal processing methods (Zhao et al. 2000).

A significant number of signal processing and reconstruction methods for THz CT are borrowed from x-ray CT. Unlike x-rays, THz waves propagate and are refracted according to the properties of a media. Because the aforementioned algorithms assume that the effects of diffraction and Fresnel loss are negligible, distortion occurs in the reconstructed image by THz CT. It is also assumed that materials consist of a target that does not absorb and scatter intensively (Wang and Zhang 2004). These limitations are hurdles to a widespread application of THz CT. THz CT is difficult to apply in biomedical applications, where tomography is widely used owing to its high absorption property.

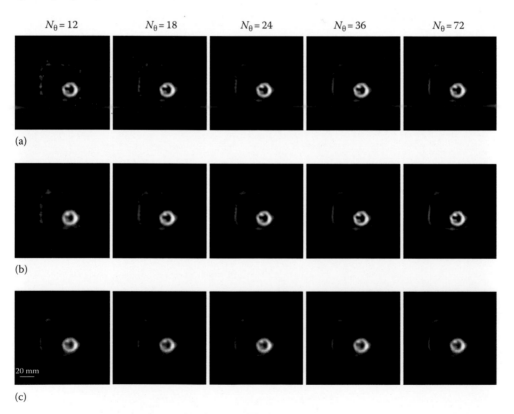

FIGURE 4.9 Reconstruction of the manufactured sample presented in Figure 4.8. Reconstructions were performed using the sinograms with 12, 18, 24, 36, and 72 projections and the BFP (a), SART (b), and OSEM (c) methods. Same scale has been used for all cross sections. (Reprinted from Recur, B. et al., *Opt. Express*, 19, 5105, 2011)

4.3.2 THz Diffraction Tomography

4.3.2.1 THz Diffraction from a Sample

THz DT is a generalized form of CT. CT is based on the geometrical model in which rays propagate in straight-line paths. There is no room for diffraction and scattering; however, a THz CT has to account for diffraction and scattering when the size of the sample's inhomogeneities approaches the wavelength. Because the wavelengths of THz are in the submillimeter range, the diffraction and scattering are common. When the sample is located across the path of a traveling planar wave (as is shown in Figure 4.10), the propagating wave is described by the scalar Helmholtz equation (Born et al. 1999):

$$\nabla^2 U(r) + k^2 n^2(r) U(r) = 0, \tag{4.15}$$

where
  $U(r)$ is the electromagnetic field
  $k = 2\pi/\lambda$ is the wave number
  $n(r)$ is the refractive index

It is expressed as

$$\nabla^2 U(r) + k^2 U(r) = -4\pi F(r) U(r), \tag{4.16}$$

where $F(r) = \dfrac{1}{4\pi} k^2 (n^2(r) - 1)$ is called the scattering potential of the medium. Equation (4.15) can then be rewritten as

$$\left(\nabla^2 + k^2\right) U^i(r) = 0 \tag{4.17}$$

$$(\nabla^2 + k^2) U^s(r) = -4\pi F(r) U^s(\mathrm{r}), \tag{4.18}$$

where $U(r)$ is the sum of the incident field $U^i(r)$ and scattered field $U^s(r)$. These differential equations can be solved by using Green's function. Yet, linear approximation methods such as the first-order Born approximation or Rytov approximation are sufficient in most cases. If the scattered field $U^s(r)$ is assumed to be small compared to $U^i(r)$, Equation (4.18) becomes

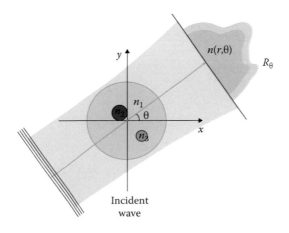

FIGURE 4.10 Schematic illustration of DT.

$$U^s(\boldsymbol{r}) = \int G(\boldsymbol{r}-\boldsymbol{r}')F(\boldsymbol{r}')U^i(\boldsymbol{r}')d^3r', \qquad (4.19)$$

which is the first-order Born approximation. The condition for using the first-order Born approximation is (Wang and Zhang 2004)

$$d\delta n < \frac{\lambda}{4}, \qquad (4.20)$$

where
d is the size of the sample
δn is the refractive index

This condition is valid only for small variations in the refractive index. The Rytov approximation assumes that the incident wave perturbation caused by the sample can be described by a change of phase in the reference wave. It is derived by using

$$U(\boldsymbol{r}) = \exp\big[\phi(r)\big] = exp\big[\phi^i(r) + \phi^s(r)\big] \qquad (4.21)$$

by combining Equations 4.17, 4.18, and 4.21

$$(\nabla^2 + k^2)U^i(\boldsymbol{r})\phi^s(\mathrm{r}) = U^i(\boldsymbol{r})\Big[\big(\nabla\phi^s(r)\big)^2 + F(r)\Big]. \qquad (4.22)$$

If the scattered complex phase $\phi^s(\mathrm{r})$ is small, the solution is

$$U^i(\boldsymbol{r})\phi^s(\boldsymbol{r}) = \int G(\boldsymbol{r}-\boldsymbol{r}')F(\boldsymbol{r}')U^i(\boldsymbol{r}')d^3r'. \qquad (4.23)$$

Further, the complex phase of the scattered field is given by

$$\phi^s(\boldsymbol{r}) = \frac{1}{U^i(\boldsymbol{r})}\int G(\boldsymbol{r}-\boldsymbol{r}')F(\boldsymbol{r}')U^i(\boldsymbol{r}')d^3r'. \qquad (4.24)$$

Although the Rytov approximation is more accurate than the first-order Born approximation, it requires the phase of the scattered field to vary slowly, relative to one wavelength (Wang and Zhang 2004).

4.3.2.2 THz Diffraction Tomography System

THz DT operates by sending a planar wave through the sample. The THz wave travels and scatters off from the sample surface. The scattered wave gives us information about the sample from which it was scattered. The THz DT system is based on the THz one-shot imaging system with a planar beam generating part and a sample position controller. The expanded THz waves illuminate the sample, and the waves are scattered from the sample. The scattered beam is measured by the nonlinear crystal that has a large area. In this case, the probe beam is also expanded using the telescope lens system to modulate the polarization with THz waves via EO crystal. The analyzer and CCD camera detect modulated probe beams through EO sampling. The sample can be moved and rotated by controlling the stages to obtain the sinogram.

THz DT measures the time-domain signal, whereas most of conventional DT systems operate in the frequency domain. For a clear qualification of the DT system, Young's double slit interference experiment was performed by Wang and Zhang, and the separation of constructive peaks versus the frequency showed a very good agreement with the theoretical predictions (Wang and Zhang 2004). The scattered waves were analyzed using the first-order Born approximation and the first-order Rytov approximation.

4.3.2.3 Quality and Limitations of THz Diffraction Tomography

Figure 4.11a and b shows the optical image and the geometrical scheme of the sample, respectively. The sample consisted of three rectangular polyethylene cylinders, each of which had a refractive index of 1.5 in the THz range. The various projections were measured by rotating the sample, and the 3D images were reconstructed using the first-order Born approximation and the first-order Rytov approximation. Wang and Zhang found that the first-order Rytov approximation provided a better result, which is presented in Figure 4.11c, and the image quality was increased followed by the frequency increment. However, because of system's SNR, the quality of the image could not become any higher for frequencies higher than 0.4 THz. Thus, the development of high-power imaging systems is highly desirable. The low output power of a commonly used source is a preemptive problem in need of a solution. Because the linear approximations permit only small changes in the refractive index, more research needs to be done to develop suitable reconstruction algorithms.

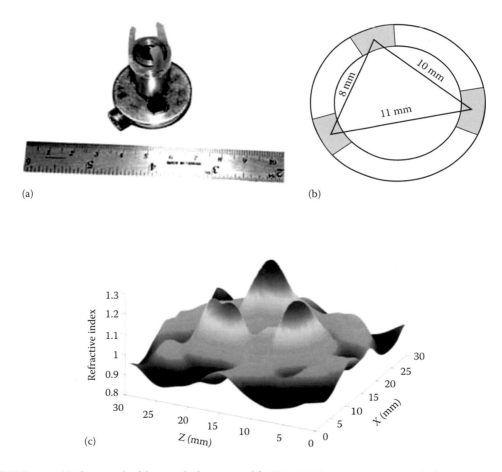

FIGURE 4.11 (a) Photograph of the sample that was used for THz DT. Test structure consisted of three rectangular polyethylene cylinders. (b) The sample's geometry. The rectangular cylinders had dimensions of 2.0 mm × 1.5 mm, 3.5 mm × 1.5 mm, and 2.5 mm × 1.5 mm (clockwise from top). (c) A reconstructed 3D image of the polyethylene cylinders. Each horizontal slice was reconstructed independently from DT and combined to form a 3D image. (Reprinted from Wang, S. and Zhang, X., *J. Phys. D Appl. Phys.*, 37, R1, 2004.)

4.3.3 Novel Technique for THz Tomography

4.3.3.1 THz Fresnel Lens Tomography

A Fresnel lens is a type of compact lens that was originally developed to reduce the weight of the lens. The design allows the construction of a lens that has a large aperture and short focal length without the mass and volume of material that would be required by a conventionally designed lens. Wang et al. demonstrated the use of the THz tomography system by using these frequency-dependent characteristics (Wang and Zhang 2003). A Fresnel lens was constructed by machining a dielectric material in a series of concentric circles of varying depth. The object's distance z of a Fresnel lens is given by

$$z = \frac{f_v z'}{z' - f_v} = \frac{r_p^2 z' v}{2cz' - r_p^2 v},$$ (4.25)

where
  r_p^2s the Fresnel zone period with the dimensions of area
  c is the speed of light
  f_v is the focal length
  z' is the image distance

Therefore, at each illumination frequency, the image formed at z' corresponds to a focused image of a plane through the target at a different depth, z.

Wang et al. performed an experiment in which three plastic sheets with different patterns and size were placed along the THz beam path, and their distances to the lens (corresponding to z) were 4, 6, and 16 cm, respectively. Inverted images of patterns in the sensor plane at a distance $z' = 5.7$ cm were measured at frequencies of 0.9, 1.1, and 1.4 THz, respectively (Figure 4.12).

Although this demonstration used broadband THz radiation as the imaging beam, and no spectroscopic information was acquired, such a tomography imaging concept is also applicable to a tunable narrowband imaging beam and can be applied in other frequency ranges, including the visible range.

4.3.3.2 Three-Dimensional THz Holography

A large number of THz imaging technologies have been developed in recent years. However, many of the proposed techniques avoid broad illumination of the object because these systems consist of a low-power THz source and only one detector, and thus they need to avoid coherent effects that may degrade the quality of an image. The 3D THz holography is an exception to this point and constitutes a novel extension of THz tomography research which is used in the time reversal of the Fresnel–Kirchhoff equation, which describes the diffraction of the reconstructed wave at the microstructure of the hologram. To generate holograms, both the intensity and the phase distribution must be recorded. Because of the coherent nature of THz-TDS, the simple THz technique has been adopted for use in holography.

In the 3D THz holography system that is shown in Figure 4.13a, THz pulses interact with the target, which diffracts the incident pulse. Then, the scattered pulses arrive at the detector and are measured by using the one-shot imaging method. The measured 2D THz image is then reconstructed as a 3D sample holographic image through the reconstruction process. Using the reconstruction algorithm based on the windowed Fourier transform, extremely rapid 3D imaging (Wang et al. 2003) has been achieved. As shown in Figure 4.13c, for the simplest case in which the polarization is neglected and the THz

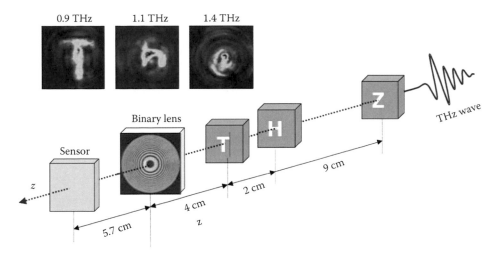

FIGURE 4.12 Schematic illustration of the tomographic imaging method that uses a Fresnel lens. Targets at various locations along the beam's propagation path are uniquely imaged on the same imaging sensor plane with different frequencies of the imaging beam. (Reprinted from Wang, S. and Zhang, X., *J. Phys. D Appl. Phys.*, 37, R1, 2004.)

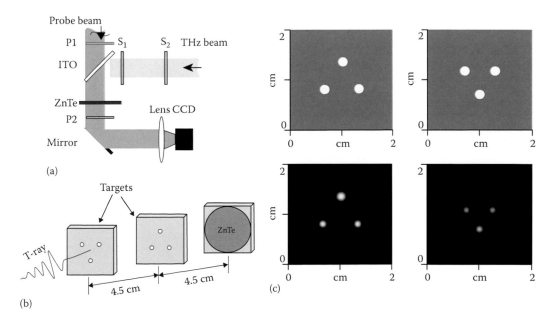

FIGURE 4.13 (a) Schematic presentation of the experimental setup for 3D THz digital holography. (b) The arrangement of targets. The diameter of each hole is 1.8 mm and the distances between holes are 6 mm. (c) Schematic of the target samples planes and their reconstructed hologram. (Reprinted from Wang, S. and Zhang, X., *J. Phys. D Appl. Phys.*, 37, R1, 2004.)

source region is excluded, it is easy to show that the THz pulse satisfies the scalar Helmholtz equation and Rayleigh–Sommerfeld diffraction formula (Wang et al. 2003):

$$\Gamma(P_0) = \frac{i}{\lambda} \iint_S \Gamma(P_1) \frac{\exp(-ikr_{01})}{r_{01}} \cos(\overline{n}, r_{01}) ds, \tag{4.26}$$

where

s is a reconstruction surface at which the diffracted THz pulses are measured

P_0 and P_1 are a point on the target and a point in the reconstruction plane, respectively

The parameter r_{01} is the distance between the points P_0 and P_1, and $\overline{n}$ is the reconstruction plane normal. Equation 4.26 is the basis for numerical hologram reconstruction, which leads to not only the distribution of the scattering center but also its reconstruction.

Furthermore, the development of pseudoholograms using CW techniques (Gorodetsky and Bespalov 2008; Mahon et al. 2006; Tamminen et al. 2010) and THz time-domain techniques (Gorodetsky and Bespalov 2010) has also been reported.

References

Abbott, D. and X. C. Zhang. 2007. Special issue on T-ray imaging, sensing, and retection. *Proceedings of the IEEE* 95: 1509–1513.

Bolduc, M., L. Marchese, B. Tremblay et al. 2010. Video-rate THz imaging using a microbolometer-based camera. Presented at *35th International Conference on Infrared Millimeter and Terahertz Waves*, pp. 1–2.

Boreiko, R. T. and A. L. Betz. 2009. Heterodyne spectroscopy of the 63 μm OI line in M42. *The Astrophysical Journal Letters* 464: L83.

Born, M., E. Wolf, and A. B. Bhatia. 1999. *Principles of Optics: Electromagnetic Theory of Propagation, Interference and Diffraction of Light*. Cambridge, U.K.: Cambridge University Press.

Chan, W. L., J. Deibel, and D. M. Mittleman. 2007. Imaging with terahertz radiation. *Reports on Progress in Physics* 70: 1325.

Cho, S. H., S. H. Lee, C. Nam-Gung et al. 2011. Fast terahertz reflection tomography using block-based compressed sensing. *Optics Express* 19: 16401–16409.

Choi, M. K., A. Bettermann, and D. W. Van Der Weide. 2004. Potential for detection of explosive and biological hazards with electronic terahertz systems. *Philosophical Transactions of the Royal Society of London. Series A: Mathematical, Physical and Engineering Sciences* 362: 337–349.

Chua, H. S., J. Obradovic, A. D. Haigh et al. 2005. Terahertz time-domain spectroscopy of crushed wheat grain. Presented at *IEEE MTT-S International Microwave Symposium Digest*, p. 4.

Duvillaret, L., F. Garet, and J. L. Coutaz. 1996. A reliable method for extraction of material parameters in terahertz time-domain spectroscopy. *IEEE Journal of Selected Topics in Quantum Electronics* 2: 739–746.

Englert, C. R., B. Schimpf, M. Birk et al. 2000. The 2. 5 THz heterodyne spectrometer THOMAS— Measurement of OH in the middle atmosphere and comparison with photochemical model results. *Journal of Geophysical Research* 105: 22.

Federici, J. F. 2012. Review of moisture and liquid detection and mapping using terahertz imaging. *Journal of Infrared, Millimeter and Terahertz Waves* 33: 1–30.

Ferguson, B., S. Wang, D. Gray, D. Abbot, and X. C. Zhang. 2002. T-ray computed tomography. *Optics Letters* 27: 1312–1314.

Fitzgerald, A. J., V. P. Wallace, M. Jimenez-Linan et al. 2006. Terahertz pulsed imaging of human breast tumors1. *Radiology* 239: 533–540.

Gorenflo, S., U. Tauer, I. Hinkov et al. 2006. Dielectric properties of oil–water complexes using terahertz transmission spectroscopy. *Chemical Physics Letters* 421: 494–498.

Gorodetsky, A. A. and V. G. Bespalov 2008. THz computational holography process and optimization. Presented at *Integrated Optoelectronic Devices*, p. 68930F9.

Gorodetsky, A. A. and V. G. Bespalov. 2010. THz pulse time-domain holography. Presented at *OPTO*, p. 760107.

Gowen, A. A., C. P. O'Donnell, C. Esquerre, and G. Downey. 2010. Influence of polymer packaging films on hyperspectral imaging data in the visible near-infrared (450950 nm) wavelength range. *Applied Spectroscopy* 64: 304–312.

Hecht, E. 1998. Hecht optics. *Addison Wesley* 997: 213–214.

Herman, G. T. 1995. Image reconstruction from projections. *Real-Time Imaging* 1: 3–18.

Ho, L., R. Müller, K. C. Gordon et al. 2009. Terahertz pulsed imaging as an analytical tool for sustained-release tablet film coating. *European Journal of Pharmaceutics and Biopharmaceutics* 71: 117–123.

Ho, L., M. Pepper, and P. Taday. 2008. Terahertz spectroscopy: Signatures and fingerprints. *Nature Photonics* 2: 541.

Hoshina, H., A. Hayashi, N. Miyoshi, F. Miyamaru, and C. Otani. 2009. Terahertz pulsed imaging of frozen biological tissues. *Applied Physics Letters* 94: 123901.

Hu, B. and M. Nuss. 1995. Imaging with terahertz waves. *Optics Letters* 20: 1716–1718.

Hua, Y. and H. Zhang. 2010. Qualitative and quantitative detection of pesticides with terahertz time-domain spectroscopy. *IEEE Transactions on Microwave Theory and Techniques* 58: 2064–2070.

Huber, A. J., F. Keilmann, J. Wittborn, J. Aizpurua, and R. Hillenbrand. 2008. Terahertz near-field nanoscopy of mobile carriers in single semiconductor nanodevices. *Nano Letters* 8: 3766–3770.

Hunsche, S., M. Koch, I. Brener, and M. Nuss. 1998. THz near-field imaging. *Optics Communications* 150: 22–26.

Ji, Y. B., E. S. Lee, S. H. Kim, J. H. Son, and T. I. Jeon. 2009. A miniaturized fiber-coupled terahertz endoscope system. *Optics Express* 17: 17082–17087.

Jiang, Z. and X. C. Zhang. 1999. 2D measurement and spatio-temporal coupling of few-cycle THz pulses. *Optics Express* 5: 243–248.

Jördens, C. and M. Koch. 2008. Detection of foreign bodies in chocolate with pulsed terahertz spectroscopy. *Optical Engineering* 47: 037003.

Kawase, K., Y. Ogawa, Y. Watanabe, and H. Inoue. 2003. Non-destructive terahertz imaging of illicit drugs using spectral fingerprints. *Optics Express* 11: 2549–2554.

Kim, G. J., S. G. Jeon, J. I. Kim, and Y. S. Jin. 2008. High speed scanning of terahertz pulse by a rotary optical delay line. *Review of Scientific Instruments* 79: 106102.

Kiwa, T., M. Tonouchi, M. Yamashita, and K. Kawase. 2003. Laser terahertz-emission microscope for inspecting electrical faults in integrated circuits. *Optics Letters* 28: 2058–2060.

Lee, A. W. M., B. S. Williams, S. Kumar, Q. Hu, and J. L. Reno. 2006. Real-time imaging using a 4.3-THz quantum cascade laser and a 320/spl times/240 microbolometer focal-plane array. *IEEE Photonics Technology Letters* 18: 1415–1417.

Lee, Y. K., S. W. Choi, S. T. Han, D. H. Woo, and H. S. Chun. 2012. Detection of foreign bodies in foods using continuous wave terahertz imaging. *Journal of Food Protection* 75: 179–183.

Li, B., W. Cao, S. Mathanker, W. Zhang, and N. Wang 2010. Preliminary study on quality evaluation of pecans with terahertz time-domain spectroscopy. Presented at *Proceedings of SPIE 7854*, p. 78543V.

Mahon, R. J., J. A. Murphy, and W. Lanigan. 2006. Digital holography at millimetre wavelengths. *Optics Communications* 260: 469–473.

Mantsch, H. H. and D. Naumann. 2010. Terahertz spectroscopy: The renaissance of far infrared spectroscopy. *Journal of Molecular Structure* 964: 1–4.

Mees, J., S. Crewell, H. Nett et al. 1995. ASUR-an airborne SIS receiver for atmospheric measurements of trace gases at 625 to 760 GHz. *IEEE Transactions on Microwave Theory and Techniques* 43: 2543–2548.

Mittleman, D., J. Cunningham, M. Nuss, and M. Geva. 1997a. Noncontact semiconductor wafer characterization with the terahertz Hall effect. *Applied Physics Letters* 71: 16–18.

Mittleman, D. M., M. Gupta, R. Neelamani et al. 1999. Recent advances in terahertz imaging. *Applied Physics B: Lasers and Optics* 68: 1085–1094.

Mittleman, D. M., S. Hunsche, L. Boivin, and M. C. Nuss. 1997b. T-ray tomography. *Optics Letters* 22: 904–906.

Nguyen, K. L., M. L. Johns, L. F. Gladden et al. 2005. Three-dimensional imaging with a terahertz quantum cascade laser. Presented at *High Frequency Postgraduate Student Colloquium*, pp. 101–104.

Park, J. Y., H. J. Choi, K. S. Cho, K. R. Kim, and J. H. Son. 2011. Terahertz spectroscopic imaging of a rabbit VX2 hepatoma model. *Journal of Applied Physics* 109: 064704.

Pickett, H. M. 2006. Microwave limb sounder THz module on aura. *IEEE Transactions on Geoscience and Remote Sensing* 44: 1122–1130.

Png, G. M., R. Flook, B. W. H. Ng, and D. Abbott. 2009. Terahertz spectroscopy of snap-frozen human brain tissue: An initial study. *Electronics Letters* 45: 343–345.

Recur, B., J. P. Guillet, I. Manek-Hönninger et al. 2012. Propagation beam consideration for 3D THz computed tomography. *Optics Express* 20: 5817–5829.

Recur, B., A. Younus, S. Salort et al. 2011. Investigation on reconstruction methods applied to 3D terahertz computed tomography. *Optics express* 19: 5105–5117.

Redo-Sanchez, A., G. Salvatella, R. Galceran et al. 2011. Assessment of terahertz spectroscopy to detect antibiotic residues in food and feed matrices. *Analyst* 136: 1733–1738.

Roeser, H. P., R. Wattenbach, E. J. Durwen, and G. V. Schultz. 1986. A high resolution heterodyne spectrometer from 100 microns to 1000 microns and the detection of CO (J = 7-6), CO (J = 6 5) and (C-13) O (J = 3-2). *Astronomy and Astrophysics* 165: 287–299.

Shen, Y. C., T. Lo, P. F. Taday et al. 2005a. Detection and identification of explosives using terahertz pulsed spectroscopic imaging. *Applied Physics Letters* 86: 241116.

Shen, Y. C., P. F. Taday, D. A. Newnham, M. C. Kemp, and M. Pepper 2005b. 3D chemical mapping using terahertz pulsed imaging. Presented at *Integrated Optoelectronic Devices*, pp. 24–31.

Tamminen, A., J. Ala-Laurinaho, and A. V. Räisänen 2010. Indirect holographic imaging: Evaluation of image quality at 310 GHz. Presented at *Proceedings of SPIE 7670*, pp. A1-A11.

Toft, P. A. and J. A. Sørensen. 1996. The radon transform-theory and implementation. Tekniske Universitet, Doctoral Dissertation.

Tonouchi, M. 2007. Cutting-edge terahertz technology. *Nature Photonics* 1: 97–105.

Usami, M., T. Iwamoto, R. Fukasawa et al. 2002. Development of a THz spectroscopic imaging system. *Physics in Medicine and Biology* 47: 3749.

van Exter, M. and D. Grischkowsky. 1990. Carrier dynamics of electrons and holes in moderately doped silicon. *Physical Review B* 41: 12140–12149.

Wachter, M., M. Nagel, and H. Kurz. 2009. Tapered photoconductive terahertz field probe tip with subwavelength spatial resolution. *Applied Physics Letters* 95: 041112.

Wang, S., B. Ferguson, D. Abbott, and X. C. Zhang. 2003. T-ray imaging and tomography. *Journal of Biological Physics* 29: 247–256.

Wang, S. and X. C. Zhang. 2003. Terahertz wave tomographic imaging with a Fresnel lens. *Chinese Optics Letters* 1: 53–55.

Wang, S. and X. Zhang. 2004. Pulsed terahertz tomography. *Journal of Physics D: Applied Physics* 37: R1.

Woodward, R. M., B. E. Cole, V. P. Wallace et al. 2002. Terahertz pulse imaging in reflection geometry of human skin cancer and skin tissue. *Physics in Medicine and Biology* 47: 3853.

Wu, Q., T. Hewitt, and X.-C. Zhang. 1996. Two-dimensional electro-optic imaging of THz beams. *Applied Physics Letters* 69: 1026–1028.

Xu, J. and X. C. Zhang. 2004. Circular involute stage. *Optics Letters* 29: 2082–2084.

Zeitler, J. A., P. F. Taday, D. A. Newnham et al. 2010. Terahertz pulsed spectroscopy and imaging in the pharmaceutical setting—A review. *Journal of Pharmacy and Pharmacology* 59: 209–223.

Zhang, X. 2002. T-ray computed tomography. *Hot Topic LEOS Newsletter*, pp. 1–4.

Zhao, S., D. D. Robeltson, G. Wang, B. Whiting, and K. T. Bae. 2000. X-ray CT metal artifact reduction using wavelets: An application for imaging total hip prostheses. *IEEE Transactions on Medical Imaging* 19: 1238–1247.

5

Compact Solid-State Electronic Terahertz Devices and Circuits

Jae-Sung Rieh
Korea University

Daekeun Yoon
Korea University

Jongwon Yun
Korea University

5.1 Introduction

The terahertz (THz) frequency band is attracting increasing interests from various sciences and engineering research communities these days. The band, which is roughly defined as 0.1–10 THz (some prefer the lower boundary defined as 0.3 THz to align with that of submillimeter), is attractive in two ways. Firstly, it is a relatively less developed territory, less exploited compared to the neighboring bands of microwave and optical spectra. The lack of proper devices to generate, detect, and control the THz signals at the band as well as the high attenuation rate of the THz wave at Earth's atmosphere can be mentioned as the reasons for the band to remain underdeveloped. Once considered as challenges, these factors now provide fertile opportunities for researchers seeking after new findings. Secondly, probably more important from a practical point of view, the band offers various unique properties that can be utilized for various useful applications: (i) THz wave shows high selectivity over different materials in its transparency, particularly exhibiting low penetration rate through water-containing substances; (ii) the band corresponds to resonance frequency of various molecules, enabling their spectroscopic detection when existing; (iii) THz wave is safe when irradiated on humans as it does not cause harmful photoionization on biological tissues; and (iv) when compared to the microwave band, the THz band provides shorter wavelength and higher frequency.

For these properties, the THz band has a great potential for various useful applications, which includes security and medical imaging, spectroscopy, sensors, product quality inspection, and broadband communication (Kim and Rieh 2012b; Rieh et al. 2011). Some of these have already been widely investigated, while others are still waiting for more practical deployment. In order to implement systems for these applications, the development of reliable and high-performance THz components is indispensible. While many different types of devices are required to compose THz systems, high-power THz source and low-noise THz detector are the two most important and also challenging ones to develop.

There are two different approaches to realize THz sources and detectors: one downward from optics and the other upward from electronics. The optical THz techniques, such as those based on femtosecond lasers and photomixing, have already been widespread for various applications such as time-domain spectroscopy and imaging. The electrical approach can be again categorized into two groups: one based on vacuum devices and the other based on solid-state devices. Various types of vacuum devices, such as klystron, magnetron, backward wave oscillator (BWO), traveling-wave tube (TWT), gyrotron, as well as free-electron laser (FEL), have been developed and are well suited for high-output-power generation. The solid-state electronic approach is based on solid-state devices such as diodes and transistors and mostly implemented with semiconductor technologies.

While both optical and vacuum electronic methods are relatively well established and quite mature, they are typically based on setup that is bulky, costly, and power consuming. They may work greatly for laboratory and special-purpose apparatus but hardly match the requirements for commercial products. On the other hand, the solid-state approach benefits from its small form factor, low power consumption, and low cost, which will be highly favored when commercial products are seriously considered. For bio-medical application as an example, the products deployed to hospital environment need to be competent in price while compact in volume if mobility is required. If handheld medical devices are considered for extreme mobility, low power consumption will be additionally desired. With all of these considered, solid-state solution best meets the requirements and thus will be highly demanded for THz generation and detection in practical biomedical applications.

This chapter is intended to describe the basic principles and current status of solid-state electronic devices for THz generation and detection. Diode-based and transistor-based approaches will be discussed in Sections 5.2 and 5.3, respectively. A brief introduction to packaging issues will be provided in Section 5.4, followed by a conclusion.

5.2 Diode-Based Approaches

Two-terminal devices, as represented by diodes, have been employed for both THz generation and detection for a relatively long time compared to transistors, mainly because diodes can operate at higher frequencies. There are various types of diodes that have been used for THz applications, and selected ones, both for generation and detection, will be described in this section.

5.2.1 Diode-Based THz Sources

THz generation with diodes exploits the negative resistance appearing across the diode terminals arising from various causes, such as transit time delay, inter-valley carrier scattering, and energy-level alignment. In this section, four types of widely accepted diodes for THz generation are discussed.

5.2.1.1 Gunn Diode

Gunn diodes make use of the Gunn effect to acquire the negative resistance (Eisele and Kamoua 2004). Many III–V compound semiconductors widely employed for electronics, such as GaAs, InP, and GaN, share a common property in the band structure. For them, the main valley in the E–k diagram is neighbored by a satellite valley in proximity, which exhibits slightly higher energy level than that of the main valley as depicted in Figure 5.1. As a result, when the electrons in the main valley are excited by external causes, such as external field from bias voltage, some electrons are transferred to the satellite valley when the acquired energy is large enough to overcome the barrier between the valleys. If the satellite valley shows a larger curvature near the bottom, which is typically the case, the electrons there will exhibit lower mobility than those residing in the main valley. As a result, the average mobility will be degraded

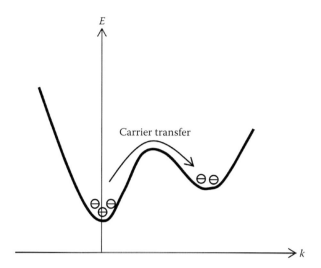

FIGURE 5.1   *E–k* diagram of semiconductors that show the Gunn effect. When electrons are transferred from the main valley to the satellite valley due to an external cause, the average mobility is degraded.

when such transfer of electrons occurs. For diodes based on this type of semiconductors, such transfer occurs when bias voltage across the diode exceeds a critical value, which reveals as a sudden drop in the current flowing through the device. This leads to a negative differential resistance established across the device. By properly terminating the two electrodes of the device, typically with a resonance circuit, oscillation is attained. The intrinsic factors such as device dimension and doping profile affect the oscillation frequency, which can reach a few hundred of GHz or higher.

There have been a great number of reports on Gunn diodes, some of which are shown in Figure 5.2a in terms of output power and oscillation frequency for various materials and different harmonic outputs. It shows that the output power may go up to nearly 20 dBm for 100–200 GHz range, and the frequency reaches around 500 GHz with output power of –10 to –15 dBm (Eisele 2010). It is worthwhile to note that Gunn diodes are widely adopted for commercial purposes and packaged products are available for various power and frequency ranges.

5.2.1.2 IMPATT and TUNNETT Diodes

IMPATT (impact ionization avalanche transit time) diodes attain the negative resistance from time delays caused by the avalanche multiplication and carrier transit across the device (Ino et al. 1976). An IMPATT diode is composed of a highly doped injection region containing a p–n junction, neighbored by a lowly doped drift region. When a large reverse bias is applied across the diode, carriers will be generated by the avalanche multiplication in the injection region and then drifted toward the terminal through the drift region, establishing a diode current. As the avalanche multiplication is basically a chain reaction of impact ionization, there will be a finite delay before enough number of carriers is generated. Additionally, there will be additional time delay for the generated carriers to transit across the drift region to reach the terminal. Hence, when external bias is abruptly applied, there will be a time delay caused by the avalanche and transit before significant current appears at the terminal. If the external bias is periodic, and the time delay exactly matches a half cycle, there will be a phase difference of 180° between voltage and current, leading to a negative resistance across the diode terminals. This negative resistance can be exploited to trigger oscillation from the diode. The oscillation frequency is dictated by the total delay and can be partly controlled by adjusting the length of the drift region and thus the transit time. On the other hand, the avalanche delay is a material property and can hardly be controlled once the material is fixed

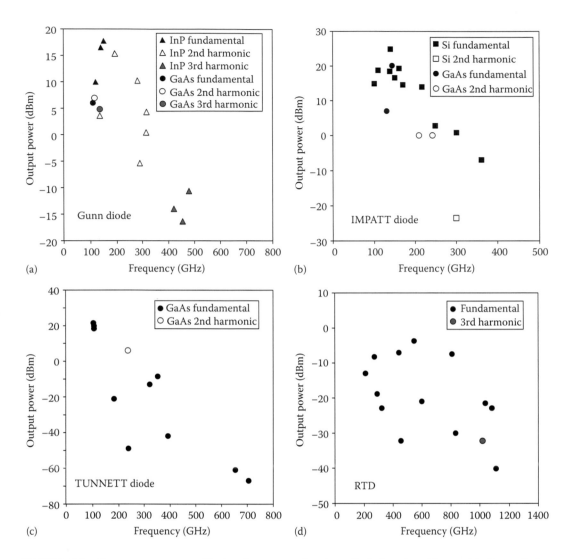

FIGURE 5.2 Output power vs. oscillation frequency of various diode-based oscillators: (a) Gunn diodes, (b) IMPATT diodes, (c) TUNNETT diodes, and (d) RTD.

for the diode. It is interesting to note that the avalanche delay is smaller for Si than GaAs, allowing Si IMPATT diodes faster than GaAs IMPATT diodes.

Instead of attempting to achieve a smaller avalanche delay for faster device operation, one may improve oscillation frequency by inducing breakdown current from a totally different mechanism that causes faster carrier generation. This is the idea behind the TUNNETT (tunneling transit time) diode, which is a variant of IMPATT diode (Nishizawa et al. 1978). It is well known that there are two mechanisms to cause breakdown in a p–n junction: avalanche multiplication and tunneling. For lower doping concentration near the junction, the former mechanism dominates, while the latter becomes dominant as the doping concentration increases. For carrier injection into the drift region, TUNNETT diodes make use of the tunneling current as opposed to the avalanche current used for IMPATT diodes. Generation of tunneling current involves a shorter time delay than for avalanche current, which results in faster operation achieved with TUNNETT diodes. It is further helped by the suppressed dispersion effect available with the tunneling mechanism. Consequently, TUNNETT diodes show higher operation frequency and lower noise level than IMPATT diodes.

Figure 5.2b shows the performance of IMPATT diodes reported in open literature in terms of output power and oscillation frequency. It shows the output power exceeding 20 dBm for 100–200 GHz range, and the maximum frequency goes up close to 400 GHz (Ino et al. 1976). On the other hand, GaAs TUNNETT diode operating beyond 700 GHz with fundamental mode has been reported as shown in Figure 5.2c (Nishizawa et al. 2008). However, the output power of TUNNETT is lower than that of IMPATT diodes and rather rapidly reduced with increasing frequency.

5.2.1.3 Resonant Tunneling Diode

Resonant tunneling diodes (RTDs) achieve the negative resistance from aligning energy levels in a quantum well with those outside (Asada et al. 2008). When semiconductor layers with different energy bandgaps are stacked, a quantum well can be formed in the conduction band (and valence band) and discrete energy levels will be created inside. Depending on the bias applied across the stack of the layers, the energy levels inside the quantum well can be aligned (or misaligned) with the conduction band edge of the adjacent semiconductor region across the barrier, which will lead to increase (or reduction) of the tunneling current. Figure 5.3 illustrates the cases with misaligned and aligned energy levels. Therefore, as the bias voltage is swept from low to high, there may exist a local reduction of current level if the increase of bias tends to further misalign the energy levels in that range. This will cause a negative differential resistance across the two terminals located on both sides of the stack, which can be exploited to generate oscillation if proper bias voltage is applied. The oscillation frequency is affected by the electron residence time in the quantum well and the transit time toward the terminal, while device parasitics also affect the device speed.

The oscillation frequency of RTD is known to be the highest among the diode-based oscillators. According to Figure 5.2d that shows the accumulated data points recently reported for RTDs, operation at 1111 GHz has been realized with fundamental mode oscillation (Cojocari et al. 2012). The output power of RTDs is relatively smaller than those of Gunn or IMPATT diodes at low frequency range, though.

Figure 5.4 compares the performance of all four types of diodes discussed earlier in a single plot. It is clearly shown that RTDs exhibit the highest frequency. It is also interesting to note that the roll-off of the output power with increasing frequency is not so pronounced for RTDs when compared to the other devices. TUNNETT diodes show high operation frequency, but their output power rapidly drops with

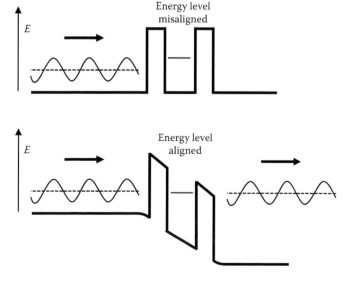

FIGURE 5.3 Band diagram of a typical RTD. The resonant tunneling occurs when the energy level in the quantum well is aligned with energy levels outside the well.

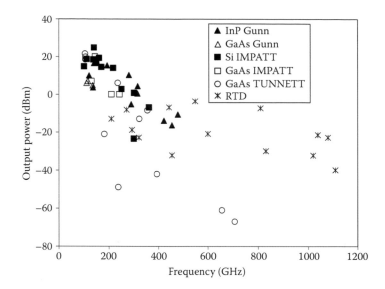

FIGURE 5.4 Performance comparison of various diode-based oscillators.

increasing frequency. For lower-frequency region below around 500 GHz, however, Gunn and IMPATT diodes would be favored for high-power operation, with slightly higher power with IMPATT diodes.

5.2.2 Diode-Based THz Detectors

There are various ways successfully developed and employed to detect electromagnetic waves. Firstly, photocurrent generation under illumination can be used for detection. This is an effective scheme for the detection of optical range of spectrum as is used for photodiodes, but not quite suitable for THz detection as prohibitively small carrier energy transition is required for this band. Secondly, thermal effect caused by THz radiation can be used for detection. This method is known to work nicely for THz detection. Detectors based on this mechanism include bolometers, pyroelectric detectors, and Golay cells, which detect the change in resistance, charge storage across a capacitor, and volume of contained gas, respectively, over temperature variation. Thirdly, rectification of alternating current (ac) signal (originally converted from EM wave by antenna) into direct current (dc) voltage can be used for detection, as in the case with the square-law detectors. This method has been routinely adopted for standard electronics and is making its way toward THz region as the device cutoff frequency improves. Finally, heterodyne technique can be used, in which signal is down-converted by a mixer and detected at lower frequency with rather a conventional ac detector. This relaxes the operation frequency requirement for the detector at the back end of the receiver chain. In fact, most of the diode-based THz detectors, such as SBD (Schottky barrier diode), HEB (hot-electron bolometer), and SIS (superconductor–insulator–superconductor) mixer, are basically operating as mixers based on the heterodyne technique, and this will be discussed in this section.

5.2.2.1 SIS Mixer

A SIS mixer is composed of a thin insulating layer sandwiched by two superconducting layers, for which Nb, NbN, and NbTiN are frequently used. When cooled down below the critical temperature, electrons form Cooper pairs inside the superconducting layer, and the density of states surges near the edge of conduction and valence bands. When a bias voltage is applied across the diode, the relative position between the bandgaps of the two superconductor layers is shifted as shown in Figure 5.5. If the bias becomes nearly equal to the bandgap, the valence band of one side and conduction band

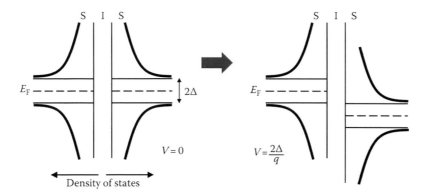

FIGURE 5.5 The band diagram of a SIS mixer. When the valence band of the left side and the conduction band of the right side of the superconductor layers are aligned, an abrupt increase in the tunneling current occurs.

of the other side of the superconductor layers would be aligned, leading to an abrupt increase in the tunneling current. This causes a strong nonlinearity in the I–V relation of the terminal current. If local oscillation (LO) radiation is added, photon-assisted tunneling takes place, and the I–V relation will show a steplike profile with the nonlinearity maintained. When a signal is applied to the device, mixing occurs via the nonlinearity and the signal is down-converted. It is noted that the required LO power is quite low for SIS mixers, around the range of a few μW. More importantly, SIS mixer is known for extremely low noise level, the noise temperature measured well below 1000 K around 1 THz for typical cases as shown in Figure 5.6a. For this reason, SIS mixers are widely deployed for astronomical heterodyne receivers where very high sensitivity is critical to capture the faint signals coming from the space. The upper frequency limit is relatively low, though, compared to other types of diode mixers to be described in Sections 5.2.2.2 and 5.2.2.3, and the measurements have been reported up to around 1.2 THz (Karpov et al. 2007).

5.2.2.2 Superconductor Hot-Electron Bolometer

Although any bolometer can be operated as a mixer, superconductor HEBs are particularly favored for THz detection owing to its large bandwidth reaching up to several THz. A typical superconductor HEB is composed of a superconductor microbridge formed across two metal pads, where Nb, NbN, and NbTiN are commonly used as the superconducting material as in the case of SIS mixers. When the bridge is heated with external radiation, the temperature is locally increased, which induces a transition from superconducting state to the normal state for the region radiated. As a result, resistance of the bridge suddenly increases, which affects the current level flowing through the device. Because the electrons are heated up faster than the phonons in the bridge due to the different heat capacity, the device is called "hot-electron" bolometer. Although the change in the resistance can be exploited as direct detection, the device is typically operated as a mixer by radiating LO power in the presence of the radio frequency (RF) signal through the bridge, leading to the generation of intermediate frequency (IF) signal for final detection. The main advantage of HEBs over SIS mixers is the higher operation frequency. As can be seen from Figure 5.6b, the operation frequency can reach over 5 THz (Zhang et al. 2010). It is noted though that the noise level of HEB mixers is slightly higher than that of SIS mixers, the noise temperature being typically measured around 1000 K near 1 THz.

5.2.2.3 Schottky Barrier Diode

An SBD is basically a metal–semiconductor junction diode. Being a majority carrier device, it inherently exhibits faster switching operation than p–n junction diodes owing to smaller reverse recovery time. As a semiconductor material, GaAs is typically selected in favor of its higher

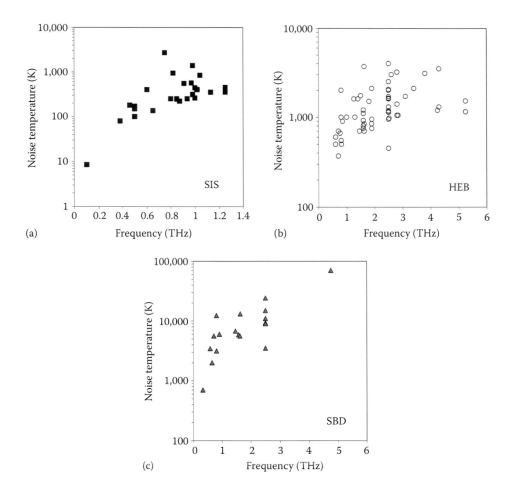

FIGURE 5.6 Output noise temperature vs. operation frequency of various diode-based mixers as a detector: (a) SIS mixers, (b) HEB mixers, and (c) SBD mixers.

operation speed, but Si is also used especially when compatibility with CMOS process is desired. The highest cutoff frequency reported so far for GaAs SBDs is 11.1 THz (Thomas et al. 2005), while that for Si SBDs is around 4 THz (Chen et al. 2010). To increase the operation frequency, the control of capacitance and parasitic resistance is critical, making the layout optimization of the device a critical step in device design. Because SBDs show nonlinear I–V characteristics as is the case for typical semiconductor junction diodes, mixing can be achieved with proper bias voltage applied. Compared to HEB or SIS mixers, the noise temperature is rather high, being measured around several thousands of K as can be seen in Figure 5.6c. However, its operation frequency is much higher than SIS mixers and similar to HEB mixers. More importantly, SBDs can be operated at room temperature, and they are compatible with conventional semiconductor process technology. These two features make SBD mixers highly desirable option for commercial products, together with transistor-based mixers to be discussed shortly.

Figure 5.7 compares the performance of all three types of diode mixers. It is clearly shown that SIS mixers show the lowest noise level, while HEB mixers would be most favored for high operation frequency. SBD mixers show similar operation frequency to HEB mixers but apparently with higher noise level due to room-temperature operation.

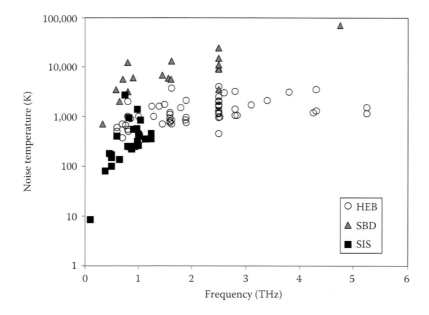

FIGURE 5.7 Performance comparison of various diode-based mixers as detectors.

5.3 Transistor-Based Approaches

While solid-state diodes can be favorably used for both THz sources and detectors as described earlier, there is one inherent critical limit for them: they are passive devices and thus provide no gain. In contrast, transistors are active devices and gain can be provided. The availability of gain is highly favored for both sources and detectors. For source side, the output power from the oscillator can be further boosted up by placing a power amplifier at the output. Also, much more various oscillator topologies can be realized because each transistor serves as a building block and can be connected in various ways. For detector side, a low-noise amplifier (LNA) preceding the other blocks will significantly lower the overall noise level of the receiver. Also, active mixers can be realized, whose positive conversion gain will suppress the noise from the following stages.

Historically, one main barrier for the transistors to enter the THz arena has been their limited operation frequency. However, continuous scaling and material innovation of semiconductor process technology over the past decades has led to transistors now operating well beyond 100 GHz and up to 1 THz with some kinds (Lai et al. 2007; Plouchart 2011; Rieh et al. 2004; Urteaga et al. 2011). It would be useful to take an overview of modern transistor technologies, and a brief comparison between them will be provided before the transistor-based circuits for THz source and detector applications are described.

5.3.1 Comparison of Transistor Technologies

Modern transistor technologies can be categorized into two groups depending on the types of semiconductor used: III–V technologies and Si-based technologies. They will be described separately in Sections 5.3.1.1 and 5.3.1.2.

5.3.1.1 III–V Transistor Technologies

Compound III–V semiconductors such GaAs and InP, as well as various alloys based on these semiconductors, are known to have superior electron transport characteristics compared to the main stream Si.

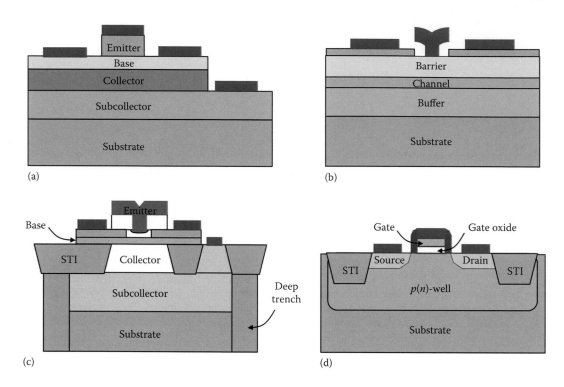

FIGURE 5.8 Simplified cross sections of various types of transistors: (a) HBT, (b) HEMT, (c) SiGe HBT, and (d) Si MOSFET.

This will lead to faster devices when employed for transistors. Additionally, transistors based on GaAs or InP tend to show higher breakdown voltage than Si transistors owing to the larger bandgap, allowing higher supply voltage and larger signal swing. This will lead to higher power operation. If large bandgap semiconductor such as GaN is employed, the output power even much higher can be achieved. III–V technologies are beneficial for passive devices such as transmission lines, inductors, and capacitors in high-frequency applications as well. GaAs or InP substrate, typically used for III–V semiconductor technologies, shows far higher resistivity than Si substrate, owing to the larger bandgap and consequent smaller intrinsic carrier concentration. This will significantly suppress the loss of those passive devices formed on the substrate, which serves as a major advantage. With these intrinsic advantages for high-frequency applications, III–V semiconductor systems have long been favored for very-high-frequency operation. There are two major types of transistors based on III–V semiconductors: HBT (heterojunction bipolar transistor) and HEMT (high-electron-mobility transistor), as will be discussed later separately. Typical cross sections of HBTs and HEMTs are illustrated in Figure 5.8, together with those of Si-based devices.

HBT is basically a bipolar transistor, the difference from the conventional ones being the larger bandgap for emitter than base. High current gain can be achieved with the larger emitter bandgap, but some of the gain can actually be traded for higher speed and lower noise, the two critically desired features for high-frequency applications, by reducing the base width and increasing the base doping. Compared to HEMTs, HBTs tend to show higher current driving capability and larger transconductance g_{m}, as are typical advantages for bipolar transistors. The fabrication steps, on the other hand, are more complicated as multiple epitaxial layers and etching steps are required because they are basically a vertical device. There are two commonly used measures for the transistor operation frequency, f_{T} and f_{max}, which refer to the frequency point where current gain and power gain, respectively, becomes unity. While both parameters are widely used, it is generally accepted that f_{max} is a more relevant parameter when RF application is concerned. As of today, the highest reported f_{max} for HBT is 1 THz (Urteaga et al. 2011) as reported by Teledyne based on their recently developed 0.25 μm InP HBT technology. Although this number is not to say that circuits based on this device can properly operate at 1 THz with enough gain, it is a good indication of transistor technology gradually entering the THz arena.

HEMT is a type of FET (field-effect transistor), which employs an undoped channel region for performance enhancement. The absence of ionized dopants in the channel region would significantly improve the mobility of the carriers traveling in the channel due to reduced scattering, resulting in improvement in both speed and noise of the device. In particular, low-temperature noise level is extremely low since both phonon scattering and impurity scattering are significantly suppressed, which makes HEMT the device of choice for receiver front end when extremely low noise performance is desired. The metal gate and the channel region are separated by a barrier layer with a larger bandgap, which supplies carriers to the channel region and also provides isolation between the gate and channel (see Figure 5.8b). Its performance is strongly dependent on the gate length, and E-beam lithography is widely adopted for the gate patterning. Also, as an effort to reduce the extrinsic gate resistance, which affects both speed and noise, gate metal with T-shaped cross section is routinely employed. Compared to HBT, the fabrication steps are simpler because less lithography steps are required. HEMT is currently considered the fastest type of transistor, as indicated by the record f_{max} of 1.2 THz achieved by Northrop Grumman based on their 35 nm InP HEMT technology (Lai et al. 2007).

5.3.1.2 Si-Based Transistor Technologies

Although lots of favorable features for high-frequency applications are available for III–V technologies as discussed earlier, they suffer from drawbacks of their own. Fabrication processes for the III–V technology typically result in nonplanar structure when completed, which partially exposes active regions of devices. With the absence of perfect passivation, such exposure may result in poor surface/interface properties near the current path, enhancing leakage current and degrading the long-term device reliability. They will be further aggravated by the typically higher defect density of III–V semiconductors than that of Si. Another fact that may be mentioned as affecting the reliability is the lower thermal conductivity of GaAs and InP (0.46 and 0.68 WK^{-1} cm^{-1}) than that of Si (1.41 WK^{-1} cm^{-1}, all at T = 300 K). Lower thermal conductivity tends to impede heat dissipation through the substrate, leading to higher device junction temperature for a given power dissipation, accelerating device wear-out and degrading device reliability. Finally, the fabrication cost of the III–V technology is on the unfavorable side, mainly due to the expensive wafers with small diameter, costly epitaxial growth steps, and the need for low-throughput E-beam lithography process. Hence, Si-based technologies can be a favored option if commercial application is considered (Rieh et al. 2012).

Two major Si-based technologies are SiGe HBT (or BiCMOS) technology and Si CMOS technology. SiGe HBT is basically a variant of Si bipolar transistors, which contains a small amount of Ge in the base region for reduced bandgap. The advantages of the smaller base bandgap are similar to those of larger emitter bandgap adopted for InP HBTs, rendering SiGe HBTs exhibit higher speed and lower noise compared to conventional Si bipolar transistors. When compared with Si MOSFETs, it benefits from the advantage of being bipolar transistor, such as higher current drivability and larger g_m. Besides, unlike Si MOSFET in which the major current path is formed near the trap-rich interface, the current vertically flows away from interface region, leading to superior $1/f$ noise characteristics. Also, its performance is determined by vertical dimension, which is better controlled than lateral dimension that is affected by lithographic variation, resulting in a better device matching than Si MOSFET (see Figure 5.8c and d). The record performance of SiGe HBTs at this point in terms of f_{max} is 500 GHz, developed by IHP based on its 0.25 μm technology (Heinemann et al. 2010).

Si MOSFET is the most prevailing device of today for various electronic applications. It is true that Si MOSFET has not been considered as a serious contender for high-frequency application until recently because of its inferior electron transport characteristics compared to III–V semiconductors. However, the continued scaling of Si MOSFETs, driven by the strong need for high-speed digital applications, together with the highly developed fabrication process technology of Si, has made Si MOSFET a competitive device even for high-frequency applications. The operation speed of the recently developed Si MOSFETs in terms of f_{max} exceeds 400 GHz (Plouchart 2011), allowing circuits based on the devices well beyond 100 GHz. When combined with the well-known merits of Si MOSFET technologies, such as easy integration with other circuit blocks including digital parts, low manufacturing cost, and highly developed design environment, Si MOSFETs are expected to emerge as a competent candidate for the THz applications.

5.3.2 Transistor-Based THz Source Components

Oscillators are used for signal generation and thus serve as a core circuit block for THz sources. There are other circuit blocks that can be integrated together with oscillators to complete a THz source. For example, when additional boost up of output power is needed, a power amplifier may follow an oscillator. When a further raise of output frequency is desired, a frequency multiplier may be added. In this section, an overview of oscillators, power amplifiers, and frequency multipliers based on transistor technology is presented.

5.3.2.1 Oscillator

While diode-based oscillators utilize the negative resistance obtained from the device-level characteristics of diodes, transistor oscillators attain the negative resistance from the circuit-level property of transistor-based networks. Terminating the nodes showing a negative resistance with a proper resonance circuit composed of inductors and capacitors or equivalent transmission-line components completes an oscillator, its oscillation frequency being determined by the resonance frequency of the resonator. Widely adopted for THz oscillators based on such configuration are LC cross-coupled oscillators and Colpitts oscillators, while some prefer the latter for its lower-phase-noise property. The operation of ring oscillators is based on rather a different mechanism, where a 180° phase inversion through a series of cascaded inverters leads to positive feedback and resultant oscillation. Although ring oscillators benefit from their small area due to the absence of a resonance circuit, they typically show high phase noise and thus rarely adopted for mm-wave and higher bands. One of the main advantages of transistor-based oscillators is their relatively large tuning range, a required feature for voltage-controlled oscillator (VCO), which can be accomplished from the use of varactors in the resonator circuit and/or the bias level control.

A gain barely larger than unity may be sufficient to trigger oscillation in transistor oscillators, making it possible for them to oscillate near f_{max} of the device used. As a result, compared to amplifiers that require device f_{max} at least two to three times of their operation frequency for sufficient gain, oscillators can work at much higher frequencies for a given process technology. Moreover, their operation frequency can be further boosted up by taking their harmonics as the output signal instead of the fundamental signal. For this purpose, the push–push technique is widely adopted when the second harmonic of an oscillator is to be extracted. At the common node of a differential pair that serves as the oscillator core, only even-mode signals appear because odd-mode signals, including the fundamental mode, are suppressed. By taking the common node as the output with an additional simple filtering circuitry, a push–push oscillator provides the second harmonic of the oscillator core as the output signal. For higher operation frequencies, triple-push technique can be employed, too, where three symmetric oscillator cores are connected and the output signal is extracted from the common node where only third harmonic (and its multiples) appears. In a similar way, n-push oscillators can be implemented by taking the nth harmonic of the oscillator cores, although the output power level will be lowered as n is increased.

As a matter of fact, lowered output power for increased frequency is a general tendency of electronic signal sources. Hence, THz oscillators typically suffer from low output power more severely than lower-frequency oscillators. As high output power is desired for most THz applications, including imaging and communication, there is a need to boost up its output power. An addition of a power amplifier is one way as will be described in the next section, but it is limited by the operation frequency of amplifiers. An alternative way to increase the output power is the power combining over multiple oscillators. Since individual oscillators tend to oscillate at random phase, simply combining the powers from oscillators would not work out to increase the power. Phase synchronization between the oscillators is required for this purpose. Various methods have been proposed for locking of the output of oscillators for this purpose (Sengupta and Hajimiri 2012; Stephan 1986), which will result in power increase when combined.

Figure 5.9 shows the recently reported data for transistor oscillators operating beyond 100 GHz. The numbers beside the data point represent the number of harmonics, while c indicates that power combining is practiced. The highest fundamental oscillation frequency reported so far is 573 GHz based on InP HBTs (Seo et al. 2011), while Si CMOS oscillator operating up to 553 GHz has been reported based on fourth harmonic (Shim et al. 2011). On a separate device, an output power near 0 dBm has been achieved up to 300 GHz based on Si CMOS technology without additional power boosting with amplifier (Tousi et al. 2012).

5.3.2.2 Power Amplifier

One of the main advantages of employing transistors over diodes is the availability of amplification, for both sources and detectors. For the source side, amplifiers are employed to boost the output power further up after signal is generated from an oscillator or any other method. One restriction though for this method is that it can cover only up to the operation frequency of the amplifier itself. Nevertheless, amplifier operation frequency keeps increasing owing the recent rapid progress in semiconductor technologies, pushing up the extent of frequency range that can benefit from the presence of amplifiers.

With the presence of the amplifier at the last stage, the output power of the preceding source is enhanced by a factor as large as the gain of the amplifier. However, the gain begins to drop if the input power becomes significantly large because of the nonlinearity of the amplifier. As a result, the output power becomes saturated if input power is too large, which is a classic problem of the amplifier nonlinearity. The saturated power tends to be smaller for amplifiers intended for high-frequency operation,

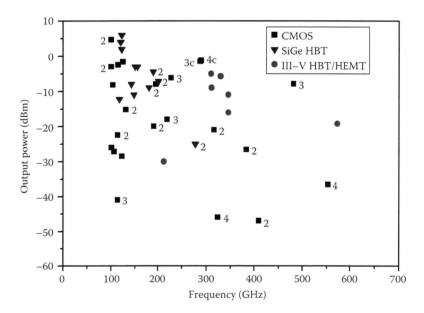

FIGURE 5.9 Output power of various transistor-based oscillators shown as a function of oscillation frequency. The numbers beside each data point indicate the number of harmonics. If a number is not shown, the oscillation is based on fundamental mode. The letter c indicates the data point that involves power combining technique.

because the breakdown voltage of the device developed for maximum operation frequency is typically reduced in favor of speed, leading to a smaller power-handling capacity.

Although there are various techniques to improve the linearity of the amplifiers through device-level optimization, another approach widely employed to overcome the saturated output power limit is to combine the output powers from multiple amplifiers connected in parallel. More specifically, the input signal is divided and streamed into multiple power amplifiers via power dividers, and then amplified in parallel, and consequently combined through a power combiner, leading to large power at the output. The output power will be theoretically N times of the saturation power if N amplifiers are connected in parallel. The power combiner is basically a reciprocal of a power divider, and thus, the same configuration can be used. It is critical, though, that the length of the path along the divider and the combiner needs to be identical for each power amplifier since phase should be coherent to obtain maximum output.

Many different power combining methods have been proposed for this purpose. Binary power combiner with T-junction is a simple yet effective way for power combining with coherent phase but suffers from poor isolation between the input branches. Wilkinson power combiner provides a symmetric combination with excellent isolation but requires rather a large area due to the need for quarter-wavelength section if frequency is not high enough (Wu et al. 2006). Transformer-based combiner can simultaneously provide power combining and impedance matching, which is an attractive feature since the impedance matching is another major issue for power amplifiers and power combining (Tai et al. 2012). Note that it is generally the case that output impedance becomes smaller and drifts away from 50 Ω as the number of devices and amplifiers connected in parallel increases, leading to a severe mismatch problem. Spatial power combining method has also been extensively studied (Harvey et al. 2000), where the powers from multiple signal sources are combined after being radiated into the free space through individual antennas.

Figure 5.10 shows the recently reported results from power amplifiers based on various technologies. The numbers beside each data point indicate the number of power combining if used. An impressive result was recently reported operating up to 650 GHz with output power of 4.8 dBm with two-way power

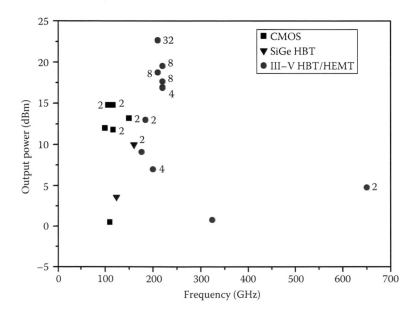

FIGURE 5.10 Saturation power vs. operation frequency recently reported for power amplifiers based on various technologies. The numbers beside each data point indicate the number of power combining. If a number is not shown, no apparent power combining is used for the data point.

combining, fabricated in InP HEMT technology from Northrop Grumman Company (Radisic et al. 2012a). Another notable data point is the output power of 22.7 dBm obtained at 210 GHz, also based on InP HEMT technology with 32-way power combining (Radisic et al. 2012b). When integrated with an oscillator source, they can serve as independent high-power THz sources.

5.3.2.3 Active Multiplier

One way to boost the oscillation frequency of oscillators is to use the harmonics of the generated signals based on push–push or higher-order push techniques as mentioned earlier. An alternative way to elevate the oscillation frequency is to employ frequency multipliers located after the oscillator. Of course, the multipliers can also be used in combination with the harmonic technique.

When a high-order multiplication is implemented by cascading multiple stages of multipliers, it is common to insert amplifiers between the multipliers to recover the power level that is degraded for every multiplication stage. For such a multiplier chain composed of cascaded multipliers and amplifiers, the last stage can be either a multiplier or power amplifier depending on the specific application considered. When high output power is a priority, a power amplifier will be at the last stage, while a multiplier will be located at the end if high-frequency operation is desired. For THz applications, the latter is usually the case, the last multiplier stage further boosting up the frequency beyond the operation frequency limit of amplifier for a given process technology.

Although passive devices such as diodes can be still used as the core device for a multiplier, transistors are usually adopted to implement active multipliers when transistor technology is available for the THz system. Advantages of active multipliers over passive ones are the potentially smaller conversion loss and larger flexibility in choosing bias condition for optimal operation as the transistor is a three-terminal device. Several approaches are commonly adopted for transistor-based multipliers, some of which are introduced as follows. One popular method is to exploit the nonlinearity of transistor operation, a similar approach as the diode-based multipliers (Lewark et al. 2011). After going through a nonlinear stage, the signal will show harmonic components in addition to the fundamental one. With proper filtering, only wanted harmonic will survive at the output. Another approach is to extract even-mode harmonics at the common node of a transistor differential pair, a similar technique as the one used for push–push oscillators (Kallfass et al. 2011). Doublers and quadruplers or even higher even-mode multipliers can be realized in this way. A method based on the principle of self-mixing with a mixer is also possible (Valenta et al. 2013). If the input signal is injected into the two input nodes of a mixer simultaneously, the input signal is self-mixed and a square term appears at the output node. As the square term leads to a frequency twice of the input frequency, such configuration can be used as a doubler. Finally, injection-locked frequency multipliers can be mentioned (Monaco et al. 2010). If an input signal is injected into an oscillator with a free-running frequency that is close to the integer multiple of the input signal frequency, the oscillator will be locked to the input, and the output frequency will be exactly the integer multiple of the input frequency, leading to a frequency multiplication.

Figure 5.11 shows the performance of the recently reported active multipliers based on both Si and III–V technologies. A source with output frequency as high as 820 GHz has been reported based on a 45× multiplier chain in SiGe technology (Ojefors et al. 2011), which is composed of two triplers and one quintupler, together with two amplifiers immersed in the chain. Although the power emerging at the multiplier output is generally smaller for higher frequency, an impressive output power around 0 dBm is reportedly obtained over 300 GHz (Lewark et al. 2012).

5.3.3 Transistor-Based THz Integrated Detector Components

Transistor is a versatile device, and it can provide various functions required to improve the detection performance, which is a favored aspect of transistor-based detection over diode-based one. There are two types of integrated detector architectures based on transistors, direct and heterodyne. Conceptual

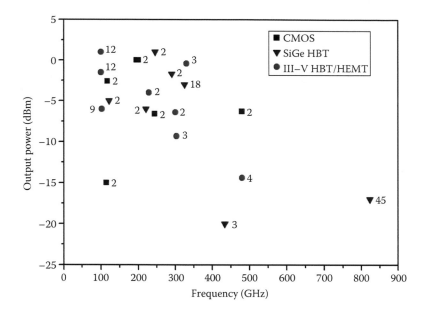

FIGURE 5.11   Output power vs. operation frequency recently reported for frequency multiplier (chain) based on various technologies. The numbers beside each data point indicate the total multiplication factor.

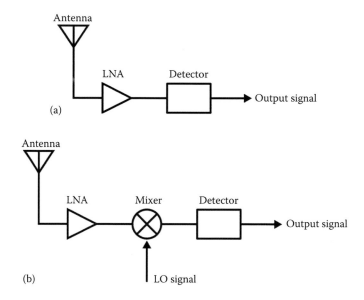

FIGURE 5.12   Typical block diagrams for (a) direct and (b) heterodyne integrated detector.

block diagrams for the direct and the heterodyne integrated detectors are shown in Figure 5.12. For the direct case, detection is made at RF level, while with the heterodyne case, detection is made at IF level after down-conversion. All the circuit blocks can be implemented with transistors for either case. In this section, an overview of the circuit components constituting the integrated detectors is provided. Although LO is also a core part of heterodyne architecture, it will not be discussed in this section as it can be realized by the source previously described.

5.3.3.1 Low-Noise Amplifier

The noise level of a system is determined by the circuit located at the first stage of the chain, a principle that also applies to the integrated detectors. If amplifiers are not available for integrated detectors, the forefront circuit needs to be either a square-law detector or mixer, whose noise level is on the high side, unfortunately. However, the presence of LNA at the front end will dramatically lower the system noise level. In fact, as will be seen shortly, noise-equivalent power (NEP) can be lowered by a factor of orders of magnitude by placing an LNA at the front end. While the effect can be compared to the case of cooling down the system to cryogenic temperature, which is commonly adopted for noise reduction in diode-based systems, no special setup such as cryostat or liquid nitrogen/helium is required.

The topology and design methodology for LNAs at THz range are not much different from those for typical microwave LNAs, and they in fact benefit from the smaller size of matching elements when operation frequency is high. However, even a tiny variation in the dimension of interconnects and passive components would result in a significant mismatch and performance degradation.

There have been increasing number of reports on LNAs operating beyond 100 GHz based on both Si and III–V technologies, but few showed actual measured data on noise figure (NF), which can be ascribed to the challenge in noise measurement at such high-frequency region. Figure 5.13 shows reported NF of LNAs operating in this region, which includes both measured and simulated values. A couple of data points are based on the actual measurement. An LNA based on SiGe HBT exhibited measured NF of 6.8 dB at 130 GHz with associate peak gain of 24.3 dB (Hou et al. 2012), while an amplifier based on InP HBT exhibited measured NF of 17.1 dB at 380 GHz with associated peak gain of 22 dB (Hacker et al. 2012). Other simulated NF with SiGe HBT LNAs ranges between 7 and 14 dB for operation frequency of 130–240 GHz, although the accuracy of the noise modeling at that high-frequency range remains questionable.

5.3.3.2 Mixer

As briefly mentioned earlier, the availability of transistors enables active mixers, which may exhibit a positive conversion gain. Once positive gain is obtained, the noise from the following stages will be suppressed by the mixer stage, which is in contrast with the case of passive mixers where such noise is actually magnified by the conversion loss. Moreover, positive conversion gain will contribute to the total gain

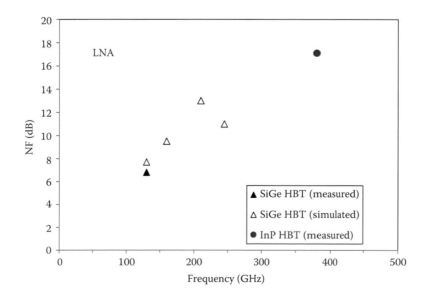

FIGURE 5.13   NF of LNAs reported for frequency beyond 100 GHz.

of the heterodyne detector, enhancing the responsivity of the system. It is true that a positive conversion gain cannot be always guaranteed for active mixers, especially for very high frequency, but still the conversion loss of active mixers of such cases will be smaller than the case of passive mixers.

Gilbert cell mixer is a popular choice for various microwave applications, and there have been attempts to use Gilbert cell mixers for the THz band, but excessively large dc power may be needed for operation (Pfeiffer et al. 2010; Schmalz et al. 2010). Alternative approaches based on nonlinearity of devices led to much reduced power dissipation while maintaining the positive conversion gain (Kim and Rieh 2012a; Kim et al. 2012c). One notable tendency for mixers operating at the THz region is the more popular adoption of the subharmonic mixers, where integer multiple of injected LO signal is mixed with the input RF signal. Although conversion gain and noise level are not as good as those of the fundamental mixers, subharmonic mixers relax the requirement for the LO frequency, which is an attractive feature for high-frequency applications (Schmalz et al. 2010).

5.3.3.3 Detector

While various types of THz detectors have been developed via various methods including diodes as discussed earlier, it is also possible to realize the THz detector circuit block based on transistors. Transistor-based detector (distinguished in this chapter from the *integrated detector*, which includes detector and other circuit blocks integrated) can be divided into two groups: active-mode and passive-mode detectors. In the active-mode detector, which can be implemented with either FETs or HBTs, the circuit is based on a differential pair with common node tied together (Zhou et al. 2011). Due to the device nonlinearity, the output signal of each branch of the differential pair will include the square terms, and they are added at the common node, while odd-harmonic terms, including the linear fundamental term, will be cancelled out. From the square terms, dc component can be extracted by filtering out the higher-order terms, which is supposed to be proportional to the input power. Thus, square-law-type detection is obtained.

The passive-mode detector is based on resistive mixer operation of FETs (Maas 1987) and, therefore, can be realized by FETs only. For detection, the FET is operated in the triode region without dedicated drain bias (Al Hadi et al. 2011). The input signal is applied to the gate electrode only, but it is coupled to the drain through a capacitance between the gate and drain. The capacitance can be a parasitic gate–drain capacitance or intentionally added external capacitance to reinforce the effect. Hence, the input signal is self-mixed at the FET due to the nonlinearity, resulting in a dc voltage appearing at the output that is proportional to the input power. In this way, square-law-type detection is achieved. It can be shown that such self-mixing occurs even for non-quasi-static FET operation mode, which is essentially equivalent to the plasma wave operation of FETs (Ojefors et al. 2009). It is noted that transistor-based detectors can operate beyond the device cutoff frequency as gain is not necessarily required for detectors unlike the amplifiers or oscillators.

5.3.3.4 Integrated Detector

With proper connection of the circuits described earlier, optionally together with LO if needed, an integrated detector can be constructed. Since such integrated detectors are based on standard circuit blocks, they can be implemented based on standard semiconductor process technology. Further, they can be integrated with other part of electronics on a single chip, which is a principal advantage of transistor-based approach over other more traditional THz methods.

There are increasing number of integrated detectors reported, based on both Si and III–V technologies, which have been fueled by the recent advancement in semiconductor technology and consequent device speed enhancement. Figure 5.14 shows some of the recently reported NEP values obtained from transistor-based THz integrated detectors from various topologies. As indicated by the plot, detection frequency of integrated detectors based on Si CMOS goes up to 1 THz while those based on III–V HEMT further reaches around 3 THz. Also shown for comparison are selected data points from other types of detectors operating at room temperature, which indicate that the transistor-based integrated

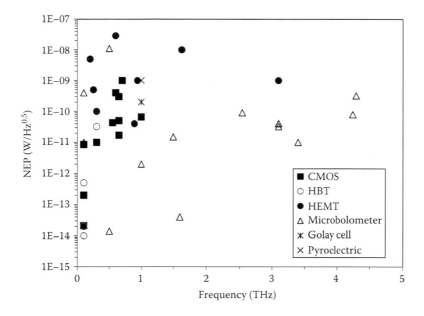

FIGURE 5.14 NEP vs. detection frequency of integrated detectors based on various technologies. Also shown for comparison are the microbolometers and typical values for Golay cell and pyroelectric detector. All data points are obtained at room temperature.

detectors show comparable or better NEP performance when compared to the traditional detectors such as Golay cells and pyroelectric detectors. It is noted that microbolometers show lower noise level while the operation frequency climbs up well beyond 4 GHz.

A clear trend is that the inclusion of an LNA reduces NEP by multiple orders of magnitude, although it is limited up to the operation frequency range of LNAs. As the operation frequency of amplifiers keeps moving upward, the range that benefits from the inclusion of LNA will be further extended in the future. Such a trend strongly indicates that transistor-based integrated detectors are well poised to become a major player in the THz applications.

5.4 Packaging Issues

The solid-state THz components are typically implemented in a planar form, in most cases on semiconductor substrates. Although on-wafer probing can be used for the characterization of the solid-state THz components up to the frequency supported by commercial RF probes, which is 750 GHz as of today, packaging is required for the practical applications with real systems. In this section, some of the practical packaging issues related to the solid-state THz components are reviewed.

5.4.1 Waveguides

Waveguides such as microstrip lines and coplanar waveguides (CPWs) are widely used for the planar implementation of the THz components on semiconductor chips. For practical applications, however, nonplanar waveguides are indispensible because the routinely employed metal packages require such nonplanar waveguide inside and the connection between the packages also needs to be made by nonplanar waveguides. Although various waveguides, such as coaxial, dielectric, parallel-plate waveguides, can be used to meet this need, the most widely used is the metallic rectangular waveguide. Rectangular waveguide is basically a hollow pipe made of metal with a rectangular cross section, the aspect ratio of which is 2 for the standard case.

TABLE 5.1 Frequency Bands and Rectangular Waveguide Standards

Frequency Band	Frequency Range (GHz)	Waveguide Standard	Internal Dimension $a \times b$ (In.)	Internal Dimension $a \times b$ (mm)	f_{low} (GHz)	f_{high} (GHz)
W-band	75–110	WR10	0.1 × 0.05	2.54 × 1.27	59.1	118.1
D-band	110–170	WR6.5	0.065 × 0.0325	1.651 × 0.8255	90.9	181.7
G-band	140–220	WR5.1	0.051 × 0.0255	1.2954 × 0.6477	115.8	231.6
H-band	220–325	WR3.4	0.034 × 0.017	0.8636 × 0.4318	173.7	347.4
Y-band	325–500	WR2.0	0.02 × 0.01	0.508 × 0.254	295.3	590.6
—	500–725	WR1.5	0.015 × 0.0075	0.381 × 0.1905	393.7	787.4
—	725–1100	WR1.0	0.01 × 0.005	0.254 × 0.127	590.6	1181.1

One outstanding feature of the rectangular waveguide compared to other types of waveguides such as coaxial, parallel-plate, microstrip lines, and CPW is the fact that it is composed of one piece of conductor. A main consequence of such structure is that transverse electromagnetic (TEM) mode is not allowed inside the waveguide. Transverse electric (TE) and transverse magnetic (TM) modes may survive, while TE_{10} mode is the one preferred as the dominant mode in many practical cases. For each mode to survive inside the waveguide, the relevant dimension needs to be large enough to allow at least one-half of the wavelength for a given frequency. The cutoff frequency is defined as the frequency below which the mode is not supported. For TE_{mn} mode, the cutoff frequency is given as a function of width and height of the cross section as follows:

$$f_c = \frac{c}{2}\sqrt{\left(\frac{m}{a}\right)^2 + \left(\frac{n}{b}\right)^2}, \quad m = 1,2,3,\ldots \quad n = 0,1,2,\ldots, \tag{5.1}$$

where a and b denote the width and height of the waveguide, respectively, and c is the speed of light.

With the cutoff frequency and the condition for the single-mode operation (TE_{10} mode), a specific dimension of waveguide cross section is recommended for each of the frequency band. Table 5.1 summarizes the frequency bands that belong to the THz region, corresponding waveguide standards, and their respective dimensions. Note that the waveguide standard is named after the width of the broad side presented in the unit of 10 mils. For TE_{10} single-mode operation, the theoretical lower frequency limit (= cutoff frequency) and upper frequency limit are obtained if $(m,n) = (1,0)$ and $(m,n) = (2,0)$ are substituted in Equation 5.1, respectively. Their explicit forms are repeated as follows for convenience:

$$f_{low} = \frac{c}{2a}, \quad f_{high} = \frac{c}{a} = 2f_{low}. \tag{5.2}$$

The calculated values of f_{low} and f_{high} in this way are also shown in Table 5.1. It is noted that the theoretically obtained usable frequency window is slightly wider than the span of the corresponding band.

5.4.2 Waveguide Transition

When two different types of waveguides are used and they have to be connected, waveguide transition is needed for a smooth link between the two. For the packaging of planar THz circuits considered here, a waveguide transition is needed between the rectangular waveguide and microstrip line or CPW line. Typically desired characteristics for waveguide transitions are broad bandwidth and low insertion loss, while low mechanical complexity is also favored from fabrication point of view. There have been a variety of transition structures developed for this purpose in microwave and mm-wave bands, some of which may be well adopted for THz region as well.

Various types of transition methods between microstrip line and rectangular waveguide have been reported as follows. The fin-line transition is composed of the signal line and ground plane of the microstrip line on either side of a thin substrate tapered in a fin-shaped pattern in an antipodal way in the transition region (Lavedan 1977). For the ridge waveguide transition, the central portion in the top-side conductor plane of a rectangular waveguide is modified to form a ridge, which is gradually lowered to be connected to the signal line of the microstrip line along the transition region, optionally combined with trough-shaped bottom-side conductor plane of the rectangular waveguide (Moochalla 1984). The E-plane probe transition refers to a structure where the microstrip signal line is protruded into the interior of a rectangular waveguide and suspended along the E-plane, the radiation from which forms a TE_{10} mode in the waveguide (Shih et al. 1988). These transition structures are known to show broadband and low-loss characteristics but involve rather high level of mechanical complexity. Another type of transition based on quasi-Yagi antenna shows broadband and easy integration with the microstrip line but requires high permittivity substrate (Kaneda et al. 1999). It is noted that the E-plane probe approach has been successfully adopted for THz region and widely used for various THz applications, one example being the transition monolithically integrated with a THz mixer (Siegel et al. 1999).

Similar transition structures have been developed for CPW to rectangular waveguide transition. The ridge waveguide (Ponchak et al. 1990), fin-line transition (Bellantoni et al. 1989), and quasi-Yagi antenna (Kaneda et al. 2000) have been employed for CPW to waveguide transitions. Another type based on aperture-coupled approach has also been reported (Simon et al. 1998). As a more recent effort, a transition based on dipole antenna was successfully adopted for InP THz circuits that can accommodate a wide MMIC chip, which showed an insertion loss less than 1 dB near 350 GHz range (Leong et al. 2009).

5.4.3 On-Chip Antennas

In most THz applications, long-range THz wave transmission is achieved through free space rather than waveguides. Guided transmission of THz waveguides, as in the case with the rectangular waveguides described earlier, is useful, but its adoption is limited to local short-range case due to the loss and cost that may be encountered if the path becomes excessively long. From this perspective, antennas are an indispensible device for THz solid-state electronic systems, which requires transition between chips and free space.

An external antenna, such as a horn antenna, can be used to serve this purpose, which is typically linked to the chip through a waveguide and a waveguide transition. While commonly adopted, such link configuration requires a package that is rather bulky (compared to chip size) and costly. An alternative way is to implement an on-chip antenna integrated with other circuits on the same chip, which directly converts the electronic signal on the chip to electromagnetic wave in free space without need for additional transition and waveguide (Rebeiz 1992). This approach is more attractive as frequency increases, since the antenna dimension is reduced for higher frequency, which THz electronics can benefit from.

Various types of on-chip antennas can be considered for THz applications, whose typical layouts are depicted in Figure 5.15. Dipole antenna is probably the most common general-purpose antenna and can also be nicely implemented on chips owing to its 2D nature. One advantage of dipole antennas formed on semiconductor chip is that they do not require backside metal plating or, if ground plane is available on the front side of the semiconductor chip, their performance is not limited by the narrow gap between the ground plane and the signal line. A downside of the on-chip dipole antenna though is that the radiation is mostly directed toward substrate due to the larger dielectric constant of the substrate, leading to a significant substrate loss due to finite resistivity and substrate modes if substrate is not thin enough. Several methods can be considered to suppress the loss, including simply thinning the wafer, forming a cavity below the antenna, or placing the chip on top of a Si lens.

There are variants of the dipole antenna designed for planar structure, which can be adopted for on-chip antennas as well. The bow-tie antenna is a 2D version of the biconical antenna, which is intended for improved bandwidth (Schuster et al. 2011). The folded dipole antenna can be also employed when

FIGURE 5.15 Layout of various on-chip antennas: (a) dipole, (b) bow-tie, (c) folded dipole, (d) patch, (e) log-periodic, and (f) log-spiral.

impedance transformation is additionally needed (Al Hadi et al. 2011). Other types of antennas can be also used for on-chip applications. The patch antenna, also called microstrip antenna, can be easily adopted for on-chip antennas. Despite its narrowband characteristics, it is widely adopted for mm-wave and THz applications for easy implementation (Han et al. 2012). For very broadband applications, frequency-independent antennas, such as log-periodic (DuHamel and Isbell 1957) or equiangular spiral antennas (Dyson 1959), can also be adopted, which yield linearly and circularly polarized patterns, respectively.

5.5 Summary

In this chapter, various types of THz sources and detectors based on solid-state diode or transistors have been described along with some relevant package issues. These solid-state electronic THz components are expected to enable compact THz systems with low cost and low power dissipation, which will be highly favored for commercial mobile THz devices in biomedical environment. It is true that the solid-state THz electronics are relatively young technologies and yet to be matured. For exactly the same reason, however, it is an attractive field for researches and retains a great potential for future commercialization.

References

Al Hadi, R., H. Sherry, J. Grzyb et al. 2011. A broadband 0.6 to 1 THz CMOS imaging detector with an integrated lens. Presented at *2011 IEEE MTT-S International Microwave Symposium Digest (MTT)*, Baltimore, MD, pp. 1–4.

Asada, M., S. Suzuki, and N. Kishimoto. 2008. Resonant tunneling diodes for sub-terahertz and terahertz oscillators. *Japanese Journal of Applied Physics* 47: 4375.

Bellantoni, J. V., R. C. Compton, and H. M. Levy. 1989. A new W-band coplanar waveguide test fixture. Presented at *IEEE MTT-S International Microwave Symposium Digest*, Long Beach, CA, pp. 1203–1204.

Chen, S.-M., Y.-K. Fang, F. Juang et al. 2010. Terahertz Schottky barrier diodes with various isolation designs for advanced radio frequency applications. *Thin Solid Films* 519: 471–474.

Cojocari, O., C. Sydlo, M. Feiginov, and P. Meissner. 2012. RTD-based THz-MIC by film-diode technology. Presented at *2012 IEEE MTT-S International Microwave Symposium Digest (MTT)*, Montreal, QC, pp. 1–3.

DuHamel, R. and D. Isbell. 1957. Broadband logarithmically periodic antenna structures. Presented at *IRE International Convention Record*, New York, vol. 5, pp. 119–128.

Dyson, J. 1959. The equiangular spiral antenna. *IRE Transactions on Antennas and Propagation* 7: 181–187.

Eisele, H. 2010. 480 GHz oscillator with an InP Gunn device. *Electronics Letters* 46: 422–423.

Eisele, H. and R. Kamoua. 2004. Submillimeter-wave InP Gunn devices. *IEEE Transactions on Microwave Theory and Techniques* 52: 2371–2378.

Hacker, J., M. Urteaga, R. Lin et al. 2012. 400 GHz HBT differential amplifier using unbalanced feed networks. *IEEE Microwave and Wireless Components Letters* 22: 536–538.

Han, R., Y. Zhang, Y. G. Kim et al. 2012. 280 GHz and 860 GHz image sensors using Schottky-barrier diodes in 0.13 μm digital CMOS. Presented at *2012 IEEE International Solid-State Circuits Conference Digest of Technical Papers (ISSCC)*, San Francisco, CA, pp. 254–256.

Harvey, J., E. R. Brown, D. B. Rutledge, and R. A. York. 2000. Spatial power combining for high-power transmitters. *IEEE Microwave Magazine* 1: 48–59.

Heinemann, B., R. Barth, D. Bolze et al. 2010. SiGe HBT technology with fT/fmax of 300 GHz/500 GHz and 2.0 ps CML gate delay. Presented at *IEDM Technical Digest IEEE International Electron Devices Meeting*, San Francisco, CA, pp. 688–691.

Hou, D., Y.-Z. Xiong, W.-L. Goh, W. Hong, and M. Madihian. 2012. A D-band cascode amplifier with 24.3 dB gain and 7.7 dBm output power in 0.13 μm SiGe BiCMOS technology. *IEEE Microwave and Wireless Components Letters* 22: 191–193.

Ino, M., T. Ishibashi, and M. Ohmori. 1976. CW oscillation with p+-p-n+ silicon IMPATT diodes in 200 GHz and 300 GHz bands. *Electronics Letters* 12: 148–149.

Kallfass, I., A. Tessmann, H. Massler et al. 2011. Balanced active frequency multipliers for W-band signal sources. Presented at *2011 European Microwave Integrated Circuits Conference (EuMIC)*, Manchester, U.K., pp. 101–104.

Kaneda, N., Y. Qian, and T. Itoh. 1999. A broad-band microstrip-to-waveguide transition using quasi-Yagi antenna. *IEEE Transactions on Microwave Theory and Techniques* 47: 2562–2567.

Kaneda, N., Y. Qian, and T. Itoh. 2000. A broadband CPW-to-waveguide transition using quasi-Yagi antenna. Presented at *2000 IEEE MTT-S International Microwave Symposium Digest*, Boston, MA, vol. 2, pp. 617–620.

Karpov, A., D. Miller, J. Stern et al. 2007. Development of 1 THz SIS mixer for SOFIA. Presented at *Eighteenth International Symposium on Space Terahertz Technology*, vol. 1, p. 50.

Kim, D.-H. and J.-S. Rieh. 2012a. A 135 GHz Differential Active Star Mixer in SiGe BiCMOS Technology. *IEEE Microwave and Wireless Components Letters* 22: 409–411.

Kim, D.-H. and J.-S. Rieh. 2012b. CMOS 138 GHz low-power active mixer with branch-line coupler. *Electronics Letters* 48: 554–555.

Kim, M., J.-S. Rieh, and S. Jeon. 2012c. Recent progress in terahertz monolithic integrated circuits. Presented at *Circuits and Systems (ISCAS), 2012 IEEE International Symposium on*, Seoul, South Korea, pp. 746–749.

Lai, R., X. Mei, W. Deal et al. 2007. Sub 50 nm InP HEMT device with fmax greater than 1 THz. Presented at *Electron Devices Meeting, 2007. IEDM 2007. IEEE International*, pp. 609–611.

Lavedan, L. 1977. Design of waveguide-to-microstrip transitions specially suited to millimetre-wave applications. *Electronics Letters* 13: 604–605.

Leong, K., W. R. Deal, V. Radisic et al. 2009. A 340–380 GHz integrated CB-CPW-to-waveguide transition for sub millimeter-wave MMIC Packaging. *IEEE Microwave and Wireless Components Letters* 19: 413–415.

Lewark, U., A. Tessmann, H. Massler et al. 2011. 300 GHz active frequency-tripler MMICs. Presented at *2011 European Microwave Integrated Circuits Conference (EuMIC)*, Manchester, U.K., pp. 236–239.

Lewark, U., A. Tessmann, S. Wagner et al. 2012. 255 to 330 GHz active frequency tripler MMIC. Presented at *Integrated Nonlinear Microwave and Millimetre-Wave Circuits (INMMIC), 2012 Workshop on*, Dublin, Republic of Ireland, pp. 1–3.

Maas, S. A. 1987. A GaAs MESFET mixer with very low intermodulation. *IEEE Transactions on Microwave Theory and Techniques* 35: 425–429.

Monaco, E., M. Pozzoni, F. Svelto, and A. Mazzanti. 2010. Injection-locked CMOS frequency doublers for mu-wave and mm-wave applications. *IEEE Journal of Solid-State Circuits* 45: 1565–1574.

Moochalla, S. S. 1984. Ridge waveguide used in microstrip transition. *Microwaves* 23: 149–152.

Nishizawa, J., K. Motoya, and Y. Okuno. 1978. GaAs TUNNETT diodes. *IEEE Transactions on Microwave Theory and Techniques* 26: 1029–1035.

Nishizawa, J., P. Płotka, T. Kurabayashi, and H. Makabe. 2008. 706-GHz GaAs CW fundamental-mode TUNNETT diodes fabricated with molecular layer epitaxy. Presented at *Physica Status Solidi (c)*, vol. 5, pp. 2802–2804.

Ojefors, E., J. Grzyb, Y. Zhao et al. 2011. A 820 GHz SiGe chipset for terahertz active imaging applications. Presented at *2010 IEEE International Solid-State Circuits Conference Digest of Technical Papers (ISSCC)*, San Francisco, CA, pp. 224–226.

Ojefors, E., U. R. Pfeiffer, A. Lisauskas, and H. G. Roskos. 2009. A 0.65 THz focal-plane array in a quarter-micron CMOS process technology. *IEEE Journal of Solid-State Circuits* 44: 1968–1976.

Pfeiffer, U. R., E. Ojefors, and Y. Zhao. 2010. A SiGe quadrature transmitter and receiver chipset for emerging high-frequency applications at 160 GHz. Presented at *2010 IEEE International Solid-State Circuits Conference Digest of Technical Papers (ISSCC)*, San Francisco, CA, pp. 416–417.

Plouchart, J.-O. 2011. Applications of SOI technologies to communication. Presented at *2011 IEEE Compound Semiconductor Integrated Circuit Symposium (CSICS)*, Waikoloa, HI, pp. 1–4.

Ponchak, G. E. and R. N. Simons. 1990. A new rectangular waveguide to coplanar waveguide transition. Presented at *IEEE MTT-S International Microwave Symposium Digest*, Dallas, TX, pp. 491–492.

Radisic, V., K. M. Leong, X. Mei et al. 2012a. Power amplification at 0.65 THz using InP HEMTs. *IEEE Transactions on Microwave Theory and Techniques* 60: 724–729.

Radisic, V., K. M. Leong, S. Sarkozy et al. 2012b. 220-GHz solid-state power amplifier modules. *IEEE Journal of Solid-State Circuits* 47: 2291–2297.

Rebeiz, G. M. 1992. Millimeter-wave and terahertz integrated circuit antennas. *Proceedings of the IEEE* 80: 1748–1770.

Rieh, J.-S., B. Jagannathan, D. R. Greenberg et al. 2004. SiGe heterojunction bipolar transistors and circuits toward terahertz communication applications. *IEEE Transactions on Microwave Theory and Techniques* 52: 2390–2408.

Rieh, J.-S., S. Jeon, and M. Kim. 2011. An overview of integrated THz electronics for communication applications. Presented at *IEEE 54th International Midwest Symposium on Circuits and Systems (MWSCAS)*, Seoul, South Korea, pp. 1–4.

Rieh, J.-S., Y. H. Oh, D. K. Yoon et al. 2012. An overview of challenges and current status of Si-based terahertz monolithic integrated circuits. Presented at *2012 IEEE 11th International Conference on Solid-State and Integrated Circuit Technology (ICSICT)*, Xi'an, China, pp. 1–4.

Schmalz, K., W. Winkler, J. Borngräber et al. 2010. A subharmonic receiver in SiGe technology for 122 GHz sensor applications. *IEEE journal of Solid-State Circuits* 45: 1644–1656.

Schuster, F., H. Videlier, A. Dupret et al. 2011. A broadband THz imager in a low-cost CMOS technology. Presented at *2010 IEEE International Solid-State Circuits Conference Digest of Technical Papers (ISSCC)*, San Francisco, CA, pp. 42–43.

Sengupta, K. and A. Hajimiri. 2012. A 0.28 THz power-generation and beam-steering array in CMOS based on distributed active radiators. *IEEE Journal of Solid-State Circuits* 47: 1–19.

Seo, M., M. Urteaga, J. Hacker et al. 2011. InP HBT IC technology for terahertz frequencies: Fundamental oscillators up to 0.57 THz. *IEEE Journal of Solid-State Circuits* 46: 2203–2214.

Shih, Y.-C., T.-N. Ton, and L. Q. Bui. 1988. Waveguide-to-microstrip transitions for millimeter-wave applications. Presented at *IEEE MTT-S International Microwave Symposium Digest*, New York, pp. 473–475.

Shim, D., D. Koukis, D. Arenas, D. Tanner, and K. Kenneth. 2011. 553-GHz signal generation in CMOS using a quadruple-push oscillator. Presented at *2011 Symposium on VLSI Circuits (VLSIC)*, Honolulu, HL, pp. 154–155.

Siegel, P. H., R. P. Smith, M. Graidis, and S. C. Martin. 1999. 2.5-THz GaAs monolithic membrane-diode mixer. *IEEE Transactions on Microwave Theory and Techniques* 47: 596–604.

Simon, W., M. Werthen, and I. Wolff. 1998. A novel coplanar transmission line to rectangular waveguide transition. Presented at *IEEE MTT-S International Microwave Symposium Digest*, Baltimore, MD, vol. 1, pp. 257–260.

Stephan, K. D. 1986. Inter-injection-locked oscillators for power combining and phased arrays. *IEEE Transactions on Microwave Theory and Techniques* 34: 1017–1025.

Tai, W., H. Xu, A. Ravi et al. 2012. A transformer-combined 31.5 dBm outphasing power amplifier in 45 nm LP CMOS with dynamic power control for back-off power efficiency enhancement. *IEEE Journal of Solid-State Circuits* 47: 1646–1658.

Thomas, B., A. Maestrini, and G. Beaudin. 2005. A low-noise fixed-tuned 300–360-GHz sub-harmonic mixer using planar Schottky diodes. *IEEE Microwave and Wireless Components Letters* 15: 865–867.

Tousi, Y. M., O. Momeni, and E. Afshari. 2012. A 283-to-296 GHz VCO with 0.76 mW peak output power in 65 nm CMOS. Presented at *2010 IEEE International Solid-State Circuits Conference Digest of Technical Papers (ISSCC)*, pp. 258–260.

Urteaga, M., M. Seo, J. Hacker et al. 2011. InP HBTs for THz frequency integrated circuits. Presented at *23rd International Conference on Indium Phosphide and Related Materials and Compound Semiconductor Week (CSW/IPRM)*, Berlin, Germany, pp. 1–4.

Valenta, V., A. C. Ulusoy, A. Trasser, and H. Schumacher. 2013. Wideband 110 GHz frequency quadrupler for an FMCW imager in 0.13-μm SiGe: C BiCMOS process. Presented at *2013 IEEE 13th Topical Meeting on Silicon Monolithic Integrated Circuits in RF Systems (SiRF)*, Austin, TX, pp. 9–11.

Wu, I., Z. Sun, H. Yilmaz, and M. Berroth. 2006. A dual-frequency Wilkinson power divider. *IEEE Transactions on Microwave Theory and Techniques* 54: 278–284.

Zhang, W., P. Khosropanah, J. Gao et al. 2010. Quantum noise in a terahertz hot electron bolometer mixer. *Applied Physics Letters* 96: 111113.

Zhou, L., C.-C. Wang, Z. Chen, and P. Heydari. 2011. A W-band CMOS receiver chipset for millimeter-wave radiometer systems. *IEEE Journal of Solid-State Circuits* 46: 378–391.

6

Waveguides for Terahertz Endoscopy

Tae-In Jeon
Korea Maritime and Ocean University

Daniel Grischkowsky
Oklahoma State University

6.1 Introduction

In recent years, the use of terahertz (THz) radiation has been the subject of much research in many areas such as biomedical imaging. The attenuation of THz radiation for the human body and the signal-to-noise ratio (SNR) of THz radiation are important parameters used to determine the detectable penetration depth of THz radiation. If the THz signal has an SNR of 1000 at 1 THz frequency, the penetration depths for adipose tissue and skin tissue are only 3 mm and 500 µm, respectively (Pickwell-MacPherson 2010). Because of the very low penetration depth for the human body, transmission imaging can only be carried out in vitro. Any in vivo studies are limited to reflection imaging for the skin tissues (Ji et al. 2009; Parrott et al. 2011; Woodward et al. 2002a,b). Therefore, a THz endoscope is required to make a diagnosis for physical checkups of organs such as the stomach or colon.

Similar to an optical endoscope, a THz endoscope needs a flexible tube, a THz delivery system (waveguides), a lens system (silicon or optical lenses), an eyepiece (monitor), and an additional channel to allow entry of medical instruments. Among the requirements for a THz endoscope, the flexible tube and THz delivery system are very difficult to implement because most THz waveguides are not flexible and have significant propagation loss. Because of these two problems, researchers have been developing THz endoscopes that have miniaturized THz transmitters (Tx) and receiver (Rx) modules that are attached at the end of optical fibers.

In Section 6.2, the advantages and disadvantages of several THz waveguides are presented. In Section 6.3, an optical fiber-coupled THz endoscope system will be introduced with laser pulse compensation for positive group velocity dispersion (GVD). Section 6.3.1 describes the principles of the laser pulse dispersion and the methods of laser pulse compression. Section 6.3.2 describes the commercial THz spectroscopy system using an optical fiber; Section 6.3.3 presents a homemade THz system using an optical fiber.

6.2 Review of Terahertz Waveguides

6.2.1 Metal Circular Waveguides

The first THz waveguide study used a metal submillimeter-diameter circular tube (McGowan et al. 1999). In order to efficiently couple the freely propagating THz radiation into the metal tube, two hyper-hemispherical silicon lenses were placed at the input and the output ends of the tube, as shown in Figure 6.1 (McGowan et al. 1999).

The measured THz pulse, transmitted through the 24 mm long 240 μm diameter stainless-steel tube, and its amplitude spectrum are shown in Figure 6.2 (McGowan et al. 1999). Because of the strong GVD of the waveguide, the 0.5 ps input THz pulse was broadened to approximately 70 ps, at which the lower frequencies traveled more slowly than the higher frequencies. Figure 6.2b shows the corresponding amplitude spectrum for the broadened 70 ps THz pulse. The low-frequency cutoff at 0.76 THz for the TE_{11} mode can be easily observed. Also, the spectrum presents a multimode component (TM_{11}) at around 1.7 THz.

The theoretical calculated field absorption and phase and group velocities for the circular waveguide are shown in Figure 6.3 (Gallot et al. 2000). The absorption coefficient of each mode is effectively infinite below the cutoff frequency and slowly approaches a constant value with increasing of the frequency.

For example, the absorption coefficient of the TE_{11} mode approaches 0.4/cm with increasing frequency. For a THz pulse transmitted through a 200 cm long circular waveguide, which is approximately the required length from the THz source to the endoscope tip, the amplitude attenuation with the propagation distance L is simply described by exp[−αL]. If the input pulse height of 1 nA is propagated for the distance, it is predicted that $1\ nA \times exp[-80] = 1.8 \times 10^{-35}\ nA \approx 0\ A$. Consequently, the input THz pulse cannot propagate through the waveguide. If the difference of the group velocity between the high (C) and the low frequencies (0.5C) is 0.5C, where C is the speed of light, the THz pulse is broadened up to 6.7 ns, which is 96 times longer than the value shown in Figure 6.2a. Therefore, it is required that there be a very small absorption coefficient and almost no GVD (transverse electromagnetic mode [TEM] mode) in order to use the waveguides with the THz endoscope.

6.2.2 Metal Wire Waveguides

Single metal wires have been studied for absorption and dispersion using millimeter waves (Goubau 1950; Sommerfeld 1952; Wentworth et al. 1961). Because of the very high conductivity of metal, weakly

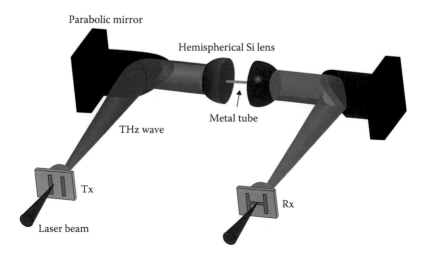

FIGURE 6.1   Schematic diagram of the optoelectronic THz time-domain spectroscopy system.

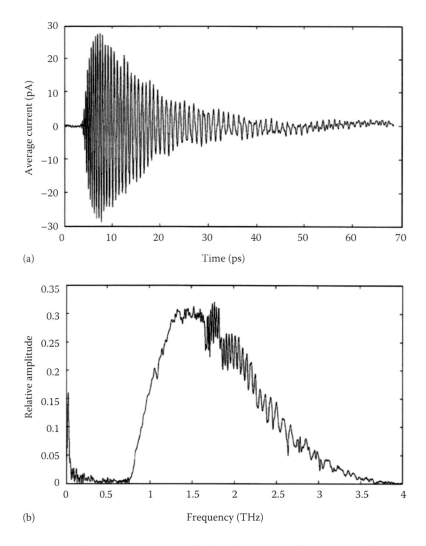

FIGURE 6.2 (a) Measured THz pulse transmitted through a 24 mm long, 240 μm diameter stainless-steel waveguide. (b) Amplitude spectra (McGowan et al. 1999). (Reproduced from McGowan, R.W., Gallot, G., and Grischkowsky, D., Propagation of ultrawideband short pulses of terahertz radiation through submillimeter-diameter circular waveguides, *Opt. Lett.*, 24, 1431–1433, 1999. Copyright 1999, With permission of Optical Society of America.)

guided electromagnetic waves propagate along the surface of the wires. Except for the TEM mode, all other modes have very high absorption coefficients and vanish almost immediately when the electromagnetic waves are coupled to the wires. Therefore, only the TEM mode waves propagate along the wires. Recently, the absorption and dispersion of propagated THz pulses on single wires have been studied (Jeon et al. 2005; Wang and Mittleman 2004).

Figure 6.4 provides a schematic diagram of the experimental setup (Jeon et al. 2005). The THz pulses are optoelectronically generated and detected by silicon-on-sapphire (SOS) chips. Commercial copper wire of 0.53 mm diameter, which is supported by two tightly fitting Teflon disks of 3 mm thickness, is directly connected to the SOS chips. The 80 μm diameter tip of the copper wire is placed near the specially designed Tx chip for optimal THz coupling.

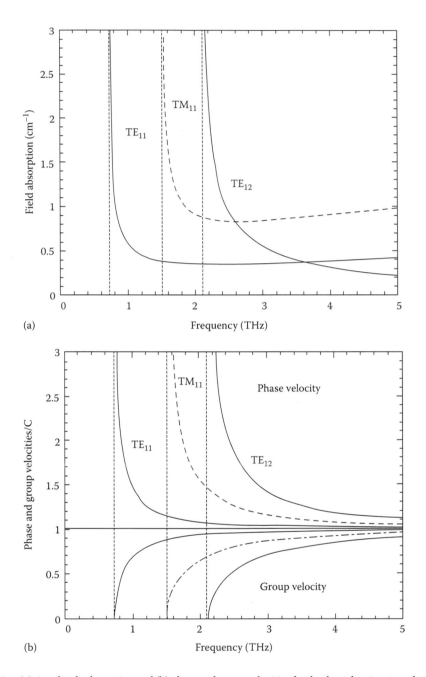

FIGURE 6.3   (a) Amplitude absorption and (b) phase and group velocities for the three dominant modes in a 240 μm diameter stainless-steel waveguide (Gallot et al. 2000). (Reproduced from Gallot, G., Jamison, S. P., McGowan, R. W., and Grischkowsky, D., Terahertz waveguides, *J. Opt. Soc. Am. B*, 17, 851–863, 2000. Copyright 2000, With permission of Optical Society of America.)

The measured propagated THz pulses and their spectra, guided by the different lengths of wires, are shown in Figure 6.5a and b. Because of the very small amplitude absorption and the low GVD of the propagated THz pulses, the THz pulses are almost identical. Figure 6.5c shows the amplitude absorption coefficient of the single metal wire. Dots and solid lines indicate the absorption values

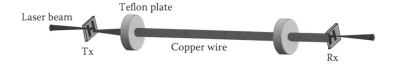

FIGURE 6.4 Schematic diagram of metal wire waveguide system.

found by measurement and in theory; these values are very small compared to those of the metal circular tubes, as shown in Figure 6.2a.

The maximum amplitude of the spectra for the 104 cm long copper wire is around 0.15 THz; the corresponding amplitude absorption is, in theory, 0.001/cm. If the metal wire waveguide is applied to the 200 cm long THz endoscope, the amplitude attenuation with the propagation distance is exp(−0.001/cm × 200 cm), giving an amplitude transmission of 0.82. This result is quite acceptable for use in a THz endoscope. For example, when a 1 nA average current of THz pulse is coupled to the single metal wire used in an endoscope, the output current of the THz pulse at the end of the 200 cm long endoscope is 820 pA, which is a sufficient THz signal for the endoscope.

However, the single metal wire has an electromagnetic loss problem with flexibility. When the single metal wire is bent, most THz waves are radiated into the air at the bend because of the very small skin depth of the metal, which is due to the very high conductivity. Figure 6.6 shows the effects of single wire bending on the guiding property of the metal wire (Wang and Mittleman 2004). When the curve depth is 2.9 cm, the main peak drops to below 1/5 of the original amplitude for the straight wire. Because of the very high bending loss, the single metal wire is not an acceptable waveguide for a THz endoscope.

6.2.3 Parallel-Plate Waveguides

The conditions of good waveguides for THz endoscopes are no cutoff frequency, low loss, low GVD, and flexibility. Metal tubes and metal wires do not satisfy these conditions. In fact, only one waveguide satisfies the aforementioned conditions, and that is a flexible parallel-plate waveguide (PPWG) made out of two thin metal strips. When THz pulses are propagated along an air gap, which is the air space between two parallel strips, the output THz pulses are in the TEM mode with low loss and low GVD. Also, when the waveguides have a curved shape, the output THz pulse is still in the TEM mode, with small loss and low GVD.

Figure 6.7 shows the experimental setup for a PPWG having a curved shape (Mendis and Grischkowsky 2001); in this PPWG, the propagation paths have lengths of 125 mm (r = 11.5 mm; radius of the curve) and 250 mm (r′ = 27.5 mm). The cross-sectional dimensions of the PPWG are 90 μm × 15 mm between the plates. Because 100 μm thick copper strips are used, the PPWG can be made into a circular shape. Two cylindrical silicon lenses are used to make a THz line focused at the air gap of the PPWG.

The propagated output THz pulses show a good SNR in the TEM mode. The full width half maximum (FWHM) is 0.22 ps for the reference pulse. This value is obtained by bringing the two cylindrical lenses into contact in their confocal position with no waveguide in place. The values are 0.25 ps for the L = 125 mm pulse and 0.39 ps for the L = 250 mm pulse. The FWHM clearly shows that the TEM mode propagates through the PPWG. The peak amplitude of the L = 250 mm pulse is reduced about 0.05 times compared to that of the reference pulse; however, the L = 250 mm pulse still has good SNR.

When the PPWG is straight with a 500 μm or 90 μm air gaps, the calculated absorption coefficients are as shown in Figure 6.8 (Marcuvitz 1986). The absorption is very small. For example, the coefficient for the 500 μm air gap is 0.0097/cm at 0.5 THz frequency. When the PPWG is applied to a 200 cm long THz endoscope using a 1 nA THz input signal, the output THz signal is 144 pA, which is a strong signal to apply to an endoscope. And while the PPWG has good guiding properties, low loss, and low GVD, the dimensions of the PPWG are too large for it to be used with the THz endoscope. The diameter of the endoscope should be less than 10 mm; however, the PPWG endoscope cannot satisfy this diameter condition.

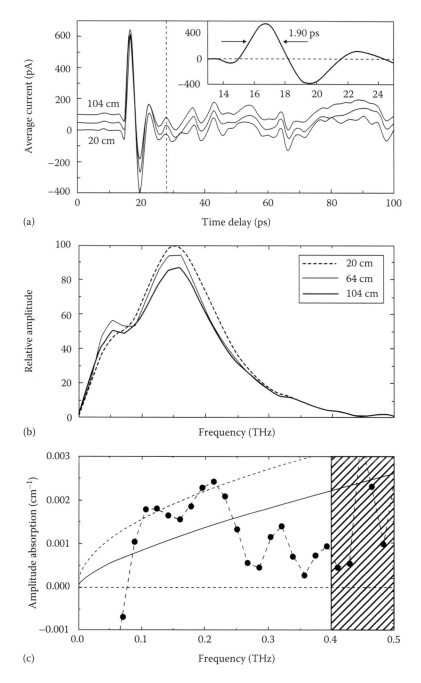

FIGURE 6.5 (a) Transmitted THz pulses for 20, 64, and 104 cm long copper wires. The pulses have been shifted for clarity. The inset shows the enlargement of the main peak of the 20 cm pulse. (b) The spectra of the pulses are truncated at the dashed line shown in (a). (c) Amplitude absorption coefficients. Dots: measured from 104 and 20 cm copper wires, the measurement in the shaded area is considered not accurate; solid line: Sommerfeld theory for a 0.52 mm diameter copper wire; dashed line: theory for TEM mode of the corresponding air-filled coaxial waveguide (Jeon et al. 2005). (Reprinted with permission from Jeon, T.-I., Zhang, J., and Grischkowsky, D., THz Sommerfeld wave propagation on a single metal wire, *App. Phys. Lett.*, 86, 161904, 2005. Copyright 2005, American Institute of Physics.)

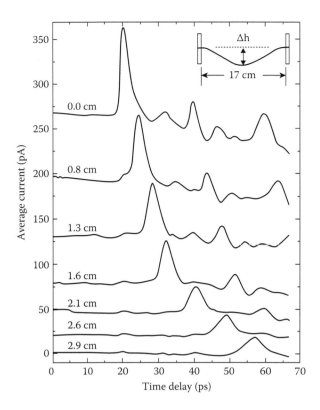

FIGURE 6.6 Transmitted THz pulses for the indicated curve depths Δh. The inset shows the curved wire (Jeon et al. 2005). (Reprinted with permission from Jeon, T.-I., Zhang, J., and Grischkowsky, D., THz Sommerfeld wave propagation on a single metal wire, *App. Phys. Lett.*, 86, 161904, 2005. Copyright 2005, American Institute of Physics.)

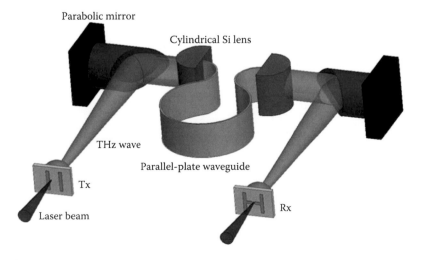

FIGURE 6.7 Schematic diagram of PPWG waveguide system.

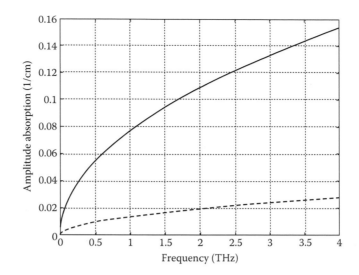

FIGURE 6.8   Calculated amplitude absorption for the flexible PPWW with a 90 μm (solid line) and 500 μm (dashed line) air gaps.

6.3 Terahertz Endoscope

Because THz beam-guided waveguide systems do not satisfy the required conditions of endoscopes, laser beam-guided endoscope systems are an acceptable alternative method. In order to make a laser beam-guided THz endoscope, laser pulse compression and miniaturized THz Tx and Rx modules are required.

6.3.1 Laser Pulse Dispersion and Compression

6.3.1.1 Laser Pulse Dispersion

The THz pulse width is proportional to the laser pulse width. Whenever a laser pulse is incident on a semiconductor surface, the photocurrent, J(t), is generated with fall time longer than rise time, as shown in Figure 6.10. The duration of the photocurrent depends on the laser pulse width and the lifetime of the photocarriers. Moreover, the THz pulse is proportional to the time derivative of the transient photocurrent, as in $E_{THz}(t) \propto \partial J(t)/\partial t$. Figure 6.9 shows the amplitudes of the laser pulse, photocurrent, and

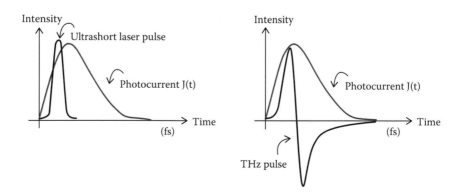

FIGURE 6.9   Temporal intensity of the laser pulse, photocurrent, and radiated THz pulse.

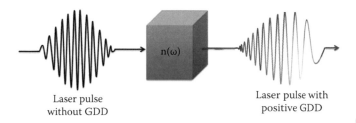

Laser pulse
without GDD

Laser pulse with
positive GDD

FIGURE 6.10 Concept of positive group-delay dispersion.

radiated THz pulse (Duvillaret et al. 2001). Therefore, the THz pulse width is proportional to the laser pulse width. In order to generate a very short THz pulse width, subhundred femtosecond laser pulse widths are required.

When making a laser pulse-guided THz endoscope, an optical fiber is a very good guiding material. Moreover, optical fibers are flexible and have diameters of only a few hundred microns. However, when subps laser pulses are propagated through such fibers, the laser pulses are broadened because of the positive GVD. The long wavelength component of the laser pulse travels faster than the short wavelength component in an optical medium, as shown in Figure 6.10.

Figure 6.11 provides a schematic diagram of a fiber-coupled laser pulse measurement setup; Figure 6.12 provides a comparison of laser pulse widths measured by an autocorrelator after Ti:sapphire laser, after beam isolator, and after 2 m long single-mode optical fiber. The 68 fs laser pulse width coming from the Ti:sapphire laser is broadened to 4.2 ps at the end of the 2 m long optical fiber; this represents a broadening of 62 times for the laser pulse width. Meanwhile, the broadened laser pulse width should be compressed after the optical fiber or should be compensated for before reaching the optical fiber in order to generate and detect THz pulses.

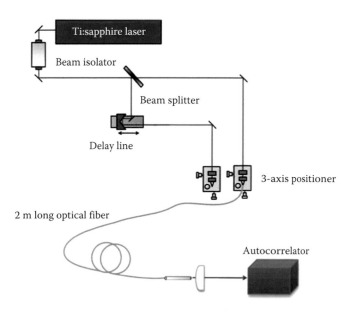

FIGURE 6.11 Schematic diagram of fiber-coupled laser pulse measurement setup.

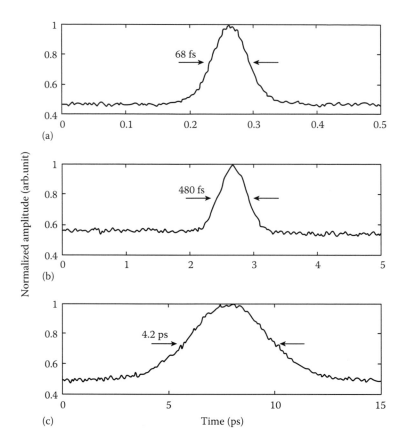

FIGURE 6.12 Measured laser pulse widths (a) after Ti:sapphire laser, (b) after beam isolator, and (c) after 2 m long single-mode optical fiber.

6.3.1.2 Laser Pulse Compression

Methods used to compensate for the positive GVD involve using a chirped mirror pair, a prism pair, and a grating pair. The beam path of the short wavelength component travels a shorter path than that of the long wavelength component, which makes negative GVD. Therefore, when the negative GVD and positive GVD are properly combined, the laser pulses recover their original pulse width.

6.3.1.2.1 Chirped Mirror

Usually, chirped mirrors are made with alternating layers of SiO_2 and TiO_2. Each layer has different depths, and these layers gradually thicken as they get closer to the substrate, as shown in Figure 6.13 (Keller and Gallmann 1997). The shorter wavelength of the pulse, which is represented by blue lines, reflects at the front surface; the longer wavelength of the pulse, which is represented by red lines, reflects from the deeper layers of the mirror. A chirped mirror is designed to have negative GVD: a laser pulse reflected from the deeper layers of the mirror travels a longer distance than does a laser pulse that reflects off the surface layers. Therefore, the shorter wavelength arrives first and the longer wavelength arrives later. Usually, chirped mirrors use ultrashort laser pulses that have pulse widths of less than 10 fs.

Whenever the laser pulses are bounced off the surface of the chirped mirror, the pulses have negative GVD. The laser pulses can bounce back and forth between two or three mirrors many times to compensate for the pulse dispersion effects, as shown in Figure 6.14.

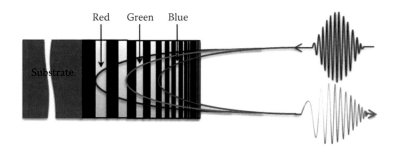

FIGURE 6.13 Schematic diagram of the chirped mirror.

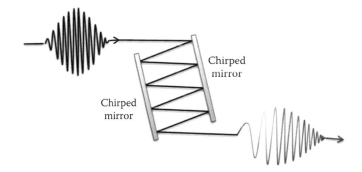

FIGURE 6.14 Schematic diagram of laser pulse bouncing between two chirped mirrors.

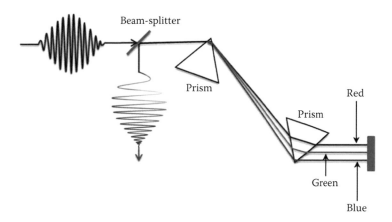

FIGURE 6.15 Schematic diagram of negative GVD using prism pair.

6.3.1.2.2 *Prism Pair*

A prism pair compressor consists of two prisms and a mirror. When a laser pulse passes through the prism, the longer wavelengths of the laser pulse have smaller refraction angles than do the shorter wavelengths of the laser pulse as shown in Figure 6.15. In contrast to the shorter wavelengths, the longer wavelengths travel to the inner part of the second prism. Since the speed of light inside the prism is much lower than that in air, the longer wavelengths are delayed more than the shorter wavelengths. Also, the reflected laser pulses from the flat mirror use the same beam path to return to the beam splitter. This situation makes a negative GVD. Prism pairs are usually used to compensate for the dispersion inside a Ti:sapphire mode-locked laser (Figure 6.16).

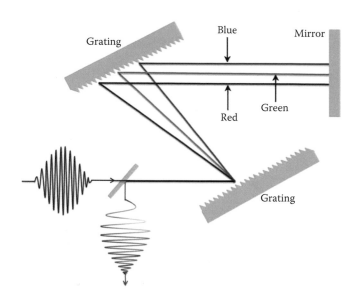

FIGURE 6.16 Schematic diagram of negative GVD using grating pair.

6.3.1.2.3 Grating Pair

Because a THz endoscope system guided by an optical fiber has a large positive GVD, a large negative GVD is required for compensation. The grating pair has a large negative GVD compared to that of the chirped mirror and the prism pair. The incident laser pulse to the grating pair is diffracted depending on the frequency of the laser pulse, as shown in Figure 6.17. The high-frequency component (short wavelength) travels a short path, and the low-frequency component (long wavelength) travels a long path. The magnitude of the negative GVD depends on the distance between the two grating pairs and can be adjusted for perfect compensation. Although the diffraction loss of the grating pair is large, a strongly negative GVD is the result. This strongly negative GVD is a good for a several-meter-long optical fiber that has strongly positive GVD.

6.3.2 Commercial THz Spectroscope System

Many THz systems that use laser pulses guided by optical fibers have been applied to THz time-domain spectroscopy (Crooker 2002; Ellrich et al. 2011; Lee et al. 2007; Vieweg et al. 2007). Although Tx and Rx modules were too big to use in the THz endoscope, the modules can be freely moved in any direction because of the optical fiber. Recently, several companies have produced commercial fiber-coupled THz spectroscopy systems. Figure 6.17 provides system images and product specifications of Fico from Zomega, T-Ray 4000 from Picometrix, and TPS Spectra 3000 from Teraview. The Spectra 3000 can perform THz imaging using a medical imaging probe that is attached to the system as shown in Figure 6.18. The medical imaging probe can perform scan 2D reflected imaging using Risley prism pairs (Figure 6.19).

Recently, THz pulsed imaging in vivo using the medical imaging probe TSP Spectra 3000 has been performed (Pickwell and Wallace 2006; Sy et al. 2010). In vivo THz pulse response from a number of different skin positions were measured, with data as given in Figure 6.20. With the use of optical fiber-guided laser pulses, the probe can freely move into any position on the skin. Although five sites of the body, including the forehead, cheek, chin, dorsal forearm, and palm were measured over 5 days, the refracted THz pulses from the five sites were quite similar. The error bars represent the statistical deviation of 50 measurements taken from 10 subjects during the 5-day period. These results indicate that the THz endoscope using an optical fiber-guided laser pulse is a very stable and reliable system.

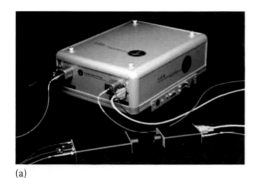

Product specifications	
Spectral range	0.1–3 THz
Freq. resolution	11 GHz
Time resolution	20 fs
Waveform rate	200–500 Hz
SNR (measures at peak, collimated transmission)	>40 dB – 500 Hz >60 dB – 1 Hz (averaged)
Geometry	Transmission Normal reflection Pitch-catch (Nonnormal reflection)
Configuration	Separated emitter and receiver heads
Software	User-extensible software via plug-in architecture

(a)

(b)

T-Ray™ 4000 system

Parameter	Specifications	Units	Comments
Bandwidth	0.02 to 2	THz	3 THz option available
Polarization extinction ratio	> 20:1		
Signal-to-noise ratio	> 70	dB	Frequency
Rapid scan range	320	ps	
Rapid scan rate	100	Hz	
Long scan range	2.8	ns	
Power requirements	110/220	VAC	50–60 Hz, 4 A
Size	1.75 × 19.5 × 7	inches	W × D × H
Weight	55	lbs	

(c)

Technical specification

THz pulsed spectrometer

Terahertz source	Laser-gated photo-conductive semiconductor emitter
Terahertz detector	Laser-gated photo-conductive semiconductor receiver
Laser	Ti:Sapphire ultrashort pulsed laser
Spectral range	0.06 THz–3 THz (2 cm^{-1}–100 cm^{-1})
Dynamic range	>4 OD @ 0.9 THz (30 cm^{-1}) in transmission
Spectral resolution	0.0075 THz (0.25 cm^{-1})
Rapid scan	30 scans/second with 1.2 cm^{-1} spectral resolution
A/D converter	16 bit

Signal-to-noise	THz	Signal/dB	SNR
	1 min rapidscan acquisition @ 1.2 cm^{-1} resolution		
Transmission	0.15	65	5000
	0.3	70	6500
	0.91	70	11000
	1.52	60	5000
	2.58	43	700
Air	0.15	50	1500
	0.3	58	2700
	0.91	57	1800
	1.52	49	1000
	2.58	30	150

FIGURE 6.17 Commercial THz spectroscopy system. (The parameters/specifications are from manual data sheet at the systems.) (a) Fico from Zomega. (Reprinted from Zomega, *Products*, Zomega Terahertz Corporation, New York, 2013, http://www.z-thz.com/index.php?option=com_content&view=article&id=51&Itemid=59.) (b) T-Ray 4000 from Picometrix. (Reprinted from PICOMETRIC, *T-Ray 4000 TD-THz System*, Picometrix's Corporation, Ann Arbor, MI, 2013, http://www.picometrix.com/documents/pdf/T-Ray4000%20DS1.pdf.) (c) TPS spectra 3000 from Teraview. (Reprinted from TeraView, *Terahetz Equipment*, TeraView's Corporation, Cambridge, U.K., 2013, http://www.teraview.com/products/terahertz-pulsed-spectra-3000/index.html.)

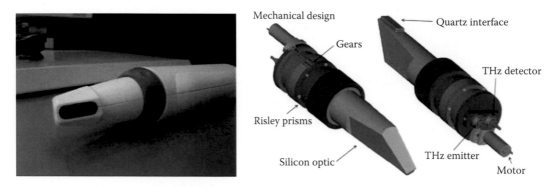

FIGURE 6.18 Medical imaging probe of TPS Spectra 3000. (Reprinted from TeraView, *Terahetz Equipment*, TeraView's Corporation, Cambridge, U.K., 2013, http://www.teraview.com/products/terahertz-pulsed-spectra-3000/index.html.)

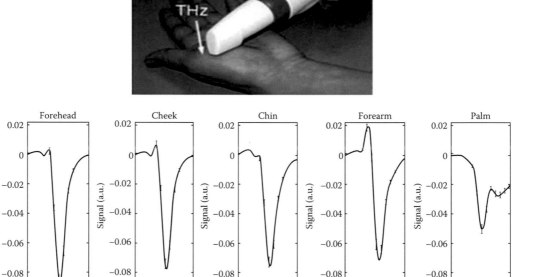

FIGURE 6.19 Average THz pulse response recorded using the handheld THz probe for the five different skin sites (Pickwell and Wallace 2006). (Reproduced with permission from Pickwell, E. and Wallace, V., Biomedical applications of terahertz technology, *J. Phys. D: Appl. Phys.*, 39: R301, 2006. Copyright 2006, Institute of Physics.)

6.3.3 Applications of Fiber-Coupled THz Endoscope System

6.3.3.1 Fiber-Coupled THz Spectroscope System

Fiber-coupled antennas for ultrafast coherent THz spectroscopy were introduced by Crooker (2002). A 20 m length of fiber was used. Because of the long optical fiber, the laser pulses were precompensated for by the grating pair. A fiber ferrule was directly put in contact with the Tx or Rx chips, as shown in Figure 6.20. The miniature fiber-coupled THz Tx and Rx modules were tested under conditions of

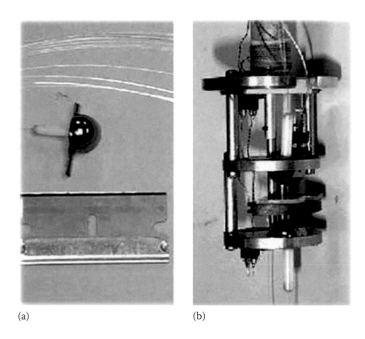

(a)   (b)

FIGURE 6.20 Photographs of (a) a fiber-coupled THz antenna and (b) THz probe (Crooker 2002). (Reproduced with permission from Crooker, S.A., Fiber-coupled antennas for ultrafast coherent terahertz spectroscopy in low temperatures and high magnetic fields, *Rev. Sci. Instr.*, 73, 3258–3264, 2002. Copyright 2002, American Institute of Physics.)

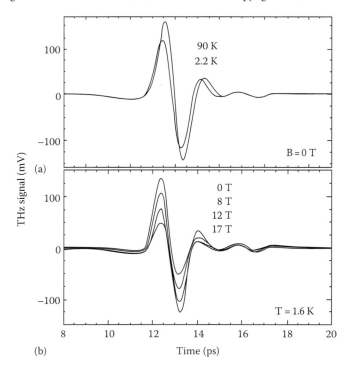

FIGURE 6.21 The measured THz pulses from an in situ fiber-coupled Rx and Rx pair. (a) Temperature variation. (b) Magnetic field variation (Crooker 2002). (Reproduced with permission from Crooker, S.A., *Rev. Sci. Instr.*, Fiber-coupled antennas for ultrafast coherent terahertz spectroscopy in low temperatures and high magnetic fields, 73, 3258–3264, 2002. Copyright 2002, American Institute of Physics.)

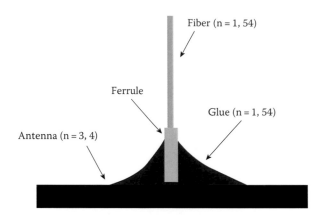

FIGURE 6.22 Schematic of the fiber-coupled antenna (Vieweg et al. 2007). (Reproduced with permission from Vieweg, N. et al., Presented at *Proceedings of SPIE* 6616, 66163M, Munich, Germany, 2007. Copyright 2007, Society of Photo Optical Instrumentation Engineers.)

low temperature and high magnetic field. The amplitude of the measured THz pulses was found to vary slightly with temperature, but the THz pulse width was unchanged. However, the amplitude of the THz pulses decreased with increasing magnitude field, as shown in Figure 6.21. The temperature- and magnetic-field-dependent THz pulse variation is related to characteristic changes of the materials, including changes of the fiber, ferrule, and Tx or Rx semiconductor chips.

Recently, an advanced fiber-coupled antenna for THz spectroscopy was presented, as shown in Figure 6.22 (Vieweg et al. 2007). By gluing the fiber directly to the photoconductive switch, the Tx and Rx modules are made compact and stable. Moreover, this method protects against contamination. If there is an air gap between the fiber end and the chip, there are two reflection losses before and after the air gap. Because the refractive index of the glue and that of the optical fiber are the same, reflection loss occurs only one time between the fiber end and the chip. Also, the difference of refractive indexes between the fiber and the chip is small, leading to only small reflection loss.

6.3.3.2 In Vivo Applications Using THz Endoscope System

In order to fabricate a miniaturized THz endoscope, the diameter of the THz Tx or Rx modules should be less than 5 mm diameter. Figure 6.24 provides a schematic of the designed and miniaturized THz Tx and Rx modules; this model is 26 mm long including the fiber ferrule (Ji et al. 2009). Each module consists of a fiber ferrule, a silicon lens (4 mm diameter), an optical lens (3 mm diameter), and a Tx chip (1.8 mm × 1.9 mm) or an Rx chip (2 mm × 2.8 mm). With a 6 mm outside diameter of the silicon holder, the modules are cut on both side surfaces to reduce the width to 4 mm, as shown in Figure 6.23c. When the Tx and Rx modules are attached in parallel, their cross section has dimensions of 8 mm (2 × 4 mm) × 6 mm.

The experimental setup for the THz endoscope system with fiber-coupled Tx and Rx is schematically represented in Figure 6.24 (Ji et al. 2009). To compensate for the positive GVD occurring in the beam isolator and the optical fibers, the laser pulses are dispersion compensated using a grating pair before being injected into the 2 m long single-mode optical fiber (Figure 6.25).

When the Tx and Rx are in a straight line at a distance of 2 cm, the measured THz pulse and spectrum are as shown in Figure 6.26. The THz pulse is obtained using a single measurement without any filtering process. Since the Tx and Rx are located close to each other, the electric field coming from the dc bias of the Tx chip led to a 520 pA offset of the THz pulse. The inset shows the noise from 0 to 3 ps. The SNR approaches 12,000:1. The FWHM of the measured THz pulse is found to be 0.5 ps; the corresponding amplitude spectrum, found using a numerical Fourier transform, extended to beyond 2 THz.

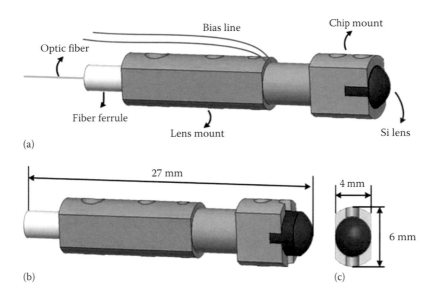

FIGURE 6.23 Schematic of the designed miniaturized THz Tx and Rx modules. (a) Tx module. (b) Rx module. (c) Cross section of the Rx module.

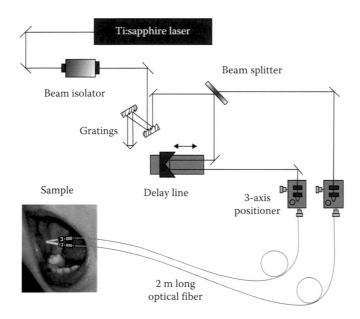

FIGURE 6.24 Experimental setup for THz endoscope (Ji et al. 2009). (Reproduced with permission from Ji, Y.B., Lee, E.S., Kim, S.-H., Son, J.-H., and Jeon, T.-I., A miniaturized fiber-coupled terahertz endoscope system, *Opt. Exp.*, 17, 17082–17087, 2009. Copyright 2009, Optical Society of America.)

Although all optical and electrical parts are fixed into the tiny module, the properties of the THz signal make this device suitable for use in a THz endoscope.

Because water is a highly absorptive material and also comprises about 70% of the human body, the THz transmission method cannot be applied to the human body. The reflection method should be used for the THz endoscope system. In other words, in order to obtain reflected THz pulses,

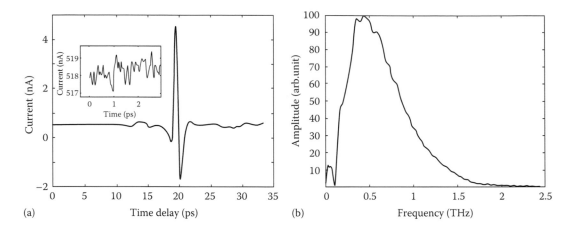

(a) Time delay (ps)   (b) Frequency (THz)

FIGURE 6.25 (a) Measured THz pulse using the miniaturized fiber-coupled THz endoscope system with Tx and Rx in a straight line at a distance of 2 cm. (b) Spectrum (Ji et al. 2009). (Reproduced with permission from Ji, Y.B., Lee, E.S., Kim, S.-H., Son, J.-H., and Jeon, T.-I., A miniaturized fiber-coupled terahertz endoscope system, *Opt. Exp.*, 17, 17082–17087, 2009. Copyright 2009, Optical Society of America.)

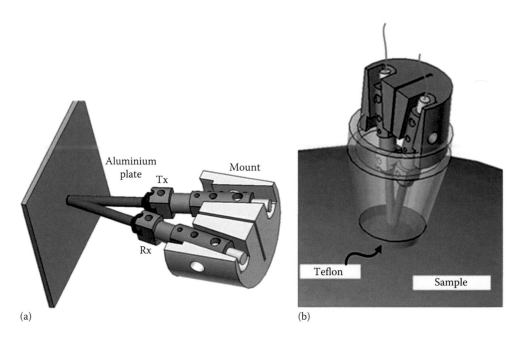

(a)   (b)

FIGURE 6.26 (a) Tx and Rx modules with an angle of 20° on a custom-made mount. (b) Teflon cap attached to Tx and Rx modules.

a new measurement system was designed and fabricated, with a schematic provided in Figure 6.26a. The Tx and Rx modules are fixed at an angle of 20° on a custom-made mount; the reflection distance between the Si lenses and the sample should always be kept at the same distance. However, it is very difficult to maintain such an exact distance for in vivo samples such as those of the stomach or colon; therefore, a Teflon cover is attached, as shown in Figure 6.26b. Due to the Teflon cover, the in vivo samples are always kept at a constant distance and a flat target surface is always maintained; this flat surface is in contact with the Teflon cap.

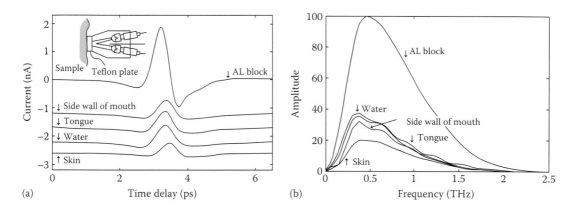

FIGURE 6.27 (a) Measured THz pulses using reflection measurement. (b) The respective spectra (Ji et al. 2009). (Reproduced with permission from Ji, Y.B., Lee, E.S., Kim, S.-H., Son, J.-H., and Jeon, T.-I., A miniaturized fiber-coupled terahertz endoscope system, *Opt. Exp.*, 17, 17082, 2009. Copyright 2009, Optical Society of America.)

After the measurement of the THz reflection from the Al surface, which is used for reference, the Al surface is replaced by samples such as those from the side wall of the mouth, the tongue, palm skin, and water (Ji et al. 2009). Figure 6.27a shows the measured reflective THz pulses; these measured pulses are numerically shifted for comparison.

The amplitude values of the reflected THz pulses from the samples are small compared to the reflected THz pulses from the Al surface because of the small reflective indexes. Figure 6.27b shows the spectra of the THz pulses. The reference spectrum extends to 2.5 THz; the relative amplitude of the other samples extends to 2.0 THz because of the lower reflection at high frequencies. The amplitude values for the side wall of the mouth, the tongue, and water are very similar; however, that of the palm skin is much lower because the refractive index of the skin is somewhat lower than that of the other samples.

Figure 6.28 shows the measured refractive index and power absorption using the frequency-dependent magnitude and phase information of the measured THz pulses. The side wall of the mouth, the tongue, and water have similar refractive indexes and power absorption values; however, the skin has much lower characteristics. The results when using THz endoscope measurement are in very good agreement with those found using THz reflection spectroscopy measurement without the fiber-coupled system (Pickwell and Wallace 2006). Using the THz reflection spectroscopy setup, Pickwell and Wallace measured the refractive index and power absorption of water, healthy skin, and human adipose (fat) tissue. The refractive index and power absorption of healthy skin are lower than those values for water, as shown in Figure 6.29. The THz reflection spectroscopy measurement and THz endoscope measurement yielded similar results.

6.4 Future Prospects

For applications of the THz endoscope, the Tx and Rx modules will need more compact size because THz endoscopes and optical endoscopes are required to work together. However, because of the silicon lenses in the Tx or Rx modules, the thickness of the THz endoscope cannot be reduced to less than two times the lens diameter. Therefore, it is necessary to combine the Tx and Rx modules, which combination will be called a transceiver module and which will have only one silicon lens. The THz endoscope system with the transceiver module will in the future be used for in vivo measurement of the stomach or the colon, as shown in Figure 6.30.

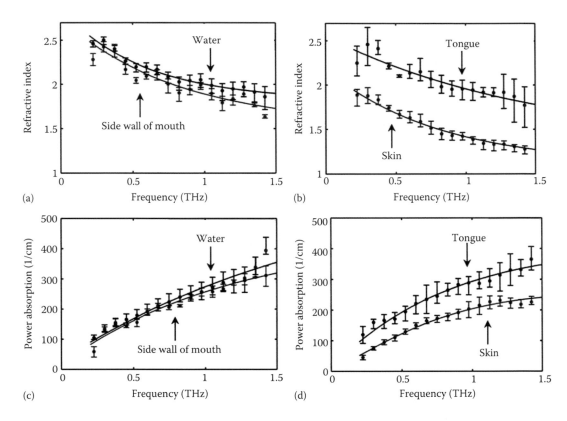

FIGURE 6.28 Measured refractive index and power absorption of samples using miniaturized fiber-coupled THz endoscope system. (a) Refractive index of water and side wall of the mouth. (b) Refractive index of tongue and skin. (c) Power absorption of water and side wall of the mouth. (d) Power absorption of tongue and skin (Ji et al. 2009). (Reproduced with permission from Ji, Y.B., Lee, E.S., Kim, S.-H., Son, J.-H., and Jeon, T.-I., A miniaturized fiber-coupled terahertz endoscope system, *Opt. Exp.*, 17, 17082–17087, 2009. Copyright 2009, Optical Society of America.)

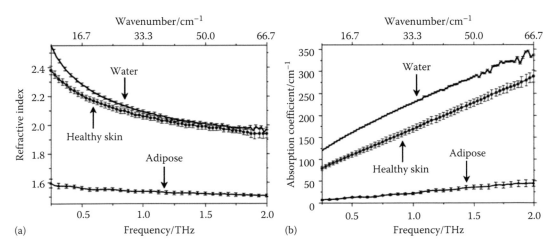

FIGURE 6.29 Measured refractive index and power absorption of samples using THz reflection spectroscopy without fiber-coupled system. (a) The refractive index and (b) power absorption coefficient of water, skin tissue, and adipose tissue (Pickwell and Wallace 2006). (Reproduced with permission from Pickwell, E. and Wallace, V., Biomedical applications of terahertz technology, *J. Phys. D: Appl. Phys.*, 39, R301, 2006. Copyright 2006, Institute of Physics.)

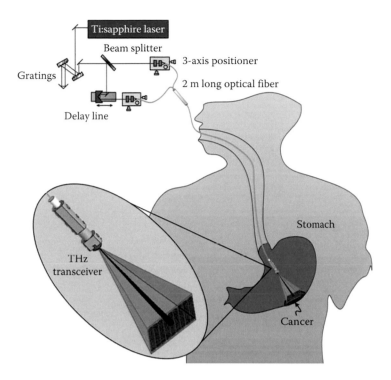

FIGURE 6.30 An application of THz endoscope with THz transceiver module.

References

Crooker, S. A. 2002. Fiber-coupled antennas for ultrafast coherent terahertz spectroscopy in low temperatures and high magnetic fields. *Review of Scientific Instruments* 73: 3258–3264.

Duvillaret, L., F. Garet, J.-F. Roux, and J.-L. Coutaz. 2001. Analytical modeling and optimization of terahertz time-domain spectroscopy experiments, using photoswitches as antennas. *IEEE Journal of Selected Topics in Quantum Electronics* 7: 615–623.

Ellrich, F., T. Weinland, D. Molter, J. Jonuscheit, and R. Beigang. 2011. Compact fiber-coupled terahertz spectroscopy system pumped at 800 nm wavelength. *Review of Scientific Instruments* 82: 053102.

Gallot, G., S. P. Jamison, R. W. McGowan, and D. Grischkowsky. 2000. Terahertz waveguides. *Journal of the Optical Society of America B* 17: 851–863.

Goubau, G. 1950. Surface waves and their application to transmission lines. *Journal of Applied Physics* 21: 1119–1128.

Jeon, T.-I., J. Zhang, and D. Grischkowsky. 2005. THz Sommerfeld wave propagation on a single metal wire. *Applied Physics Letters* 86: 161904.

Ji, Y. B., E. S. Lee, S.-H. Kim, J.-H. Son, and T.-I. Jeon. 2009. A miniaturized fiber-coupled terahertz endoscope system. *Optics Express* 17: 17082–17087.

Keller, U. and L. Gallmann. 1997. *Ultrafast Laser Physics*. Zurich, Switzerland: ETH Zurich. http://www.ulp.ethz.ch/education/ultrafastlaserphysics/3_Dispersion_compensation.pdf, accessed January 15, 2013.

Lee, Y., S. Tanaka, N. Uetake et al. 2007. Terahertz time-domain spectrometer with module heads coupled to photonic crystal fiber. *Applied Physics B* 87: 405–409.

Marcuvitz, N. 1986. *Waveguide Handbook* (IEEE Electromagnetic Waves Series). London, U.K.: Peter Peregrinus.

McGowan, R. W., G. Gallot, and D. Grischkowsky. 1999. Propagation of ultrawideband short pulses of terahertz radiation through submillimeter-diameter circular waveguides. *Optics Letters* 24: 1431–1433.

Mendis, R. and D. Grischkowsky. 2001. THz interconnect with low-loss and low-group velocity dispersion. *IEEE Microwave and Wireless Components Letters* 11: 444–446.

Parrott, E. P., S. M. Sy, T. Blu, V. P. Wallace, and E. Pickwell-MacPherson. 2011. Terahertz pulsed imaging in vivo: Measurements and processing methods. *Journal of Biomedical Optics* 16: 106010.

Pickwell, E. and V. Wallace. 2006. Biomedical applications of terahertz technology. *Journal of Physics D: Applied Physics* 39: R301.

Pickwell-MacPherson, E. 2010. Practical considerations for in vivo THz imaging. *IEEE Transactions on Terahertz Science and Technology* 3: 163–171.

PICOMETRIC. 2013. *T-Ray 4000 TD-THz System*. Ann Arbor, MI: Picometrix's Corporation. http://www.picometrix.com/documents/pdf/T-Ray4000%20DS1.pdf, accessed January 15, 2013.

Sommerfeld, A. 1952. Part II—Derivation of the phenomena from the Maxwell equations. In *Electro-dynamics*, ed. Sommerfeld, A., pp. 177–190. New York: Academic Press.

Sy, S., S. Huang, Y.-X. J. Wang et al. 2010. Terahertz spectroscopy of liver cirrhosis: Investigating the origin of contrast. *Physics in Medicine and Biology* 55: 7587.

TeraView. 2013. *Terahertz Equipment*. Cambridge, U.K.: TeraView's Corporation (http://www.teraview.com/products/terahertz-pulsed-spectra-3000/index.html).

Vieweg, N., N. Krumbholz, T. Hasek et al. 2007. Fiber-coupled THz spectroscopy for monitoring polymeric compounding processes. Presented at *Proceedings of SPIE* 6616, 66163M, Munich, Germany.

Wang, K. and D. M. Mittleman. 2004. Metal wires for terahertz wave guiding. *Nature* 432: 376–379.

Wentworth, F. L., J. C. Wiltse, and F. Sobel. 1961. Quasi-optical surface waveguide and other components for the 100- to 300-GHz region. *IRE Transactions on Microwave Theory and Techniques* 9: 512–518.

Woodward, R. M., B. E. Cole, V. P. Wallace et al. 2002a. Terahertz pulse imaging in reflection geometry of human skin cancer and skin tissue. *Physics in Medicine and Biology* 47: 3853.

Woodward, R. M., V. P. Wallace, B. E. Cole et al. 2002b. Terahertz pulse imaging in reflection geometry of skin tissue using time domain analysis techniques. presented at *Proceedings of SPIE* 4625: 160–169.

Zomega. 2013. *Products*. New York: Zomega Terahertz Corporation. http://www.z-thz.com/index.php?option=com_content&view=article&id=51&Itemid=59, accessed January 15, 2013.

II

Fundamental Biological Studies by Terahertz Waves

7

Terahertz Characteristics of Water and Liquids

Seung Jae Oh
Yonsei University

Seungjoo Haam
Yonsei University

Jin-Suck Suh
Yonsei University

7.1 Introduction

Terahertz (THz) technology has emerged as a novel biomedical method for diagnosing cancer and studying the dynamics of biomaterials (Fitzgerald et al. 2006; Kawase et al. 2003; Oh et al. 2007, 2009, 2011; Pickwell and Wallace 2006; Woodward et al. 2003). In order to further investigate biomedical sciences and technologies, studies need to be conducted on the dynamics of biological molecules, such as DNA, proteins, and lipids (Son 2009). Most biological molecules can only function in biological environments. Therefore, an understanding of biological environments is essential to develop biomedical technologies based on biological sciences. Liquids, which include water, are major components that determine the biological environment. The interaction between liquids and biological molecules can indicate the biological molecular dynamics (MD) information based on the biological environment. The intra- and intermolecular dynamics in biological molecules are caused by hydrogen bonding between water and biological molecules or between water molecules themselves. The typical timescale of the hydrogen bonding network of water at room temperature is in the picosecond range, which corresponds to a few THz (~10^{12} Hz) in frequency. The bulk dielectric relaxation and oscillatory motion of liquids occur in the THz frequency range, which is positioned between the microwave and far-infrared frequency ranges. This frequency range can also be used as a bridge to connect the spectral results of the microwave and infrared regions. For several decades, THz spectroscopy has been used to explain the dielectric relaxation, hydrogen bonding network, and ultrafast dynamics of liquids including water and molecular, polar, and nonpolar liquids.

This chapter presents some important results obtained from fundamental research on the dynamics of water and liquids, which was performed using THz spectroscopy, and briefly discusses useful applications using the properties of liquids. Water molecules are the main components and activators in a biological medium, which includes cells, proteins, and DNA. Therefore, this chapter starts with an introduction to water in Section 7.2. This section reviews the history of the THz spectroscopy of water molecules in Section 7.2.1. The THz properties of water molecules are explained for various phases, such as vapor, ice, and liquids, in Section 7.2.2. In Section 7.2.3, the THz optical properties of liquid-phase water are discussed with a special focus on the hydrogen bonding network. In Section 7.3, the THz optical properties of liquids are discussed using the polarities of liquid molecules because THz waves have a strong interaction with the transient dipole moments and show a large absorption difference between polar and nonpolar liquids. A liquid mixture composed of two types of liquids is discussed in Section 7.4. This section includes studies of electrolyte solutions, which are essential processes for maintaining life, such as metabolism and osmotic action.

7.2 Terahertz Characteristics of Water

7.2.1 Historical Review

Water is the most important liquid in biological media. The biochemical interactions between liquid water and biological molecules determine the biological environment and its activities. Despite the importance of liquid water, the anomalous physical and chemical reactions of liquid water molecules, which are caused by their hydrogen bonds, are not completely understood. Numerous studies have been performed using a variety of experimental and theoretical methods since 1892, when Röntgen first determined the structure of liquid water (Rontgen 1892). Later in 1971, Rahman et al. were the first to study water using MD simulations (Rahman and Stillinger 1971). Numerous MD simulations have been performed since then. On the other hand, experimental studies of the intermolecular resonant or dielectric relaxation processes were implemented using several methods, including microwave and Fourier transform infrared (FTIR) spectroscopy, FTIR laser, Raman scattering, and optical-heterodyne detected Raman-induced scattering (Afsar and Hasted 1977; Czumaj 1990; Mizoguchi et al. 1992; Simpson et al. 1979; Vij and Hufnagel 1989). These experimental methods indicated that water MDs occurred in the infrared band greater than 100 cm^{-1}. These dynamics included H-bond stretching, restricted translation, parallel to OH$\cdots$O, O–O stretching, and longitudinal phonon. Microwave spectroscopy is used to investigate the intermolecular dynamics in the spectral range below 10 cm^{-1}. The experimental results of Guillot et al. in the far-infrared spectrum agreed well with the MD simulation results based on the ab initio MD simulation proposed by Car et al. (Car and Parrinello 1985; Guillot et al. 1991). Barthel et al. showed that a dielectric relaxation model, such as the multiple Debye model, could be used to study the fast and slow relaxation processes of solvents, including water (Barthel and Parrinello 1991). The development of femtosecond-pulse lasers and other optoelectronics techniques enabled detection in the broad THz frequency region. After the emergence of the THz spectroscopy, studies were conducted on the spectral band gap between the microwave and infrared region that were relevant to bulk dielectric relaxation and oscillatory intermolecular motions. In 1989, Grischkowsky et al. (1989) determined the emission and detection of THz signals in free space, thus developing the THz time-domain spectroscopy (TDS) technique (Fattinger and Grischkowsky 1989). In 1990, Hu et al. developed a THz detection technique using the electrooptic sampling method (Hu et al. 1990). THz TDS enabled the study of material properties for semiconductors, dielectric materials, vapors, and liquids in the THz frequency range. In 1995, the temperature-dependent THz spectroscopy of water MDs was reported by Thrane et al. using the reflection spectroscopy technique (Thrane et al. 1995). In 1966, Kindt et al. published THz spectroscopy results of polar liquids, including water, using a transmission-type system (Kindt and Schmuttenmaer 1996). They showed the fast and slow reorientation times of hydrogen bond dynamics using multiple Debye model fitting. A year later, the fast and slow reorientation times and their interactions were verified experimentally using a mid-infrared pump–probe experiment by Woutersen et al. (1997). In 1997, Rønne et al. reported the temperature-dependent Debye relaxation time in the THz frequency region (Rønne et al. 1997, 1999, 2002). They also reported the temperature-dependent dielectric relaxation dynamics of heavy water and water from a super-cooled state to near the boiling point so that they could understand the temperature dependence, isotope shift, and relaxation time of the water. Using a simulation based on the Debye model, Pickwell et al. were able to distinguish significant differences between normal and abnormal tissues (Pickwell et al. 2004). Later, in 2008, Yada reported the dielectric relaxation time of water and heavy water using THz time-domain attenuated total reflection (ATR) spectroscopy, rather than the conventional reflection or transmission-type spectroscopy (Yada et al. 2008). This study showed that the fast relaxation component of water and heavy water was governed by the collision processes without hydrogen bonding. Recently developed high-power THz sources, which generate signals using the photoionization of air or nonlinear crystals such as $LiNbO_3$, lead to the study of the nonlinear phenomena of water molecules (Nagai et al. 2010).

7.2.2 Vapor and Ice

7.2.2.1 Water Vapor

Water vapor is the gas phase, which is typically produced from the evaporation or boiling of liquid water. The simple structure of the water vapor aids in the understanding of the molecular structure and dynamics of water. The optical properties of water vapor were investigated using various optical spectroscopy techniques, including microwave and Fourier transform spectroscopy. To precisely analyze the experimental results, information regarding the relevant line center, strength, and broadening was required. THz TDS offered the most accurate line values for the spectrum gap between the microwave and infrared regions. Martin van Exter et al. were the first to apply THz TDS to water vapor (Exter et al. 1989; Figure 7.1).

They measured the THz spectrum of water vapor with a 5 GHz resolution in the frequency range of 0.2–1.45 THz and identified the nine strongest lines. Although the line centers in the spectrum of water vapor were precisely measured, the measurements of the line strength and line broadening were affected by the temperature and local molecular environment. Cheville et al. reported the far-infrared properties and self-broadening rotational line widths of high-temperature water vapor (Cheville and Grischkowsky 1999). They reported the change in rotational line broadening from water vapors at high temperatures for the transition frequencies of 1.0–2.5 THz. To obtain the rotational transitions of water vapor at high temperatures, THz TDS of a near-stoichiometric propane–air flame was implemented at 1490 K. They showed that the experiment results could be fitted with a numerical convolution of the Lorentz line shape. The line widths for 29 pure rotational transitions were obtained. These results agreed with the predicted line strength and center frequency.

7.2.2.2 Ice

Ice is the solid phase of water. As the temperature decreases below a critical point, the water molecules form a hexagonal structure that is assembled by hydrogen bonding. Examining the structure of ice yields information regarding the hydrogen bonding network of water molecules in ice. In addition, frozen biological samples yield more static hydrogen bonding information as compared to biological samples containing liquid water. Therefore, studies of ice provide fundamental research on the intra- and intermolecular interactions between water and biomolecules. Numerous studies on ice have been performed, and dielectric relaxation, which is related to hydrogen bonds, has been reported in the broad frequency regions from microwave to ultraviolet. However, studies in the THz frequency region, related to the molecular reorientation dynamics, were only performed by a few groups owing to the lack of an effective THz wave source. Mishima et al. reported the absorption constant at 0.25–0.75 and 0.75–1.3 THz using methods such as grating monochromator and bolometer THz spectroscopy, respectively (Mishima et al. 1983). However, the imaginary part of refractive index was 30% lower than the imaginary indices, which were extracted from the measured data at 10–100 GHz by Matsuoka et al. (1996). In order to verify the fitting results and fill in the gap for THz frequencies in Matsuoka et al.'s study, Zhang et al. used temperature-dependent THz TDS (Zhang et al. 2001; Figure 7.2).

They obtained the complex THz optical constants and showed that their result was consistent with the results of other groups. The measured complex refractive constant at 1 THz was $1.793 + 0.0205i$.

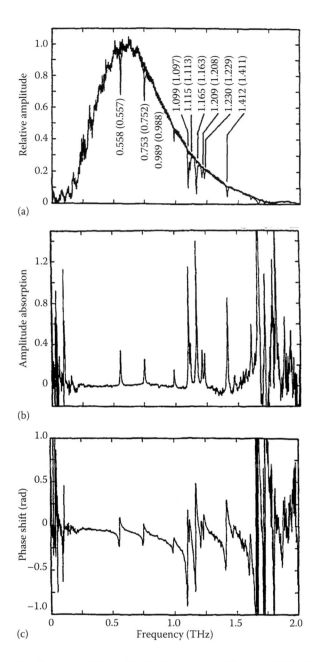

FIGURE 7.1 The (a) amplitude spectra, (b) amplitude absorption coefficient, and (c) relative phase shift of water vapor. (Reprinted with permission from Exter, M., Fattinger, C., and Grischkowsky, D., Terahertz time-domain spectroscopy of water vapor, *Opt. Lett.*, 14, 1128, 1989. Copyright 2013, American Institute of Physics.)

Its imaginary index was much lower than that of liquid water, 2.1 + 0.56i. They proposed a theoretical model that successfully described the temperature-dependent changes in the imaginary refractive index. Furthermore, the small refraction index of ice implied that the penetration depth of THz waves in tissues could be enhanced by freezing the tissues. Hoshina et al. froze porcine tissues below −33°C to decrease the large absorbance of the liquid water. They were clearly able to distinguish the differences in the THz signal between the muscle and adipose tissue and showed that the THz wave transmittance of tissues was greatly increased when the water was frozen (Hoshina et al. 2009).

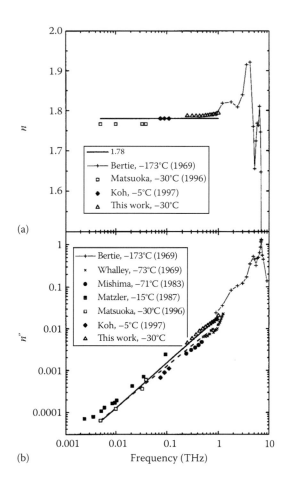

FIGURE 7.2 (a) Real and (b) imaginary refractive indices of ice in the range of 10^9–10^{13} Hz. Experimental data points were obtained by various groups and are labeled by the first author's names in the graph (Bertie et al. 1969; Koh 1997; Matsuoka et al. 1996; Matzler and Wegmuller 2000; Whalley and Labbé 1969). The solid and dashed lines in (b) are calculated using Equation 7.2 (Zhang et al. 2001). (Reprinted with permission from Zhang, C., Lee, K.-S., Zhang, X.-C., Wei, X., and Shen, Y., Optical constants of ice Ih crystal at terahertz frequencies, *Appl. Phys. Lett.*, 79, 491, 2001. Copyright, 2013, American Institute of Physics.)

7.2.3 Liquid Water

7.2.3.1 Hydrogen Bonding

A hydrogen bond is the attractive interaction between an electronegative atom and a hydrogen atom that binds the electronegative atom by a covalent bond. The hydrogen bond energy of water is around 2.6 kJ/mol, which is smaller than the energy of a covalent or ionic bond but larger than the van der Waals force (Rønne et al. 2002). Water and biomolecules such as DNA and proteins contain hydrogen bonds. The intra- and intermolecular dynamics of biological media can be represented by hydrogen bonds. The unique properties of liquid water are due to the hydrogen bonds existing between water molecules. A water monomolecule contains two electronegative lone pairs due to oxygen, each of which permits a hydrogen bond with other water molecules. Such hydrogen bonds between water molecules enable water to exhibit different phases, such as the liquid and solid phases. The unique properties of water, such as high boiling and melting temperatures and a higher viscosity than most other liquids, are because of the greater number and strength of intermolecular hydrogen bonds in them.

When solutes are dissolved in water, water hinders the intra- and intermolecular hydrogen bonding between the two solute molecules, and the donors and acceptors on solute molecules form a hydrogen bond network with other water molecules.

The hydrogen bond dynamics of water molecules can be represented by the permanent and induced dipole moments, which are produced by the intra- and intermolecular interactions. Since the average lifetime of a hydrogen bond is on the order of subpicoseconds, the reorientation dynamics of permanent and induced dipole moments are represented by the dipole relaxation model in low THz frequencies. The Debye model is commonly used to explain the dynamics of hydrogen bonds as a dipole relaxation of water molecule groups (Barthel and Buchner 1991; Møller et al. 2009). This model is based on the proposition that the dipole moment of a liquid is composed of a permanent dipole moment (P_μ) caused by the local electric field interacting with thermal motion and induced polarization (P_α). Both dipole moments are depicted by Equations 7.1 through 7.3:

$$\vec{P} = \varepsilon_0(\varepsilon - 1)\vec{E}, \tag{7.1}$$

$$\vec{P}_\mu(\omega,t) = \varepsilon_0(\varepsilon - \varepsilon_\infty)\vec{E}(t)L_{i\omega}\left[f_P^{or}(t) \right], \tag{7.2}$$

$$\vec{P}_\alpha = \varepsilon_0(\varepsilon_\infty - 1)\vec{E}, \tag{7.3}$$

where
$L_{i\omega}\left| f_P^{or}(t) \right|$ is the Laplace transform
$f_P^{or}(t)$ is the pulse-response function defined as $(1+i\omega\tau)^{-1}$

Therefore, the combination of the dielectric relaxation constants with dc conductivity can be written as

$$\hat{\varepsilon}(\omega) = \varepsilon_\infty + \sum_{j=1}^{n} \frac{\varepsilon_j - \varepsilon_{j+1}}{1 + i\omega\tau_j}, \tag{7.4}$$

where
ε_∞ and ε_s are the infinite and static dielectric constant, respectively
ε_2 is the intermediate permittivity
τ_1 and τ_2 are the slow and fast relaxation time constants, respectively

Both relaxation times are relevant to the collective reorientation motion, hydrogen bonding formation, and decomposition dynamics. Numerous studies, which have investigated the hydrogen bonding motion of water, have been implemented using this model. These studies are discussed in detail in the following section.

7.2.3.2 Molecular Dynamics of Water

The MD of liquid water can be characterized using the intra- and intermolecular dynamics due to hydrogen bonding. The interaction of water molecules, which have a high polarity, can be represented by permanent and induced dipole moments. The Debye relaxation model has been used to investigate the dielectric relaxation motion of a water molecule based on the dipole moments. Few groups have reported that the hydrogen bonding dielectric relaxation motions of water can be classified into two time components: picosecond and femtosecond timescales. Barthel et al. reported the fast and slow relaxation processes of solvents using the multiple Debye model (Barthel and Buchner 1991). They obtained the dielectric relaxation time using millimeter waves below 500 GHz. The water spectrum above 500 GHz was first measured by Thrane et al. They applied the temperature-dependent THz spectroscopy method to investigate the water MDs using reflection type TDS (Thrane et al. 1995). Their experiment showed

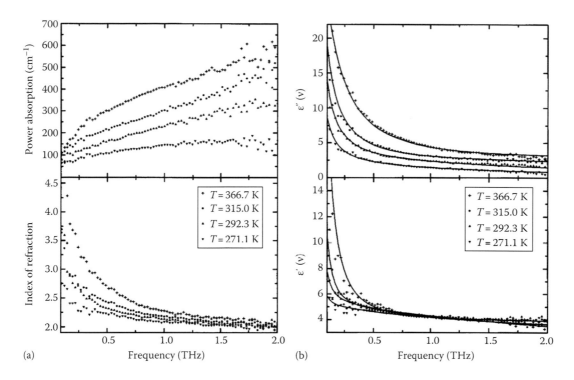

FIGURE 7.3 (a) The refractive index and power absorption of liquid water shown as a function of frequency at four different temperatures. (b) The complex dielectric constant of liquid water (points) as a function of frequency at four different temperatures. The lines show the fits of the double Debye model. (Reprinted with permission from Rønne, C., Thrane, L., and Åstrand, P.-O., Investigation of the temperature dependence of dielectric relaxation in liquid water by THz reflection spectroscopy and molecular dynamics simulation, *J. Chem. Phys.*, 107, 5319, 1997. Copyright 2013, American Institute of Physics.)

that some parts of the orientational relaxation in liquid water occurred without hydrogen bonds breaking because the extracted enthalpy of activation was smaller than the bonding energy. Kindt et al. implemented the THz transmission-type spectroscopy for polar liquids including water. They determined the fast and slow reorientation times of water molecules using multiple Debye model fitting (Kindt and Schmuttenmaer 1996). Rønne et al. later obtained the temperature-dependent Debye relaxation time in the THz frequency region (Rønne et al. 1997; Figure 7.3).

Their experimental results agreed well with those of several molecular simulation models. The fast decay time is related to the fast reorientation of single water molecules. Rønne et al. studied these phenomena using a TDS reflection setup (Rønne et al. 1999). They focused on a two-component model to explain the liquid water MDs. The model hypothesizes that two liquids phases, which are a low-density liquid (LDL) and high-density liquid (HDL), can coexist. The two-component model usually can be used to account for the many thermodynamic features of liquid water. They found that the slow relaxation time changed with temperature, although any correlation between the fast relaxation time and temperature was within the experimental uncertainty. This indicated that the slow decay time, which is included in the single Debye model, has a relationship with the structural relaxation of water molecules. The fast relaxation time did not have a relationship with the structure of liquid water. They also determined the temperature-dependent dielectric relaxation dynamics of heavy water and water from a super-cooled state to near the boiling point. They showed that the temperature dependence and isotope shift conform to the water molecular structural relaxation with slow decay times. Yada et al. attempted to add a third relaxation time, which represented the microscopic motion of water molecules (Yada et al. 2008). They used THz time-domain ATR spectroscopy rather than conventional reflection or transmission-type spectroscopy to determine the dielectric relaxation time of water and heavy water. Their study showed that the fast relaxation component of water and heavy water was governed by the collision process without hydrogen bonding. Further, they reported the complex dielectric constants of the liquid water isotopes, H_2O, D_2O, and $H_2{}^{18}O$ in the broadband region from 0.2 to 7 THz and showed the dispersion of the complex susceptibility due to the intermolecular stretching vibration around 5 THz. They were able to determine that the fastest relaxation time can be related to the rotation motion of a single water molecule.

7.3 Terahertz Characteristics of Liquids

7.3.1 Nonpolar Liquids

Liquids can be widely classified into two categories: polar and nonpolar, depending on whether their molecular structures have polarities. The dynamics of polar liquids consists of the dipole–dipole intermolecular interactions and hydrogen bonds. These molecular interactions influence the movement of individual molecules and are organized strongly through the collective, diffusive, and reorientation motions due to their polarity. The approximate value of a liquid's polarity can be represented by the dielectric constant of the liquid. The dielectric constant of water, which has a strong polar nature, is 80 at 20°C. In contrast, liquids with dielectric constants less than 15 are generally considered to be nonpolar. Nonpolar liquids such as oil, fat, benzene, and cyclohexane have almost no polarity and are water insoluble, that is, they are hydrophobic. The transient dipole moments induced by collisions in the liquid are the reason for the low absorption constant of the nonpolar liquids. In 1992, Pedersen and Keiding measured the optical properties of nonpolar liquids such as benzene, carbon tetrachloride, and cyclohexane using THz TDS (Pedersen and Keiding 1992). Carbon tetrachloride was the heaviest of the three molecules and showed that absorption peak was around 1.2 THz, which represented the slower dynamics of the induced dipole moments. The absorption coefficients of benzene were an order of magnitude larger than that of cyclohexane because the π-orbitals of benzene had a larger contribution compared to the σ-electrons of cyclohexane. To account for the dielectric response of nonpolar molecules in the THz range, they suggested two important models. The first model was based on the proposition that the response was due to collisions, which induced the transient dipole moments. The second model for benzene and carbon tetrachloride determined the absorption spectrum directly from the multipole moments of the molecules. The low absorption constants of the nonpolar liquid in the THz frequency region support the optical and electrical properties of colloidal nanoparticles or solvent electron excluding the effects of the solvent. Knoesel et al. reported the electron dynamics of the solvent electron in *n*-hexane using the optical pump–THz probe method. They obtained the electrical parameter using the Drude model and showed that the quasifree electrons in the solvent dominated the THz signal modulation due to the optical pump. One of the reported applications of nonpolar liquids to the THz technique is a variable focus lens for use in the THz range (Scherger et al. 2011). They were able to alter the focal length by changing the volume of medical white oil on the inside of the lens body.

7.3.2 Polar Liquids

7.3.2.1 Polar Protic Liquids

Polar liquids have large absorption coefficients in the THz range, which are approximately 10–100 times greater than those of nonpolar liquids. It is due to the dipole–dipole intermolecular interaction and hydrogen bonds between polar molecules. The polar liquids can be classified as protic and aprotic liquids depending on whether they exhibit hydrogen bonding. Protic liquids have intermolecular hydrogen bonding as well as ion–dipole interaction owing to their O–H or N–H bond. Thus, both cations and anions, which are, respectively, positively and negatively charged species, can be dissolved in polar

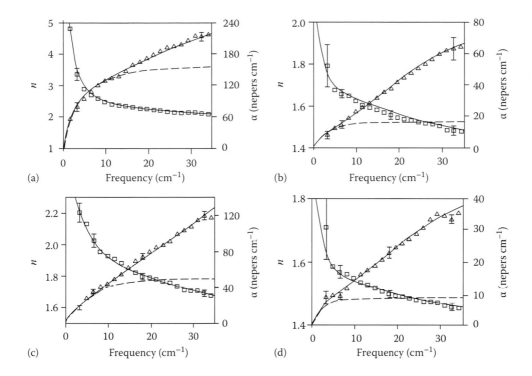

FIGURE 7.4 Comparison of the experimental and double Debye relaxation model fitted data. Solid lines are double Debye relaxation model fitted data of (a) water, (b) methanol, (c) ethanol, and (d) 1-propanol. Triangles indicate the frequency-dependent power absorption coefficient and squares indicate the index of refraction. Dashed line is R(ĩ) calculated with parameters from Barthel et al. (1990). (Reprinted with permission from Kindt, J. and Schmuttenmaer, C., Far-infrared dielectric properties of polar liquids probed by femtosecond terahertz pulse spectroscopy, *J. Phys. Chem.*, 100, 10373, 1996. Copyright 2013 American Chemical Society.)

protic liquids. Water, ethanol, and methanol are examples of protic liquids. This section presents the few reported THz spectroscopy results. In 1991, Barthel and Buchner reported that dielectric relaxation studies can be an efficient method for determining the MDs of a solution (Barthel and Buchner 1991). They showed that protic liquids can be represented by three relaxational processes using the multiple Debye relaxation model: (1) intramolecular rotation of solvent molecules as a monomer and in hydrogen bonding networks, (2) the re-formation of the perturbed solvent structure, and (3) the fast relaxation time of around 1 ps because of the hydrogen bonding. In 1996, Kindt and Schmuttenmaer studied the dielectric properties of water, methanol, ethanol, 1-propanol, and liquid ammonia using THz TDS (Kindt and Schmuttenmaer 1996; Figure 7.4).

They extended their dielectric relaxation study to the high-frequency region and compared their THz spectroscopic data with the data from the microwave study of Barthel et al. (1990). They also used the multiple Debye relaxation and Cole–Cole models to describe the liquids (Barthel et al. 1990). They determined that in the case of alcohols, extending the data set to 1 THz led to a triple Debye model that had a significantly faster relaxation time for the second and third Debye processes compared to water. Then, they measured the power absorption spectrum of liquid ammonia below −33°C. The results showed a similar curve to water. However, although the dipole moment of ammonia is lower than that of water, the absorption coefficient of water is lower than that of ammonia in the THz region. This may be due to the weaker hydrogen bonding in ammonia. In 2009, Moller et al. compared the experimental and theoretical results of water and water–ethanol mixtures in the THz frequency range (Møller et al. 2009). The link between the dielectric relaxation function at low frequencies and vibration mode in intra- and intermolecular processes at high frequencies was established. They showed that THz waves could distinguish the difference between a benign aqueous mixture and hazardous liquids, such as fuels and organic solvents. The aqueous mixtures are discussed in detail in Section 7.4.

7.3.2.2 Polar Aprotic Liquids

Polar aprotic liquids have strong dipole–dipole interactions, but no hydrogen bonding, due to the absence of O–H or N–H bonds. Therefore, in contrast to protic liquids, which strongly solvate positively and negatively charged solutes, aprotic liquids only solvate positively charged solutes. Acetone, acetonitrile, dimethyl sulfoxide (DMSO), and dimethylformamide (DMF) are examples of aprotic liquids. DMSO and DMF, which are miscible with water, are commonly used as solvents for chemical reactions. DMF in the liquid phase has a high dipole moment, and its strong dipole–dipole interactions govern the intermolecular dynamics. In 1990, Buchner and Yarwood reported the THz experimental results of DMF diluted in carbon chloride. They measured the absorption coefficient of the diluted DMF in carbon tetrachloride (Buchner and Yarwood 1990). They showed that the dilution of DMF affected the total translational and rotational band profiles in THz regions and showed that the short- and long-time spectral density were dependent on the surrounding molecular interactions, similar to other organic solvents. They also determined that at the lowest frequency, as the percentage of DMF, gradually decreased the dielectric relaxation time (τ_D) due to the reduction of dipole–dipole interaction. Their results showed that collective reorientation exhibited faster dynamics at the diluted state.

DMSO permeates the skin very easily. It is commonly used for the delivery of cosmetics and drugs via the skin. In 2012, Kim et al. investigated using drug delivery through the skin for medicine solvated in DMSO (Kim et al. 2012). They predicted the time-dependent change in the distribution and penetration of DMSO, including the medicine, using THz 2D and B-scan imaging. DMSO is usually used as a cryoprotectant. It decreases cell death during the freezing process by reducing ice formation. The study of DMSO as a cryoprotectant may be helpful for the THz spectroscopic imaging study of freezing tissues of cells, which were recently reported to suppress the strong water effects (Sim et al. 2013).

7.4 Terahertz Characteristics of Liquid Mixtures

7.4.1 Aqueous Mixtures

Most liquids in biological media typically exist as a mixture. However, studies on liquids are only now beginning to consider pure liquids. In addition, liquids and liquid mixtures have been rarely studied, despite the need for the investigation of their properties for biological sciences and technologies. This section reviews the MD studies of liquid mixtures such as aqueous and nonaqueous mixtures and electrolyte solutions, using THz spectroscopy. Conventional studies of liquid mixtures are typically performed to formulate a hypothesis for mixture behaviors as an ideal mixture model and to compare this model with the experimental results obtained for actual mixtures. Spectroscopic techniques, including Raman, infrared, far-infrared, microwave, and nuclear magnetic resonance spectroscopy, as well as MD simulations, have been used to explain the structure and dynamics of mixtures. In 1998, Venables and Schmuttenmaer reported the dynamics in an acetonitrile and water mixture using THz TDS (Venables and Schmuttenmaer 1998; Figure 7.5).

They obtained the absorption and refractive indexes of aqueous mixtures with various volume fractions. The experimentally measured complex optical constants, as a function of the volume fraction, were compared with the ideal extracted values, which were extracted using an ideal mixture equation and the THz optical constants of neat liquids. The measured optical constants did not agree with the expected values for an ideal mixture. The dielectric behavior and the relaxation time of the actual and ideal mixtures were obtained by double Debye model fitting. The dielectric constants and relaxation times of various volume fractions between actual and ideal mixtures were compared. The results

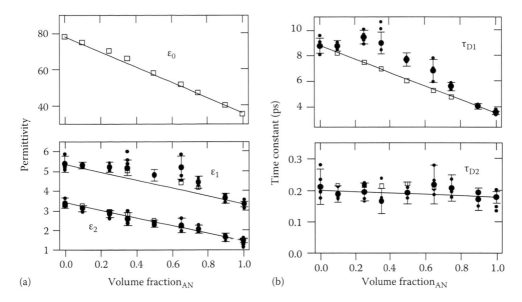

FIGURE 7.5 The results of fitting a double Debye model to the measured data. Part (a) shows the static and high-frequency permittivity as a function of the acetonitrile volume fraction (vol. fraction), and part (b) shows the Debye relaxation time constants for the two processes, also as a function of volume fraction. The large filled circles with error bars represent the fitted parameters and 1σ uncertainty. The solid lines represent an ideal mixture. The open squares represent the results of fitting the double Debye model to an ideal dataset generated by combining the absorption coefficient and index of refraction of the two neat liquids based on their volume fractions using simple equations. The faster Debye process characterized by τ_{D2} and ε_2 behaves essentially ideally, but the slower process characterized by τ_{D1} and ε_1 shows significant deviations from ideality. The relaxation time constant for the actual mixture is roughly 25% longer than for an ideal mixture when the acetonitrile volume fraction is between 25% and 65%. (Reprinted with permission from Venables, D. S. and Schmuttenmaer, C. A., Far-infrared spectra and associated dynamics in acetonitrile–water mixtures measured with femtosecond THz pulse spectroscopy, *J. Chem. Phys.*, 108, 4935, 1998. Copyright 2013, American Institute of Physics.)

indicated that these mixtures could be treated as a mixture model when the mixture was considered a uniform solution, even though the mixture contained two components. Further, they extended their work to include the case of the binary mixtures of water with acetone, acetonitrile, and methanol (Venables and Schmuttenmaer 2000a; Figure 7.6).

They reported that the absorption coefficients for the mixtures were substantially smaller compared to those for ideal mixtures. In addition, the Debye relaxation time constants for the actual mixtures were longer than those for the ideal mixtures, which was similar to the previous results. In 2009, Uffe Møller et al. conducted an ATR THz spectroscopy study of a water–ethanol mixture (Møller et al. 2009). They showed that the dielectric function, as the microscopy dynamics of water–ethanol mixtures, was relevant to their macroscopic thermodynamics properties using dielectric relaxation and vibrational models. In particular, they found that the mixing volume fraction of ethanol–water mixtures was correlated with the intermediate and fast relaxation times and enthalpy of mixtures. They also showed that the inspection of liquids inside bottles was feasible by exploiting the dielectric function of aqueous mixtures (Jepsen et al. 2008).

7.4.2 Nonaqueous Mixtures

Compared with aqueous mixtures, nonaqueous mixtures have received less attention because they barely exist in nature. Therefore, only a few THz studies of nonaqueous mixtures have been reported. In 1996, Flanders studied the frequency-dependent absorption coefficients of $CHCl_3$, CCl_4, and their mixtures by pulsed THz time-domain transmission spectroscopy and fitted the curves to the absorbance spectra of

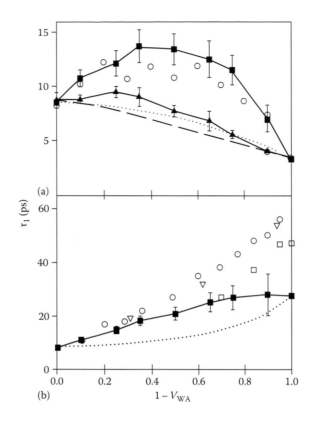

FIGURE 7.6 The slow Debye time constants of (a) the acetone/water (filled squares) and acetonitrile/water (filled triangles) mixtures and (b) the methanol/water mixtures (filled squares). The time constants based on the ideal optical constants are shown as a dotted line for the acetone/water and methanol/water systems, and as a dashed line for water/acetonitrile mixtures. Literature data for acetone/water mixtures are from Kumbharkhane at 25°C (open circles) (Kumbharkhane et al. 1996). For the methanol/water system, our data are compared to microwave data from D. Bertolini (30°C) (Bertolini et al. 1983), S. Mashimo (23°C) (Mashimo et al. 1989), and Kaatze (25°C) (Kaatze et al. 1989) shown as open squares, open circles, and open triangles, respectively. (Reprinted with permission from Venables, D. S. and Schmuttenmaer, C. A., Spectroscopy and dynamics of mixtures of water with acetone, acetonitrile, and methanol, *J. Chem. Phys.*, 113, 11222, 2000. Copyright 2013, American Institute of Physics.)

the liquid mixtures using a single Mori function (Flanders et al. 1996). These were based on the mole-fraction-weighted sums of the absorption coefficients of pure $CHCl_3$ and CCl_4. This analysis indicated that a bulk dipole reducing mechanism was caused by the clustering of $CHCl_3$ molecules around CCl_4. Therefore, this work showed that the integrated absorption coefficient for the collisionally induced absorption of $CHCl_3$–CCl_4 collisions was less than that of $CHCl_3$–$CHCl_3$ collisions by 2.6 ± 0.4 THz cm^{-1}. In 2000, Venables and Schmuttenmaer studied the structure and dynamics of nonaqueous mixtures such as acetone/methanol, acetonitrile/methanol, and acetone/acetonitrile and of dipolar liquids by infrared and THz spectroscopy (Venables and Schmuttenmaer 2000b). They compared mixtures constituting a nonassociating and an associating liquid of methanol, as well as mixtures of two nonassociating dipolar liquids. The methanol-containing mixtures provided an interesting comparison with aqueous mixtures since water is also an associating liquid. In their nonaqueous mixture, the absorption coefficients and refraction indices of the mixtures changed monotonically when changing from one neat liquid to another. The composition dependence of the absorption coefficients and refraction indices of the mixtures demonstrated that their mixtures showed an ideal behavior, except at high frequencies in acetone/methanol mixtures. The single difference between aqueous and nonaqueous mixtures is that aqueous mixtures have isosbestic points above 2.4 THz in the refractive index of acetone/methanol mixtures. They also found the absorption coefficients to be approximately 0.45, 1.35, and 2.7 THz for acetone/methanol, acetonitrile/methanol, and acetone/acetonitrile mixtures, respectively.

7.4.3 Electrolyte Solutions

Bodily fluids in tissues and cells exist in the form of an electrolyte solution, and the appropriated balance of the electrolyte concentration between the intra- and extracellular environments is a key factor for maintaining life. Consequently, understanding the dynamics of this solution is necessary to investigate certain biological phenomena as well as to diagnose and treat many diseases. THz waves have been shown to be a helpful method for investigating the dynamics of electrolyte solutions because the THz frequency response is based on two properties: the solution conductivity by electrolytes and the dynamics of these waves in terms of their freely collective, diffusive, reorientational motions in liquid. When water molecules meet ion molecules, the water molecules surround the electrolyte molecules with an opposite charge and serve as a means of hydration. Simply put, water molecules form a sheath around ions via electric gravitation. THz spectroscopy has been applied to investigate such charged molecular systems. Dodo et al. investigated the THz spectroscopy of electrolyte solutions using synchrotron orbit radiation and a Martin–Puplett-type Fourier transform spectrometer (Dodo et al. 1993). They observed the THz absorbance of various concentrations of LiCl solution and concluded that as the electrolyte concentration increases, the absorption coefficients decrease. These studies also demonstrated that the dynamics of the reduction in the absorption coefficient of the electrolyte solution could be explained as being due to the Debye relaxation of the electric dipole motion of water molecules. Zoidis et al. performed THz spectroscopy studies on the solutions of NaCl, LiCl, and HCl in water to investigate the ionic effects (Figure 7.7; Zoidis et al. 1999).

They deliberated over the ionic interactions of the electrolyte solution with respect to several parameters such as conductivity at low frequencies, water MD, M + H_2O rattling modes, and corresponding changes in the molecular structure of water as well as the collective mode for ions. They measured the THz optical constants at solute concentrations ranging from 0 to 10 M and temperatures ranging from −100°C to 80°C. A single Debye relaxation model was applied to handle and analyze the obtained data. They showed experimentally that as the electrolyte concentration increased, the difference in the THz spectral intensity was smallest for the NaCl solution even at concentrations as high as 5 M and that the increase in the ratio of THz absorbance of the LiCl solution was the highest. They explained that this was due to the network breaking and motion of the restricted H_2O molecules. Further, it was also noted that there could be a contribution due to the higher frequency of the Li + H_2O *rattling* mode caused by decreased mass. They found that the HCl solution showed a reversed effect, which was determined to be the result of increased proton polarizability. Dodo et al. (2002) studied the transport properties of ions in highly concentrated electrolyte solutions (Dodo et al. 2002). Transmission and reflection experiments were used to extract the refractive indices and absorption coefficients of LiCl liquids at various concentrations in the frequency range of 0.4–1.2 THz. They found that the behavior of ions in concentrated electrolyte solutions may be depicted as a high-density coupled plasma regardless of the ion species used. However, the collective plasma oscillations of the ions were not observed in the measured frequency range, even when the concentration of LiCl was varied between 2 and 13 mol/L, and the ion species was changed to support the different ion masses. They concluded that the plasma oscillation of ions was determined by the long-range collective interactions between ions, which are caused by short-range interactions between the ions and water molecules. Oh et al. (2007) considered the biological medium as a binary mixture, which consisted of pure water and molecular materials such as electrolytes (Oh et al. 2007). The THz optical constants of a Luria–Bertani (LB) media as biological mixtures were measured with various concentrations of ingredients. The ideal mixture using mass fractions of ingredients was used to extract the optical constants of the electrolytes. They showed that the conductivities were relevant to the electrolyte concentration using the modified multiple Debye model and by adding the dc conductivity equation. They found that slow relaxation times were shortened in the case of high electrolyte concentrations, whereas fast relaxation times

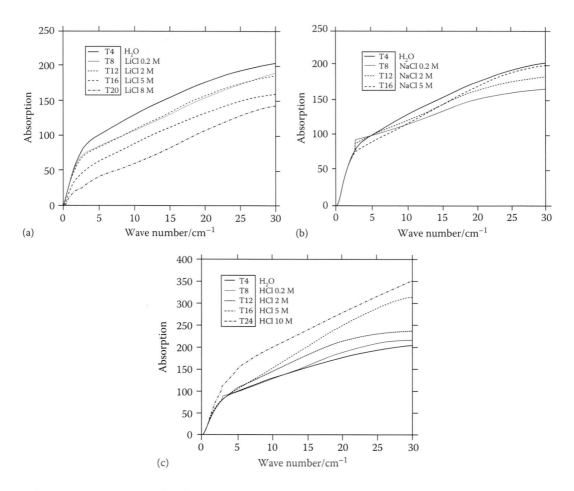

FIGURE 7.7 THz spectra of LiCl (a), NaCl (b), and HCl (c) solution in water at 20°C and at different concentrations. (Reprinted with permission from Zoidis, E., Yarwood, J., and Besnard, M., Far-infrared studies on the intermolecular dynamics of systems containing water. The influence of ionic interactions in NaCl, LiCl, and HCl solutions, *J. Phys. Chem. A*, 103, 220, 1999. Copyright 2013 American Chemical Society.)

decreased only slightly. The investigation of the THz properties of electrolyte solutions could involve basic studies in biomedical subject areas such as neurology. Masson et al. (2006) demonstrated that THz near-field microscopy allows the functional imaging of excitable living cells such as neurons and revealed that axonal Na⁺ accumulation was associated with water exchange between intra-axonal and extracellular compartments (Masson et al. 2006). Further, they detected variations in ion concentration as small as 10 μM in 20 fL of water. Finally, through the investigation of the central neural tube of *Lumbricus terrestris* worms, they showed that THz waves could be employed to probe the dynamics of electrolyte solutions in cells and tissues.

References

Afsar, M. N. and J. B. Hasted. 1977. Measurements of the optical constants of liquid H_2O and D_2O between 6 and 450 cm^{-1}. *Journal of the Optical Society of America A* 67: 902–904.

Barthel, J., K. Bachhuber, R. Buchner, and H. Hetzenauer. 1990. Dielectric spectra of some common solvents in the microwave region. Water and lower alcohols. *Chemical Physics Letters* 165: 369–373.

Barthel, J. and R. Buchner. 1991. High frequency permittivity and its use in the investigation of solution properties. *Pure and Applied Chemistry* 63: 1473–1482.

Bertie, J., H. Labbe, and E. Whalley. 1969. Absorptivity of Ice I in the range 4000–30 cm^{-1}. *The Journal of Chemical Physics* 50: 4501.

Bertolini, D., M. Cassettari, and G. Salvetti. 1983. The dielectric properties of alcohols–water solutions. I. The alcohol rich region. *The Journal of Chemical Physics* 78: 365–372.

Buchner, R. and J. Yarwood. 1990. Far-infrared studies of molecular dynamics and interactions in *N,N*-dimethylformamide. *Molecular Physics* 71: 65–77.

Car, R. and M. Parrinello. 1985. Unified approach for molecular dynamics and density-functional theory. *Physical Review Letters* 55: 2471–2474.

Cheville, R. and D. Grischkowsky. 1999. Far-infrared foreign and self-broadened rotational linewidths of high-temperature water vapor. *JOSA B* 16: 317–322.

Czumaj, Z. 1990. Absorption coefficient and refractive index measurements of water in the millimetre spectral range. *Molecular Physics* 69: 787–790.

Dodo, T., M. Sugawa, and E. Nonaka. 1993. Far infrared absorption by electrolyte solutions. *The Journal of Chemical Physics* 98: 5310–5313.

Dodo, T., M. Sugawa, E. Nonaka, and S.-i. Ikawa. 2002. Submillimeter spectroscopic study of concentrated electrolyte solutions as high density plasma. *The Journal of Chemical Physics* 116: 5701.

Fattinger, C. and D. Grischkowsky. 1989. Terahertz beams. *Applied Physics Letters* 54: 490–492.

Fitzgerald, A. J., V. P. Wallace, M. Jimenez-Linan et al. 2006. Terahertz pulsed imaging of human breast tumors1. *Radiology* 239: 533–540.

Flanders, B., R. Cheville, D. Grischkowsky, and N. Scherer. 1996. Pulsed terahertz transmission spectroscopy of liquid CHCl$_3$, CCl$_4$, and their mixtures. *The Journal of Physical Chemistry* 100: 11824–11835.

Guillot, B., Y. Guissani, and S. Bratos. 1991. A computer-simulation study of hydrophobic hydration of rare gases and of methane. I. Thermodynamic and structural properties. *The Journal of Chemical Physics* 95: 3643–3648.

Hoshina, H., A. Hayashi, N. Miyoshi, F. Miyamaru, and C. Otani. 2009. Terahertz pulsed imaging of frozen biological tissues. *Applied Physics Letters* 94: 123901.

Hu, B., X. C. Zhang, D. Auston, and P. Smith. 1990. Free-space radiation from electro-optic crystals. *Applied Physics Letters* 56: 506–508.

Jepsen, P. U. and J. K. Jensen. 2008. Characterization of aqueous alcohol solutions in bottles with THz reflection spectroscopy. *Optics Express* 16: 9318–9331.

Kaatze, U., R. Pottel, and M. Schäfer. 1989. Dielectric spectrum of dimethyl sulfoxide/water mixtures as a function of composition. *The Journal of Physical Chemistry* 93: 5623–5627.

Kawase, K., Y. Ogawa, Y. Watanabe, and H. Inoue. 2003. Non-destructive terahertz imaging of illicit drugs using spectral fingerprints. *Optics Express* 11: 2549–2554.

Kim, K. W., K.-S. Kim, H. Kim et al. 2012. Terahertz dynamic imaging of skin drug absorption. *Optics Express* 20: 9476–9484.

Kindt, J. and C. Schmuttenmaer. 1996. Far-infrared dielectric properties of polar liquids probed by femtosecond terahertz pulse spectroscopy. *The Journal of Physical Chemistry* 100: 10373–10379.

Koh, G. 1997. Dielectric properties of ice at millimeter wavelengths. *Geophysical Research Letters* 24: 2311–2313.

Kumbharkhane, A., S. Helambe, M. Lokhande, S. Doraiswamy, and S. Mehrotra. 1996. Structural study of aqueous solutions of tetrahydrofuran and acetone mixtures using dielectric relaxation technique. *Pramana* 46: 91–98.

Mashimo, S., S. Kuwabara, S. Yagihara, and K. Higasi. 1989. The dielectric relaxation of mixtures of water and primary alcohol. *The Journal of Chemical Physics* 90: 3292.

Masson, J. B., M. P. Sauviat, J. L. Martin, and G. Gallot. 2006. Ionic contrast terahertz near-field imaging of axonal water fluxes. *Proceedings of the National Academy of Sciences of the United States of America* 103: 4808–4812.

Matsuoka, T., S. Fujita, and S. Mae. 1996. Effect of temperature on dielectric properties of ice in the range 5–39 GHz. *Journal of Applied Physics* 80: 5884–5890.

Matzler, C. and U. Wegmuller. 2000. Dielectric properties of freshwater ice at microwave frequencies. *Journal of Physics D: Applied Physics* 20: 1623.

Mishima, O., D. Klug, and E. Whalley. 1983. The far-infrared spectrum of ice Ih in the range 8–25 cm. Sound waves and difference bands, with application to Saturn's rings. *The Journal of Chemical Physics* 78: 6399.

Mizoguchi, K., Y. Hori, and Y. Tominaga. 1992. Study on dynamical structure in water and heavy water by low-frequency Raman spectroscopy. *The Journal of Chemical Physics* 97: 1961–1968.

Møller, U., D. G. Cooke, K. Tanaka, and P. U. Jepsen. 2009. Terahertz reflection spectroscopy of Debye relaxation in polar liquids [Invited]. *Journal of the Optical Society of America B* 26: A113–A125.

Nagai, M. and K. Tanaka 2010. THz nonlinearity of water observed with intense THz pulses. Presented at *2010 Conference on Lasers and Electro-Optics (CLEO) and Quantum Electronics and Laser Science Conference (QELS)*, San Jose, CA, pp. 1–2.

Oh, S. J., J. Choi, I. Maeng et al. 2011. Molecular imaging with terahertz waves. *Optics Express* 19: 4009–4016.

Oh, S. J., J. Kang, I. Maeng et al. 2009. Nanoparticle-enabled terahertz imaging for cancer diagnosis. *Optics Express* 17: 3469–3475.

Oh, S. J., J. H. Son, O. Yoo, and D. H. Lee. 2007. Terahertz characteristics of electrolytes in aqueous Luria-Bertani media. *Journal of Applied Physics* 102: 074702-1–074702-5.

Pedersen, J. and S. Keiding. 1992. THz time-domain spectroscopy of nonpolar liquids. *IEEE Journal of Quantum Electronics* 28: 2518–2522.

Pickwell, E., B. Cole, A. Fitzgerald, V. Wallace, and M. Pepper. 2004. Simulation of terahertz pulse propagation in biological systems. *Applied Physics Letters* 84: 2190–2192.

Pickwell, E. and V. Wallace. 2006. Biomedical applications of terahertz technology. *Journal of Physics D: Applied Physics* 39: R301.

Rønne, C., P.-O. Åstrand, and S. R. Keiding. 1999. THz spectroscopy of liquid H_2O and D_2O. *Physical Review Letters* 82: 2888.

Rønne, C. and S. R. Keiding. 2002. Low frequency spectroscopy of liquid water using THz-time domain spectroscopy. *Journal of Molecular Liquids* 101: 199–218.

Rønne, C., L. Thrane, P.-O. Åstrand et al. 1997. Investigation of the temperature dependence of dielectric relaxation in liquid water by THz reflection spectroscopy and molecular dynamics simulation. *The Journal of Chemical Physics* 107: 5319.

Rahman, A. and F. H. Stillinger. 1971. Molecular dynamics study of liquid water. *The Journal of Chemical Physics* 55: 3336.

Rontgen, W. 1892. The structure of liquid water. *Annals of Physics* 45: 91–97.

Scherger, B., M. Scheller, C. Jansen, M. Koch, and K. Wiesauer. 2011. Terahertz lenses made by compression molding of micropowders. *Applied Optics* 50: 2256–2262.

Simpson, O. A., B. L. Bean, and S. Perkowitz. 1979. Far infrared optical constants of liquid water measured with an optically pumped laser. *Journal of the Optical Society of America A* 69: 1723–1726.

Son, J.-H. 2009. Terahertz electromagnetic interactions with biological matter and their applications. *Journal of Applied Physics* 105: 102033.

Thrane, L., R. H. Jacobsen, P. Uhd Jepsen, and S. R. Keiding. 1995. THz reflection spectroscopy of liquid water. *Chemical Physics Letters* 240: 330–333.

van Exter, M., C. Fattinger, and D. Grischkowsky. 1989. Terahertz time-domain spectroscopy of water vapor. *Optics Letters* 14: 1128–1130.

Venables, D. and C. Schmuttenmaer. 1998. Far-infrared spectra and associated dynamics in acetonitrile–water mixtures measured with femtosecond THz pulse spectroscopy. *The Journal of Chemical Physics* 108: 4935–4944.

Venables, D. S. and C. A. Schmuttenmaer. 2000a. Spectroscopy and dynamics of mixtures of water with acetone, acetonitrile, and methanol. *The Journal of Chemical Physics* 113: 11222.

Venables, D. S. and C. A. Schmuttenmaer. 2000b. Structure and dynamics of nonaqueous mixtures of dipolar liquids. II. Molecular dynamics simulations. *The Journal of Chemical Physics* 113: 3249.

Vij, J. K. and F. Hufnagel. 1989. Millimeter and submillimeter laser spectroscopy of water. *Chemical Physics Letters* 155: 153–156.

Whalley, E. and H. Labbé. 1969. Optical spectra of orientationally disordered crystals. III. Infrared spectra of the sound waves. *The Journal of Chemical Physics* 51: 3120.

Woodward, R. M., V. P. Wallace, R. J. Pye et al. 2003. Terahertz pulse imaging of ex vivo basal cell carcinoma. *Journal of Investigative Dermatology* 120: 72–78.

Woutersen, S., U. Emmerichs, and H. Bakker. 1997. Femtosecond mid-IR pump-probe spectroscopy of liquid water: Evidence for a two-component structure. *Science* 278: 658–660.

Yada, H., M. Nagai, and K. Tanaka. 2008. Origin of the fast relaxation component of water and heavy water revealed by terahertz time-domain attenuated total reflection spectroscopy. *Chemical Physics Letters* 464: 166–170.

Zhang, C., K.-S. Lee, X.-C. Zhang, X. Wei, and Y. Shen. 2001. Optical constants of ice Ih crystal at terahertz frequencies. *Applied Physics Letters* 79: 491–493.

Zoidis, E., J. Yarwood, and M. Besnard. 1999. Far-infrared studies on the intermolecular dynamics of systems containing water. The influence of ionic interactions in NaCl, LiCl, and HCl solutions. *The Journal of Physical Chemistry A* 103: 220–225.

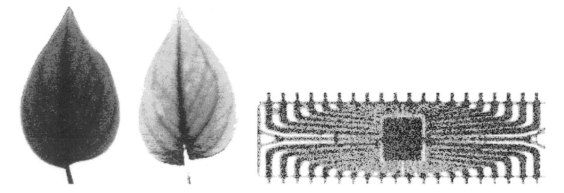

FIGURE 1.2 Demonstration of imaging with THz waves. (Reprinted from Hu, B. and Nuss, M., Imaging with terahertz waves, *Opt. Lett.*, 20, 1716–1718, 1995. With permission of Optical Society of America.)

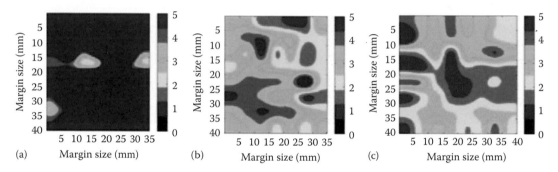

FIGURE 16.3 Diffuse reflectance spectroscopy for tumor margin assessment. Parameter maps per margin were obtained from the ratio of β-carotene (absorption parameter) and the wavelength-averaged reduced scattering coefficient (scattering parameter). Blue areas generally represent healthy tissue, whereas red areas represent tumor. (a) Pathologically confirmed negative margin. (b) Pathologically confirmed margin positive for DCIS. (c) Pathologically confirmed margin positive for IDC. (Reproduced from Wilke, L. G., J. Brown, T. M. Bydlon et al., *Am. J. Surg.*, 198, 566, 2009, with permission from Elsevier.)

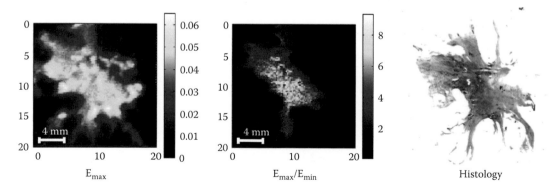

FIGURE 16.4 This figure shows two terahertz images generated using the maximum of the reflected pulse (E_{max}) and the ratio between maximum and minimum of the reflected pulse (E_{max}/E_{min}). In the E_{max} image, all the tissues are shown, tumor with surrounding adipose tissue. In E_{max}/E_{min}, only the tumor is visible and correlated well with the tumor shown in the histology image. (Reproduced from Ashworth, P. C., E. Pickwell-MacPherson, E. Provenzano et al., *Opt. Exp.*, 17, 12444, 2009. With permission of Optical Society of America.)

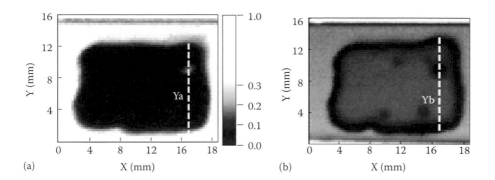

FIGURE 17.3   THz images of an oral melanoma specimen at (a) −20°C and (b) 20°C. (Reprinted with permission from Sim, Y. C., K.-M. Ahn, J. Y. Park, C. Park, and J.-H. Son, *IEEE J. Biomed. Health Inform.*, 17, 779, 2013a. Copyright © 2013 IEEE.)

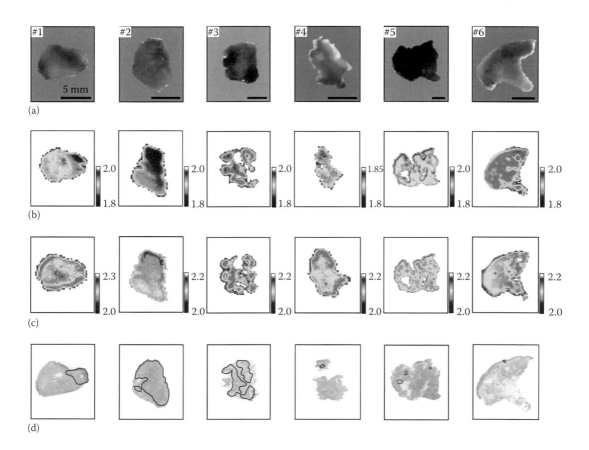

FIGURE 17.7   (a) Visible images, THz images at (b) −20°C and (c) 20°C, and (d) pathological images of oral carcinoma specimens. (Reprinted from Sim, Y. C., J. Y. Park, K.-M. Ahn, C. Park, and J.-H. Son, *Biomed. Opt. Exp.*, 4, 1413, 2013b. Copyright © 2013 by Optical Society of America. With permission of Optical Society of America.)

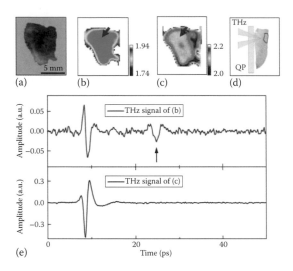

(a)   (b)   (c)   (d)

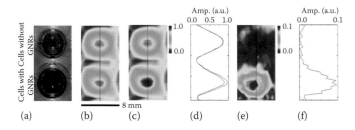

(e)

FIGURE 17.10 THz detection of an infiltrative cancerous tumor hidden within a tissue sample. (a) A visible image, and THz images at (b) –20°C and (c) 20°C of the surface of the specimen and (d) a pathological image from a vertical section. (e) Temporal THz waveforms extracted from the areas designated by the red arrows in (b) and (c). (THz: THz radiation, QP: quartz plate). (Reprinted from Sim, Y. C., J. Y. Park, K.-M. Ahn, C. Park, and J.-H. Son, *Biomed. Opt. Exp.*, 4, 1413, 2013b. Copyright © 2013 by Optical Society of America. With permission of Optical Society of America.)

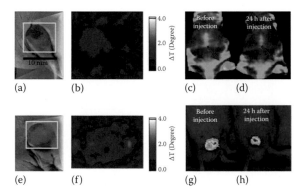

(a)   (b)   (c)   (d)   (e)   (f)

FIGURE 18.5 Images of cancer cells with and without GNRs. (a) Visible image; (b) THz image without IR illumination; (c) THz image with IR illumination; (d) amplitudes along the lines in (b) (black) and (c) (red); (e) differential image between (b) and (c); and (f) amplitude along the line in (e). (Reprinted from Oh, S. J., J. Kang, I. Maeng et al., *Opt. Exp.*, 17, 3469, 2009. Copyright © 2009 by Optical Society of America. With permission of Optical Society of America.)

(a)   (b)   (c)   (d)

(e)   (f)   (g)   (h)

FIGURE 18.9 In vivo THz molecular images with NIR absorption images for comparison. (a) Photograph of a mouse with an A431 cancer tumor; (b) THz molecular image of the mouse shown in (a); (c) and (d) NIR absorption images of the mouse shown in (a) before and 24 h after NPP injection, respectively; (e) photograph of a mouse with a smaller tumor (size = 2.1 cm³); (f) THz molecular image of the mouse shown in (e); (g) and (h) NIR absorption images of the mouse shown in (e) before and 24 h after NPP injection, respectively. ((a,b) With kind permission from Springer Science+Business Media: Oh, S. J., Y.-M. Huh, J.-S. Suh et al., *J. Infrared Millim. Terahertz Waves*, 33, 74, 2012. Copyright © 2012; (c,d) Reprinted with permission from Son, J.-H., Nanotechnology, 24, 214001, 2013. Copyright © 2013 by Institute of Physics Publishing; (e–h) Reprinted from Oh, S. J., J. Choi, I. Maeng et al., *Opt. Exp.*, 19, 4009, 2011. Copyright © 2011. With permission of Optical Society of America.)

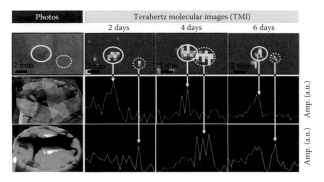

FIGURE 18.11 Tracking of MSCs transplanted into a mouse by TMI of tagged SPIOs. See text for details.

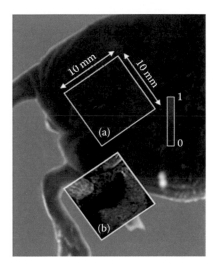

FIGURE 18.12 In vivo (a) TMI and (b) MRI images of a mouse 24 h after transfection of SPIOs into SKOV3 ovarian cancer cells. (From Park, J. Y., H. J. Choi, K.-S. Cho, K.-R. Kim, and J.-H. Son, *J. Appl. Phys.*, 109, 064704, 2011. Copyright © 2011 by IEEE. Reprinted by permission of IEEE.)

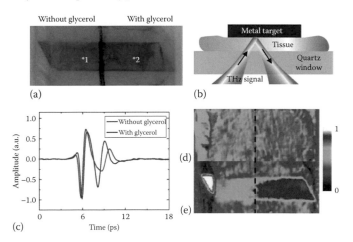

FIGURE 19.1 Enhancement of measurement depth of mouse tissue through the application of glycerol as a THz PEA. (a) Photo of a mouse's abdominal skin placed on a metal target with (*1) and without (*2) glycerol application, (b) measurement scheme, (c) temporal THz waveforms reflected by the tissue, and THz images reconstructed using (d) the first and (e) second peaks of waveforms in (c). The tissue size is 5×3 cm^2. (From Oh, S. J., Kim, S.-H., Jeong, K. et al., Measurement depth enhancement in terahertz imaging of biological tissues, *Opt. Exp.*, 21, 21299, 2013. Copyright © 2013 by Optical Society of America. Reprinted by permission of Optical Society of America.)

8

Probing Solvation Dynamics by Terahertz Absorption Spectroscopy

Benjamin Born
*Weizmann Institute
of Science*

Matthias Heyden
University of California

Simon Ebbinghaus
Ruhr-Universität Bochum

Martina Havenith
Ruhr-Universität Bochum

8.1 Introduction

Water covers more than 70% of the surface of the earth. Nearly all biological processes take place in water. It is now widely recognized that hydration water in the proximity of protein surfaces plays an essential role for the structure, stability, and dynamics of proteins. Hydrogen bond networks have been proposed to stabilize intermediates during protein folding or protein aggregation and have recently been visualized as forming extended water channels in photosystems. Yet, it is part of an ongoing scientific debate whether solvent dynamics influences, or even *enslaves*, the dynamics of biomolecules or, more speculative, whether solvent fluctuations actually contribute to biomolecular (mal)function, for example, in biological assembly, protein–substrate binding, or enzymatic turnover. Going beyond individual proteins, the active role of water in protein interactions, protein aggregation, and *conformational diseases* is currently a topic of intense and even controversial debate. Whereas in bulk water the hydrogen bonds break and reform every picosecond (ps) on average, in the vicinity of huge biological solutes like proteins and enzymes, water molecules show a change in the dynamics of the hydrogen bond network. Terahertz (THz) absorption spectroscopy is an experimental tool to detect subtle changes of nonlocal dipole fluctuations due to intermolecular vibrations, giving direct access to collective water network dynamics. THz absorption spectroscopy allows to report on long-ranged, solute-induced changes of solvent dynamics (solute-induced changes of the THz absorption were found up to 5–6 Å for disaccharides and beyond 15 Å for proteins by THz spectroscopy), which was much more far reaching than had been anticipated before. The combination of THz spectroscopy and stopped-flow methods allows the

135

mapping of changes on ps timescale hydration dynamics in real time. By performing kinetic terahertz absorption (KITA) experiments, changes in hydration dynamics can be observed during biochemical processes, that is, protein folding and enzymatically driven peptide hydrolysis. The experimental results suggest that water in biological contexts is more than a spectator but may support biological function as a mediating chemical matrix between interacting biomolecules, especially during their association, fine alignment, and binding.

The mode by which hydration properties might affect solute–solute interactions and biological function has been the topic of a recent discussion at the interface of biology, chemistry, and physics. There is a general trend to extract information about hydration dynamics from experiments and simulations under equilibrium conditions. As an additional factor, the solute concentration can be tuned. More information can be obtained by perturbing the biomolecule–water network by temperature changes and photoreactions or by utilizing site-directed mutagenesis.

The overall goal is to obtain a comprehensive picture of the role of water network dynamics for biomolecule dynamics, interactions, and function with temporal and spatial resolution. In the following, we will give an overview of the experimental techniques applied to answer these questions and elucidate the specific aspects these techniques can resolve.

For decades, nuclear magnetic resonance (NMR) spectroscopy and x-ray crystallography have been applied to study hydration dynamics of bound water molecules in the first hydration shell (Gallagher et al. 1994) and in cavities (Halle 2004) of proteins or to emphasize transport of water molecules across membranes (Ernst et al. 1995; Fujiyoshi et al. 2002). Both techniques, NMR and x-ray crystallography, provide static and dynamic information about bound water in the first hydration shell of a biomolecule up to 3 Å away from the biomolecular surface. For instance, NMR studies of carbohydrate hydration suggest that carbohydrates only influence water dynamics in the first hydration shell of the carbohydrate solute (Winther et al. 2012). In principle, x-ray crystallography resolves high-resolution structures of biomolecules and their bound water molecules. In addition to structural information, NMR also resolves hydration dynamics ranging from nanoseconds to seconds.

The fast ps dynamics of the hydrogen bond network of hydration water can be resolved by neutron scattering utilizing deuterated proteins to reveal the correlated motions of the hydration shell with the protein. By analyzing radial distribution functions of scattered particles, the observed hydration properties were different from bulk water dynamics up to the second hydration shell. The first hydration shell appeared as more rigid, whereas the second hydration shell contains water molecules with a distinct water dynamics (Frölich et al. 2009; Wood et al. 2008).

Dynamical hydration processes on the timescale of tens to hundreds of ps are monitored by dielectric relaxation spectroscopy at MHz to GHz frequencies. Model peptides with both, hydrophobic and hydrophilic, side chains have been shown to reduce dynamics of their hydration water up to a 9 Å distance from the peptide surface that corresponds to three hydration shells (Head-Gordon 1995; Murarka and Head-Gordon 2008; Nandi et al. 2000).

Subpicosecond (sub-ps) hydration dynamics are accessible by fluorescence spectroscopy using fluorescent local probes (Zhong 2009). Tryptophan fluorescence depends on hydration dynamics due to the pronounced dielectric constant of water. A red shift is caused by stabilization of dipole moments in excited states, which is detected by femtosecond pump–probe laser techniques (Zhong et al. 2002). A local relaxation process of 1–8 ps has been observed and a global process of 20–200 ps (Zhang et al. 2007). The hydration dynamics of apomyoglobin are inhomogeneous and reach beyond 10 Å, which is more than three hydration shells (Zhang et al. 2009). The water network relaxes in an inhomogeneous manner around the protein. The heterogeneity arises from site-specific, local effects caused by electric charge, polarity, or hydrophilicity. Local change translates to global hydration as demonstrated by mutations in protein cavities (Born et al. 2009b; Ebbinghaus et al. 2007; Evans and Brayer. 1990).

These findings depict that protein and water dynamics must be correlated. To depict such a correlation, Frauenfelder suggests a mechanism of solvent slaving by the protein, and in case of noncorrelated dynamics, he suggests the notion of nonslaved motions (Fenimore et al. 2002).

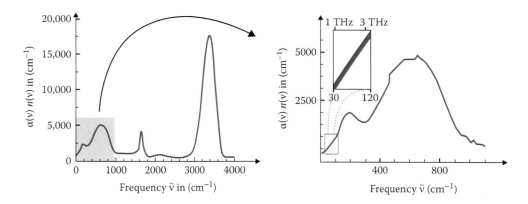

FIGURE 8.1 Computed spectrum of bulk water from the THz to the IR region (Silvestrelli et al. 1997). Left: the IR spectrum of water. Two intermolecular vibrational modes with collective character occur at 200 cm⁻¹ (translational hydrogen bond stretch vibrations) and 600 cm⁻¹ (librational motions). Two intramolecular vibrational modes are observed at 1600 cm⁻¹ (OH-deformation vibration) and 3300 cm⁻¹ (OH-stretch vibration). Right: THz spectrum of water. The absorbance of water increases linearly from 1 to 3 THz (inset). (Reprinted with kind permission from Springer Science+Business Media: *J. Infrared Millimeter Terahertz Waves*, Terahertz dance of proteins and sugars with water, 30, 2009, 1245–1254, Born, B. and Havenith, M., Copyright 2009.)

A method to directly access the coupled protein–water network dynamics is THz absorption spectroscopy. The THz region is the frequency band between the infrared (IR) and microwave region. THz radiation excites collective modes of water and of biological solutes dissolved in water (Dexheimer 2007; Leitner et al. 2006). The dielectric ps and sub-ps motions that contribute to the spectrum of aqueous solutions in the THz region are rotational and translational diffusion and libration dynamics and the large amplitude motions of biological solutes. The collective resonances of diffusional–rotational motions occur at a frequency beyond ≈1 THz; also see Figure 8.1.

Technical advance within the last decade allowed to spectroscopically access the THz region and to close the *THz gap* (Jepsen et al. 2011; Siegel 2002, 2004). Today, water films extending several tens of micrometers can be analyzed by THz spectroscopy (Bergner et al. 2005; Schmuttenmaer 2004) as well as fully hydrated biological solutes like carbohydrates, amino acids (Siegrist et al. 2006; Zhang et al. 2008), model peptides, nucleic acids, lipids, osmolytes, and proteins (Arikawa et al. 2008; Born et al. 2009a,b; Chen et al. 2005a; Dexheimer 2007; Ebbinghaus et al. 2008; Falconer and Markelz 2012; Leitner et al. 2008; Xu et al. 2006b). THz spectroscopy specifically probes long-ranged dynamics of collective networks and displays fast dynamics, which average out in many static or scattering experiments (Born et al. 2009a,b; Ebbinghaus et al. 2008; Heugen et al. 2006; Heyden et al. 2008; Leitner et al. 2008).

THz spectroscopy became a *hot topic* recently, not least because of its potential for biophysical research; an increasing number of research groups have joined the field of studying hydration dynamics of completely hydrated biological solutes (Born and Havenith 2009; Chen et al. 2005b; Ebbinghaus et al. 2007; Heugen et al. 2006; Knab et al. 2007; Markelz 2008b; Plusquellic et al. 2007; Xu et al. 2006a,b; Zhang and Durbin 2006). During the last decade, systematic THz studies and simulations aimed at deciphering the complex combinations of modes that contribute to the THz spectrum of aqueous solutions (Ebbinghaus et al. 2008; Zhang and Durbin 2006). Despite the apparent spectral simplicity, there is a noticeable influence of biological solutes on the collective hydration dynamics as several THz hydration studies of proteins discussed in this chapter revealed (Born and Havenith 2009; Ebbinghaus et al. 2008; Markelz 2008a; Whitmire et al. 2003).

We will now describe how the solute concentration affects the THz spectrum. For wetted protein films of hen egg white lysozyme (HEWL), the THz absorbance increased when the level of hydration reached ratios of minimum 23 g water per gram HEWL (Knab et al. 2006). Enhanced protein flexibility was suggested to be the major component of the increase in THz absorption. Tuning the temperature of the

HEWL–water system around the temperature of the dynamical transition of 200 K, where biological activation is observed, resulted in a significant increase of THz absorption as well as increase in the hydration level (Markelz et al. 2007). A neutron scattering study confirmed the result (Wood et al. 2008).

A different study on myoglobin (Mb) shows that the THz absorbance of the samples was correlated with the hydration level, which ranged from the dry protein powder up to 98 wt% of water. Between 1 and 42 wt%, the THz absorbance increased strongly with the water content of hydrated Mb powders, while a weaker increase was observed for completely dissolved Mb beyond 50 wt% (Zhang and Durbin 2006). Strikingly, a two-component model of interacting hard spheres serving as models for the two interacting species water and proteins was inadequate to describe the experimental results.

We will compare the study at high solute concentration with THz hydration studies in a highly diluted sample to gain further insight to coupled solute–water network dynamics.

Studying the increase of hydration up to a fully solvated sample results in an experimental challenge: since water has a high absorption coefficient, sufficiently powerful THz sources are required in order to measure dilute aqueous solutions.

At Ruhr-University Bochum, we have set up a THz spectrometer that allows precise measurements of the THz absorption coefficients using a high-power pulsed laser that can be operated between 1 and 4 THz (Bergner et al. 2005; Brundermann et al. 2000). Previous THz absorption studies at 2.4 THz of solvated sugars and proteins revealed a nonlinear behavior of the THz absorbance with increasing solute concentration (Born and Havenith 2009; Ebbinghaus et al. 2007; Heugen et al. 2006; Heyden et al. 2008), which could not be interpreted anymore by a two-component model taking into account only water and the biological solute. Only by introducing hydration water as a third component, we were able to model and interpret the nonlinearity of the THz absorbance. The maximum in THz absorption for dilute protein solutions can also be explained qualitatively within this model as detailed in Section 8.2. The maximum (Born and Havenith 2009; Ebbinghaus et al. 2007) in the low-concentration regime (99 wt% water) (Born and Havenith 2009; Ebbinghaus et al. 2007) is reached at solute concentrations at which the density of solute molecules in solution is high enough that the dynamical hydration shells of neighboring solute molecules begin to significantly overlap (Leitner et al. 2008). The occurrence of this maximum at low solute concentrations indicates the presence of large hydration shells, extending 10 Å and more from the protein surface.

For sugars that have a homogeneous surface like lactose (Heugen et al. 2006), trehalose, and glucose (Heyden et al. 2008), the experimentally obtained concentration-dependent THz absorbance could be fitted with quantitative precision to a three-component model yielding the size and absorption cross section of the dynamical hydration shell of each respective carbohydrate. The dynamical hydration shell of the monosaccharide glucose extends approximately 4 Å from the carbohydrate surface and approximately 6 and 7 Å for the disaccharides lactose and trehalose (Heyden et al. 2008). When evaluating the number of hydrogen bond contacts from the carbohydrate to the water network, a direct correlation of THz absorbance change with the available hydrogen bonds between the carbohydrate and water was found. The dynamical hydration shell around the monosaccharide glucose includes approximately 50 water molecules, whereas approximately 150 water molecules and up to 190 water molecules were determined for lactose and trehalose, respectively.

Furthermore, a KITA spectrometer has been set up in Bochum that allows to directly measure the evolution of THz absorbance during self-assembly reactions, such as protein folding, and during metalloenzyme catalysis (Grossman et al. 2011; Kim et al. 2008). In both cases, during folding and enzyme catalysis, conformational changes are preceded by fast changes in the water network dynamics indicating a coupling of protein and water network dynamics.

In parallel to the improvement of THz technology and experiments, theoretical studies have investigated the THz/far-IR spectrum of biomolecules (Ebbinghaus et al. 2007; Leitner et al. 2006, 2008; Whitmire et al. 2003). The coupled biological solute–water network dynamics have been simulated with the goal to decipher the distance dependence of solutes and hydration water (Heyden et al. 2008; Leitner et al. 2006, 2008; Schröder et al. 2006).

8.2 Concepts for Understanding THz Hydration

The experimentally detected interplay of the water network with biological solutes is complex and requires a conceptual description beyond a simple two-component model. A simplified approach to describe the experimental results of our THz hydration studies considers three components and is related to an effective medium theory (Born et al. 2009a; Elber and Karplus 1986; Leitner et al. 2008).

The first assumption is based that biomolecules alone absorb much less THz radiation than water: $\alpha_{solute} \ll \alpha_{bulk\,water}$, where α is the respective THz absorption coefficient. In a first approximation, particles are described as hard spheres that do not directly interact with the water network. When increasing the concentration of biomolecules, water is replaced by the less absorbing solute resulting in a decrease of THz absorption. This so-called THz defect (Leitner et al. 2008) can be described by a linear decrease of THz absorbance with increasing solute concentration. The actual decrease in THz absorption compared to bulk water depends on the biomolecule's internal dielectric properties.

In order to advance this simple two-component model, a third component was introduced: the water in the so-called hydration shell (Heyden et al. 2008). Collective intermolecular vibrations of water within the hydration shell are influenced by its interactions with the solutes. For sugars and proteins, we found that $\alpha_{solute} \ll \alpha_{bulk\,water} < \alpha_{hydration\,water}$ (Heyden et al. 2008).

For some proteins, it was found that the presence of a more strongly absorbing hydration shell overcompensates the loss of THz absorption caused by replacing water molecules by significantly more THz-transparent solute molecules under dilute conditions. This leads to an initial increase of THz absorption with increasing protein concentration (THz excess) (Leitner et al. 2008). The maximum in THz absorption is reached at solute concentrations, which are high enough that hydration shells of distinct solutes begin to overlap significantly. At this point, the volume fraction of the hydration shell water does no longer increase linearly with the solute concentration because additional solute molecules now replace strongly absorbing water as well as even more strongly absorbing hydration shell water of other solutes. When the solute concentration increases further, the THz absorption of the solution ultimately decreases linearly with the solute concentration, which is in agreement with previous experimental observations (Xu et al. 2006a) at higher protein concentrations. This linear decrease stems from the fact that the THz absorption of the solution is dominated by the absorption of solutes and their overlapping hydration shells. Further increase of the solute concentration hence now leads to the replacement of hydration water with the significantly more transparent solutes (see also Figure 8.2).

Simulation studies reveal that the typical hydrogen bond lifetimes of water molecules in the dynamical hydration shell of a protein can be reduced considerably; see, for example, Heyden and Havenith (2010). We found that the collective vibrations of the hydrogen bond network of the hydration shell are correlated to the dynamics of hydrogen bond rearrangements, which take place on the ps timescale ($1\,THz = 10^{12}\,Hz = 10^{-12}\,s = 1\,ps^{-1}$). This correlation foots on the fluctuations of the dipole moment of the protein–water hydrogen bond network, which are probed by THz absorption.

In pioneering studies, it could be shown that the size of the dynamical hydration shell for proteins can reach up to 20 Å from a protein surface (Born and Havenith 2009; Ebbinghaus et al. 2007), involving more than thousand water molecules. This extends well beyond static hydration radii (up to 3 Å) (Heyden et al. 2010), which are obtained from experimental observables that are sensitive to differences in the hydrogen bond network structure or local single molecule dynamics (in contrast to the collective, delocalized vibrations of the hydrogen bond network at THz frequencies). Considering that proteins like many other biomolecules are fractal in their dimension and mixed hydrophilic–hydrophobic entities, the dynamical hydration shells of proteins must be assumed to exhibit heterogeneous dynamics. As a consequence, even the three-component model, which holds very well for hydrated carbohydrates (Heyden et al. 2008), can only give a qualitative description for proteins; however, it is not sufficient to describe the observed changes in THz absorbance (Born and Havenith 2009; Matyushov 2010). This is in agreement with molecular dynamics simulations (Bagchi 2005; Heyden and Havenith 2010).

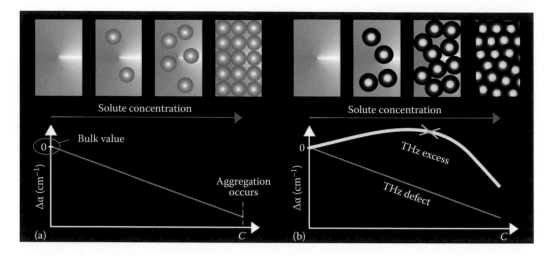

FIGURE 8.2 Concepts of THz hydration: shown are the absorbance-dependent concepts of (a) THz defect and (b) THz excess. The THz defect describes a linear change of THz absorbance with solute concentration for biomolecules (spheres) considered as hard spheres in water. The THz excess behavior is nonlinear for the concentration-dependent THz absorption coefficient, thereby taking into account dynamical hydration (dark spheres) of biomolecules (bright spheres) in water. (Reprinted with kind permission from Springer Science+Business Media: *J. Infrared Millimeter Terahertz Waves*, Terahertz dance of proteins and sugars with water, 30, 2009, 1245–1254, Born, B. and Havenith, M., Copyright 2009.)

By ab initio molecular dynamics simulations, it was possible to dissect the contributions of local and collective vibrations to the THz spectrum of bulk water (Heyden et al. 2010). The absorption of water around 2.4 THz was attributed to a collective bending vibration of hydrogen bonds, which could be analyzed in detail for intermittently stable water molecule clusters of up to eight water molecules containing the tetrahedral environment of two hydrogen-bonded water molecules (Heyden et al. 2010). These involve a correlated vibrational motion of water molecules separated by 5 Å and more.

The absorption spectrum of water and aqueous solution in the THz region can be calculated as the Fourier transform of the time autocorrelation of the total dipole moment M:

$$\alpha(\omega) = F(\omega) \int_0^\infty dt \, \exp(i\omega t)\langle M(0)M(t)\rangle,$$

with the frequency-dependent prefactor F taking into account harmonic quantum corrections (Ramírez et al. 2004); the square brackets describe the ensemble average. This equation exemplifies that vibrational absorption spectroscopy indeed probes the fluctuations of the total dipole moment of the system, as stated before.

In molecular dynamics simulations with simple nonpolarizable force fields, which still dominate simulations of biomolecular solutes in explicit solvent, accurate calculations of the THz absorption coefficient are hindered. This stems from the fact that the IR activity of hydrogen bond network vibrations in water is dominated by the mutual dynamic polarization of the involved water molecules, while changes of the involved static molecular dipole moments largely cancel. This can be overcome with recently suggested approaches to take the polarizability of water molecules into account a posteriori, when analyzing the simulation (Torii 2011).

In our studies, we found that the vibrational density of states (VDOS), which describes the spectrum of vibrations without taking into account spectroscopic cross sections, represents a useful tool to analyze solute-induced effects on the vibrations of the hydrogen bond network of water. While this approach does not explicitly take into account the collective character of these vibrations, we find that solute-induced effects on the VDOS of water molecules in the first few hydration layers correlate strongly with the experimentally observed changes in THz absorption in the solute hydration shell. The VDOS is obtained as the Fourier transform of time autocorrelation functions of the atomic velocities v:

$$\text{VDOS}(\omega) \propto \int_0^\infty dt \, \exp(i\omega t)\langle v(0)v(t)\rangle.$$

The THz excess of a protein solution compared to bulk water at 80 cm^{-1} or 2.4 THz could be attributed to a blue shift in the VDOS of the protein–hydration water relative to bulk water, increasing the mode density at this frequency. This relative frequency shift of absorbing THz modes results in a predicted decrease of absorption for the hydration water below 50 cm^{-1} (1.5 THz) and an increase above 50 cm^{-1} (1.5 THz) (Heyden et al. 2010, 2012), which is supported by experimental observations (Heyden et al. 2010; Kim et al. 2008).

Molecular dynamics simulations also show that the blue shift of vibrational frequencies in the hydrogen bond network of hydration water of numerous solutes is accompanied by a slowdown or retardation of dynamical processes on the ps timescale. These ps processes are rearrangements of the hydrogen bond network due to breaking and reforming of hydrogen bonds, rotational relaxation, and translational diffusion (Heyden et al. 2012). Therefore, it has been proposed frequently that observed changes in the THz absorption coefficient in the hydration water of a solute are intimately coupled to dynamical processes on the corresponding ps timescale.

8.3 THz Absorbance of Aqueous Protein Solutions

In Bochum, a p-Ge difference laser spectrometer (Bergner et al. 2005) has been set up to determine THz absorption spectra and the dynamical hydration radius of several biomolecules including the proteins λ-repressor (Ebbinghaus et al. 2008, 2007), antifreeze protein (AFP) (Ebbinghaus et al. 2010, 2012; Meister et al. 2013), ubiquitin (Born and Havenith 2009), and human serum albumin (HSA) (Luong et al. 2011); also see Figure 8.3.

By measuring the absorption coefficient of a sample and a reference solution in a difference THz spectrometer, even small changes in the THz absorbance of aqueous protein solutions can be detected. It is crucial that both sample and reference are kept at equal experimental condition such as temperature or humidity. The expected changes in the absorption coefficient are in the range of 1%–5% for the protein solution compared to bulk water.

In a two-component model, the absorption coefficient of the solution alpha scales linearly with protein concentration as

$$\alpha = \alpha_{\text{protein}} \frac{V_{\text{protein}}}{V} + \alpha_{\text{buffer}} \frac{V - V_{\text{protein}}}{V}$$

$$\approx \alpha_{\text{buffer}}(1 - c_{\text{protein}} \cdot \rho_{\text{protein}}),$$

where
   V is the total volume and V_{protein} is the volume of the protein
   c_{protein} is the solution concentration of the protein
   ρ_{protein} is the solution density of the protein

For small globular proteins, $\rho_{\text{protein}} \approx 1.4\,\text{g/mL}$ (Fischer et al. 2004).

The observed concentration-dependent THz absorbance is found to be nonlinear in most cases for native, folded proteins. At low protein concentrations, a linear increase is detected. The linear increase of THz absorbance is attributed to an increased net absorption coefficient of the protein and its surrounding

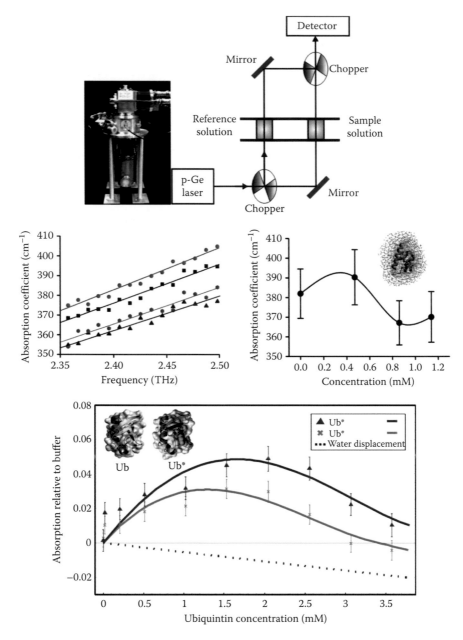

FIGURE 8.3 Top: p-Ge laser spectrometer in two-beam configuration for simultaneous detection of reference and sample absorbance. Center left: THz spectrum of a λ-repressor solution. Shown is the THz absorption coefficient of buffer (squares) and of λ-repressor solutions in buffer at concentrations of 0.47 mM (circles), 0.86 mM (triangles), and 1.14 mM (hexagons). (Reprinted with kind permission from Springer Science+Business Media: *J. Infrared Millimeter Terahertz Waves*, Terahertz dance of proteins and sugars with water, 30, 2009, 1245–1254, Born, B. and Havenith, M., Copyright 2009.) Center right: concentration-dependent THz excess derived from the same λ-repressor solutions. The inset shows the structure of the λ-repressor embedded in its dynamical hydration shell. Bottom: concentration-dependent THz absorbance of ubiquitin wild-type Ub (triangles) and pseudo-wild-type Ub* (crosses) compared to an estimated THz defect behavior for ubiquitin (dots). (Taken from Born, B. and Havenith, M., Terahertz dance of proteins and sugars with water, *J. Infrared Millimeter Terahertz Waves*, 30, 1245–1254. Reproduced by permission of The Royal Society of Chemistry)

dynamical hydration shell, as has been described in detail in the previous section. When increasing the protein concentration, a decrease of THz absorbance is detected, with a maximum that describes the concentration range at which the strongly absorbing hydration shells of distinct protein solute begin to overlap significantly. Beyond that concentration, the hydration shells will overlap, which means that adding further proteins to the solution will not further contribute to an increase of strongly absorbing hydration shell water. At very high protein concentrations, the THz absorbance decreases since the proteins have a lower absorption coefficient at 3 THz than bulk water. As described previously, at higher concentrations, additional proteins primarily replace hydration shell water in the solution, leading to a final linear decrease of the absorption coefficient as a function of protein concentration. This leads us to a three-component model, which could be used to describe this behavior quantitatively for carbohydrate solutions (Heugen et al. 2006; Heyden et al. 2008):

$$\alpha = \alpha_{\text{protein}} \frac{V_{\text{protein}}}{V} + \alpha_{\text{shell}} \frac{V_{\text{shell}}}{V} + \alpha_{\text{buffer}} \frac{V - V_{\text{protein}} - V_{\text{shell}}}{V}.$$

In addition to the volume fraction of the solutes in the solution V_{protein}/V, this model also includes the volume fraction of the hydration shells V_{shell}/V, which is a nonlinear function of the solute concentration c and depends strongly on the hydration shell thickness d. Assuming hard solute particles, which cannot overlap with each other and a specific solute shape, $[V_{\text{shell}}/V](c, d)$ can be computed efficiently and accurately in numerical simulations.

For protein solutions, a quantitative description of the concentration-dependent absorption could not be achieved based on this model. This is attributed to heterogeneous properties of the protein–water interface, a concentration-dependent coupling of internal vibrational modes of the protein and the surrounding hydration shell water hydrogen bond network, and potential effects of protein aggregation or oligomerization in solution. However, the model allows for a qualitative description of the concentration-dependent THz absorption of the protein solution. For this qualitative analysis, the concentration at which the THz excess reaches its maximum is used. Given that at this concentration hydration shells of distinct molecules overlap, the distribution of distances between the surfaces of nearest neighbor proteins in the solution at this concentration can be used to determine an approximate hydration shell thickness.

The list of hydrated proteins that have been studied with THz spectroscopy so far includes bacteriorhodopsin (Whitmire et al. 2003), bovine serum albumin (Xu et al. 2006b), cytochrome c (Chen et al. 2005a), lysozyme (Knab et al. 2006; Xu et al. 2006a), Mb (Zhang and Durbin 2006), ubiquitin (Born et al. 2009a), and the λ-repressor (λ_{6-85}^{*}) (Ebbinghaus et al. 2007, 2008), both with various mutants, the AFP (Ebbinghaus et al. 2010, 2012; Meister et al. 2013) and HSA (Luong et al. 2011). The determined dynamical hydration shells around proteins were found to extend beyond 15 Å from the protein surface corresponding to at least five shells of hydration water or a volume of ≈7500 Å³ (Born et al. 2009a; Ebbinghaus et al. 2007).

Acidic denaturation of ubiquitin and of the λ-repressor resulted in a decrease of the THz absorbance (Born et al. 2009a; Ebbinghaus et al. 2008). Intuitively, unfolded proteins exhibit more hydrophobic side chains exposed to the water network. Molecular dynamics simulations of spherical hydrophilic and hydrophobic model particles revealed a frequency shift of vibrational modes in the hydration water of hydrophilic solute–water interfaces, which leads to an increased mode density and to an increased THz absorption relative to hydration water in the vicinity of hydrophobic side chains and bulk water at 3 THz (Heyden and Havenith 2010). Consequently, partial or complete unfolding of a protein in solution, exposing previously buried amino acid side chains with distinct effects on the vibrations in the hydration shell, affects the THz absorption of the protein solution.

A recent study employed an extension of Maxwell's theory of dielectrics to develop an alternative model that describes the concentration-dependent THz absorption of solutions of amino acids, carbohydrates, and proteins, for which experimental data are available (Heugen et al. 2006; Niehues et al. 2011). The resulting model accurately describes the THz absorption of these solutions using a single parameter, a scalar prefactor of the solute–solvent interface dipole in Maxwell's theory (Heyden et al. 2012), which can be related to the change of the frequency-dependent polarity of the solution due to the presence of the solute. In addition, molecular dynamics simulations, which were used to test the robustness of the model without fitting to experimental data, demonstrated that instantaneous correlations between the dipole of the protein solute (the DNA-binding domain of the λ-repressor) and its hydration shell gradually increase as a function of the thickness of the considered hydration shell, eventually reaching a plateau at hydration shell thicknesses of 20 Å and more (Heyden et al. 2012). Hence, also this alternative route for the analysis of experimental THz absorption data, using a very distinct theoretical model to describe the THz absorption of biomolecular solutions, reveals long-ranged solute-induced effects and extended hydration shells with dielectric properties different from the bulk.

To gain further insight into the coupling of water network dynamics and protein folding, the influence of temperature-induced unfolding and refolding of the protein HSA, which is a major transporter protein in our blood plasma and regulates pH and the osmotic pressure, has been studied (Luong et al. 2011). Circular dichroism revealed that heat denaturation of HSA occurs reversibly at ≈55°C, which only affected the quaternary structure, preceding the irreversible unfolding of the protein backbone above 70°C. Along the same line, the heat-induced changes of THz absorbance at 55°C were reversible; heating the buffered HSA solution beyond 70°C altered the THz absorbance irreversibly (Luong et al. 2011). Similar as in the previous THz hydration studies of acidic protein denaturation of globular proteins, the heat-induced change in THz absorbance of the HSA protein solution could be explained by an exposure of hydrophobic amino acid residues from the protein core toward the water network upon unfolding (Born et al. 2009a).

AFPs lower the freezing point of water with respect to the melting point, enabling insects and fish to survive at subfreezing temperatures. Different classes of AFPs are known, ranging in structure from intrinsically disordered carbohydrate chains to highly repetitive β-barrels. THz spectroscopic studies show that they commonly function via a combination of short-range local water interaction and long-range coupling to the water network. THz spectroscopic studies of the disordered antifreeze glycoprotein (AFGP) revealed long-ranged effects on hydration water dynamics with a dynamical hydration radius of ≈20 Å at 20°C. At 5°C, the dynamical hydration shell increases by up to five additional layers of water to ≈35 Å, indicating a perturbation of the entire water network (Ebbinghaus et al. 2010). Complexation of the exposed cis-hydroxyl groups by borate results in a significant decrease of the antifreeze activity and a reduction in long-range water interaction. Site-directed mutagenesis studies of the α-helical AFP from the winter flounder revealed that secondary structure motives can be a prerequisite for long-range water interaction (Ebbinghaus et al. 2012).

THz spectroscopy of the hyperactive AFP from the insect *fire-colored beetle* showed that extended AFP–water network dynamics are a fundamental criterion for the extraordinarily high antifreeze activity of insect AFPs (Meister et al. 2013). The addition of the osmolyte sodium citrate increased the antifreeze activity and the size of the dynamical hydration shell underlining the importance of the long-range hydration dynamics for antifreeze activity.

8.4 KITA of Protein Folding and Enzyme Catalysis

KITA spectroscopy allows following changes in the hydrogen bond network dynamics and vibrations of water during biochemical reactions in real time; also see Figure 8.4. As a first example, the folding of ubiquitin was studied by KITA (Kim et al. 2008). In this study, time-resolved methods to probe the formation of secondary and tertiary protein structure were combined with time-resolved mapping of the hydration dynamics. Proteins are denatured at high (7 M) GuHCl concentrations. By mixing the GuHCl-denatured protein with GuHCl-free buffer solution, folding of dissolved proteins to their native form is initiated and can be followed in a time-resolved way. In our KITA experiment, folding was initiated by combining buffered solution of denatured ubiquitin with a six-times surplus of buffer solution to guarantee refolding at submolar GuHCl concentration. The folding reaction of ubiquitin was decelerated to millisecond rates by studying the refolding process at −20°C and −28°C, which required the addition of a cryoprotective agent. The KITA experiment of ubiquitin folding was monitored in 45:55 ethylene glycol/water mixtures.

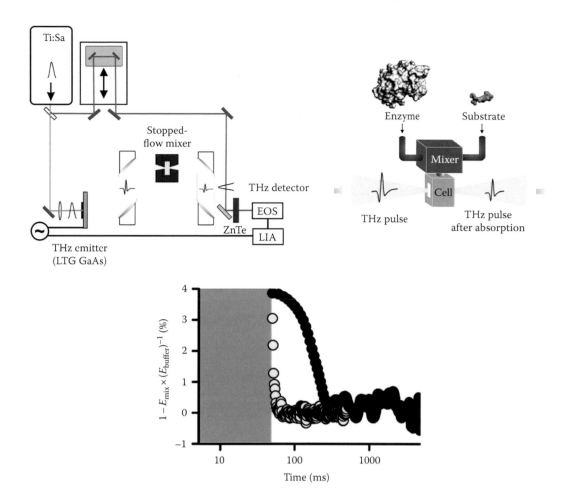

FIGURE 8.4 KITA setup. Top left: a stopped-flow mixer is placed in the focused THz beam of a THz time-domain spectrometer in transmission configuration. Top right: an enzyme and a substrate solution are placed in two reservoirs of the mixer. Rapid mixing initiates the enzymatic reaction that is detected by KITA. Bottom: typical KITA process of the hydrolysis of a short peptide substrate (bright circles) and of another peptide substrate with a slower catalytic rate (dark circles) by the metalloprotease MT1-MMP. The KITA signal in (%) displays the changes in THz intensity relative to the buffer intensity, which is equal to 0%. The mixer dead time is represented by a gray box. (Taken from Born, B. et al., Protein–water network dynamics during metalloenzyme hydrolysis observed by kinetic THz absorption [KITA]. Presented at *Proceedings of SPIE*, p. 85850E, 2013, Copyright 2013 SPIE.)

Mixing was performed in a commercial stopped-flow apparatus where the solutions of denatured protein and buffer from two separate reservoirs have been combined in an observation cell with z-cut quartz windows and a volume of approximately 40 μL.

Using a THz time-domain spectrometer, collective protein–hydration water dynamics upon refolding was monitored over a range of tens of milliseconds with THz pulses of ps duration. The electric field of the THz pulses is attenuated and phase shifted when passing the reaction cell due to absorbance and refraction of the aqueous sample solution inside the cell. THz pulses were generated in a photoconductive antenna of low-temperature-grown gallium arsenide. The THz emitter was optically pumped by femtosecond pulses of a Ti/Sa laser. For detection, the THz pulses were focused on a zinc telluride crystal by a 90°-off-axis parabolic mirror. Electrooptical detection was accomplished with a Pockels effect where a second near-IR gating pulse (at 800 nm) was delayed on a translation stage and focused on the zinc telluride detection crystal. The gating pulse was delayed by ≈0.6 mm ps⁻¹. The THz pulse had ≈4 ps duration with a full width at half maximum of 600 fs and a spectral range of ≈0.1–1 THz. The kinetic traces of protein folding were monitored at fixed positions of the mechanical delay stage.

The KITA experiment showed that the water network is dynamically rearranged around ubiquitin in less than 10 ms. The time constant for the formation of secondary structures as detected by time-resolved circular dichroism and by real-time small-angle x-ray scattering (SAXS) is found to be considerably longer: the SAXS measurement revealed that the final structural collapse of the ubiquitin backbone essential for the final protein compactness takes place on a time of 2 s. Time-resolved circular dichroism and time-resolved fluorescence studies revealed also time constants in the range of 1 s. This observation shows that changes in the hydration dynamics precede the protein folding process and are instead associated with the initial steps of protein folding.

For all mutants of ubiquitin, we observed a similar fast relaxation of the KITA response on a timescale of several milliseconds, independent of the respective core mutation or temperature. The KITA relaxation processes of all mutants are on the order of (18 ± 10) ms, which is within the dead time of the mixer (50 ms).

The combined biophysical real-time approach showed a fast millisecond phase of folding and a slow phase in the regime of seconds that is in agreement with a simulation study of the coupling of water dynamics to folding of the SH3 protein, which is of similar size as ubiquitin (Grantcharova et al. 1998). Native protein–protein contacts form during an early state of folding and shape a partially hydrated hydrophobic core in a cooperative manner. In the final folding transition, the hydrophobic protein core is shielded from the water network, and for SH3, less than 20 water molecules are expelled from the partially folded protein (Cheung et al. 2002). Since only few water molecules with hydrogen bond contacts to the protein backbone are being replaced by protein–protein hydrogen bonds in the late phase of folding of ubiquitin, KITA is not sensitive to the processes assigned to the fine alignment of the hydrophobic core.

The KITA setup was also used to monitor changes in the hydration dynamics during an enzymatic reaction, in particular the hydrolysis of a peptide substrate by the human metalloenzyme membrane type 1-matrix metalloproteinase (MT1-MMP) (Grossman et al. 2011). MT1-MMP is an extracellular matrix (ECM) zinc-dependent metalloprotease with a transmembrane domain and is involved in ECM remodelling. PLGLAR (proline, leucine, glycine, leucine, alanine, arginine) served as a model substrate, which has a high affinity to the active site of MT1-MMP. KITA was combined with other time-resolved techniques—time-resolved x-ray absorption spectroscopy and fluorescence—to follow the kinetics of the reaction at the active site and changes in the THz response of the surrounding hydration water quasi-simultaneously. The enzymatic reaction was initiated by mixing of enzyme and substrate solutions in assay buffer using a stopped-flow mixer. The (E/S) ratio was 1:20 for all experiments to operate in substrate surplus conditions beyond the diffusion limit.

Combining KITA, time-resolved x-ray absorption, time-resolved fluorescence spectroscopies, and the results of molecular dynamics simulations, we proposed the following mechanism: prior to substrate binding, a gradient of water dynamics toward the active site of MT1-MMP is established. Molecular dynamics simulations suggest typical hydrogen bond lifetimes of ≈7 ps for water molecules in the vicinity of the zinc ion at the active site and of ≈1 ps in bulk. In the absence of a substrate molecule, the catalytic zinc ion of the metalloenzyme is tetracoordinated with three histidine nitrogens and a structural water molecule as shown by x-ray absorption.

In the first nanoseconds after stopped-flow mixing, there is an unspecific binding of the substrate at the surface of MT1-MMP. This is followed by specific substrate binding to the active site. For PLGLAR at approximately 65 ms after mixing, the formation of the Michaelis enzyme–substrate complex takes place, which can be monitored as a change in the charge state of the zinc ion. This goes along with a change in the coupled protein–hydration dynamics. By direct binding of the PLGLAR substrate to the catalytic zinc ion, the coordination number of zinc increases from four to five. The charge state changes, which can be monitored by extended x-ray absorption fine structure as a frequency shift of the edge of the zinc atom. Changes in the oxidation state of the zinc metal ion during Michaelis complex formation seem to cause a change of the long-range hydration dynamics as observed by KITA.

To test the hypothesis that collective water network dynamics and enzymatic proteolysis reaction are coupled, the same experimental protocol was repeated with a peptide substrate of the sequence Mca-RPLPA-Nva-WML-Dnp-NH$_2$ (with Nva = norvaline and the FRET pair Mca = (7-methoxycoumarin-4-yl) acetyl and Dnp = dinitrophenol). In this case, the enzymatic reaction requires a longer time. In accord with this, we observed a longer time for the change in hydration dynamics. Compared to PLGAR, the typical time constant has increased from 65 to 140 ms.

By a combination of time-resolved experiments, which are able to follow the kinetic structural changes at the active site as well as hydration water dynamics, we could prove a strict correlation of both as a function of reaction time, hinting toward an active role of the hydration water properties during catalysis. It can be speculated whether the retardation of the hydrogen bonding assists in formation of a Michaelis complex. Future studies should address the more specific question of a correlation between biological function and hydration dynamics.

8.5 Discussion and Outlook

Water is the ubiquitous biological solvent occurring in every living cell where it shows distinct dynamics in the vicinity of an enormous variety of biological molecules and assemblies up to tissues. In cells, liquid–liquid demixing and liquid phase transitions are hallmarks of biological assembly and of the regulation and onset of function (Hyman and Simons 2012). Yet, the complexity of collective dynamical interactions of biomolecules with each other and with the water network makes the characterization of the contribution of water to cellular processes a scientific challenge (Ball 2012).

The analysis of hydration dynamics around isolated biomolecules is the initial step for understanding the influence of hydration water dynamics on biological systems. In this respect, THz spectroscopy is considered to be a powerful tool to study hydration dynamics. Also, systematic perturbations of an equilibrated aqueous solution by stopped-flow mixing or temperature jumps allow kinetic studies with KITA; also see Figure 8.5.

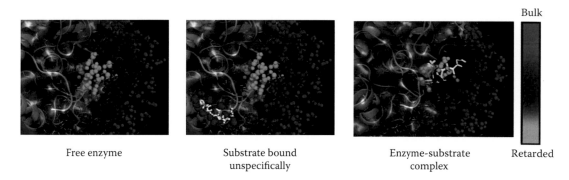

Free enzyme   Substrate bound   Enzyme-substrate   Retarded
unspecifically   complex

FIGURE 8.5 Gradient of water dynamics at the active site of the metalloprotease MT1-MMP and water network dynamics during Michaelis complex formation. Distinct hydration are present as active site hydration water, hydration water of separated enzyme and substrate, hydration water of the substrate in the Michaelis complex, and bulk water. Left: prior to substrate binding, a sharp gradient of water dynamics is observed at the active site of MT1-MMP. Water dynamics increase in direction of the bulk. Center: right after mixing, the substrate (white) binds to the surface of MT1-MMP but not to the active site. The active site gradient of hydrogen bond exchange dynamics supports substrate association toward the catalytic zinc ion via a charge-induced water retardation at the active site. Right: upon Michaelis complex formation, the hydration dynamics of the substrate gets further retarded and a mild gradient of water dynamics develops at the active site. (Reprinted by permission from Macmillan Publishers Ltd.: *Nat. Struct. Mol. Biol.*, Grossman, M., Born, B., Heyden, M. et al., 18, 1102, Figure 8.5(b), Copyright 2011.)

Water is thought to be the matrix of life (Szent-Gyorgyi 1979), that is, the medium in which biochemical reactions take place. However, the question of whether water is an active player therein remains still to be answered (Ball 2011; Chaplin 2006). A recent molecular dynamics study supports the assumption that water is an active player during noncovalent binding and Michaelis complex formation (Baron et al. 2012). The model enzyme–substrate system of the study was composed of a hydrophobic substrate and an enzyme with a nonpolar active site. A vibrational spectroscopy study in real time showed that during enzyme catalysis, active site water molecules show retarded dynamics compared to the water molecules in the bulk (Jha et al. 2012). Retarded active site water molecules of the enzyme KSI may set the stage for catalysis due to their distinct dielectric properties. Recent neutron scattering results suggest a global active effect of hydration on protein dynamics, namely, that the interior and exterior dynamics of a protein are both influenced by hydration dynamics (Wood et al. 2013).

A comprehensive notion of the different contributions of water network dynamics to biological structures and function could have an impact on drug design strategies. Natural enzymes exhibit higher rates of catalysis than most of their artificially synthesized versions (Alexandrova 2008; Ruscio et al. 2009). Implementing water molecules and their specific dipolar nature and dynamics could be a missing link for closing the gap of functionality, which exists today between artificial and natural enzymes. In antibody design, improved catalytic rates were observed already, when water molecules have been included explicitly to the design process (Acevedo 2009). Drug design strategy could benefit even further from considering the entire water network in which enzymes and antibodies of interest are embedded.

The role of water in life-related processes is still an enigma, but the further development of multimodal setups and integrated experimental and computational hydration studies with THz spectroscopy as a powerful tool for direct observation of fast and coupled biomolecule–water network dynamics might play its part in deciphering a more active participation of water as has been anticipated before.

Acknowledgment

We thank Martin Gruebele (University of Illinois) and David Leitner (University of Nevada) for their contributions to the design and analysis of the initial THz absorption studies of solvated proteins, which have been funded by a joint HFSP grant to Martina Havenith, Martin Gruebele, and David Leitner. Moran Grossman, Irit Sagi (both Weizmann Institute of Science), and Gregg B. Fields (University of Texas) are acknowledged for their close collaboration on the MMP project. Konrad Meister (RUB), Art de Vries (University of Illinois), Martin Gruebele, and Martina Havenith have been partners in the AFP project funded by the *VW Stiftung*. Rajib K. Mitra (Bose National Center) and Trung Quan Luong (RUB) have carried out the HSA measurements. We thank Erik Bründermann (RUB) for the development of the p-Ge laser spectrometer and him and Matthias Krüger (RUB) for their contributions to the initial setup of the KITA experiment. Gerhard Schwaab and Diedrich Schmidt (both RUB) are acknowledged for their development of tools to analyze the THz data. Simon Ebbinghaus acknowledges funding from the *NRW Rückkehrerprogramm*. Benjamin Born is a crossdisciplinary fellow of the *Human Frontier Science Program*.

References

Acevedo, O. 2009. Role of water in the multifaceted catalytic antibody 4B2 for allylic isomerization and Kemp elimination reactions. *The Journal of Physical Chemistry B* 113: 15372–15381.

Alexandrova, A. Y. 2008. Evolution of cell interactions with extracellular matrix during carcinogenesis. *Biochemistry* 73: 733–741.

Arikawa, T., M. Nagai, and K. Tanaka. 2008. Characterizing hydration state in solution using terahertz time-domain attenuated total reflection spectroscopy. *Chemical Physics Letters* 457: 12–17.

Bagchi, B. 2005. Water dynamics in the hydration layer around proteins and micelles. *Chemical Reviews* 105: 3197–3219.

Ball, P. 2011. More than a bystander. *Nature* 478: 467–468.

Ball, P. 2012. The importance of water. In *Astrochemistry and Astrobiology*, eds. Ian W. M. Smith, Charles S. Cockell, Sydney Leach, pp. 169–210. New York: Springer.

Baron, R., P. Setny, and F. Paesani. 2012. Water structure, dynamics, and spectral signatures: Changes upon model cavity–ligand recognition. *The Journal of Physical Chemistry B* 116: 13774–13780.

Bergner, A., U. Heugen, E. Brundermann et al. 2005. New p-Ge THz laser spectrometer for the study of solutions: THz absorption spectroscopy of water. *Review of Scientific Instruments* 76: 063110.

Born, B. and M. Havenith. 2009. Terahertz dance of proteins and sugars with water. *Journal of Infrared, Millimeter, and Terahertz Waves* 30: 1245–1254.

Born, B., M. Heyden, M. Grossman, I. Sagi, and M. Havenith 2013. Protein–water network dynamics during metalloenzyme hydrolysis observed by kinetic THz absorption (KITA). Presented at *Proceedings of SPIE*, p. 85850E.

Born, B., S. J. Kim, S. Ebbinghaus, M. Gruebele, and M. Havenith. 2009a. The terahertz dance of water with the proteins: The effect of protein flexibility on the dynamical hydration shell of ubiquitin. *Faraday Discussions* 141: 161–173.

Born, B., H. Weingärtner, E. Bründermann, and M. Havenith. 2009b. Solvation dynamics of model peptides probed by terahertz spectroscopy. Observation of the onset of collective network motions. *Journal of the American Chemical Society* 131: 3752–3755.

Brundermann, E., D. R. Chamberlin, and E. E. Haller. 2000. High duty cycle and continuous terahertz emission from germanium. *Applied Physics Letters* 76: 2991–2993.

Chaplin, M. 2006. Do we underestimate the importance of water in cell biology? *Nature Reviews Molecular Cell Biology* 7: 861–866.

Chen, J.-Y., J. R. Knab, J. Cerne, and A. G. Markelz. 2005. Large oxidation dependence observed in terahertz dielectric response for cytochrome c. *Physical Review E* 72: 040901.

Cheung, M. S., A. E. García, and J. N. Onuchic. 2002. Protein folding mediated by solvation: Water expulsion and formation of the hydrophobic core occur after the structural collapse. *Proceedings of the National Academy of Sciences* 99: 685.

Dexheimer, S. L. 2007. *Terahertz Spectroscopy: Principles and Applications*. London, U.K.: Taylor & Francis.

Ebbinghaus, S., S. J. Kim, M. Heyden et al. 2007. An extended dynamical hydration shell around proteins. *Proceedings of the National Academy of Sciences* 104: 20749–20752.

Ebbinghaus, S., S. J. Kim, M. Heyden et al. 2008. Protein sequence-and pH-dependent hydration probed by terahertz spectroscopy. *Journal of the American Chemical Society* 130: 2374–2375.

Ebbinghaus, S., K. Meister, B. Born et al. 2010. Antifreeze glycoprotein activity correlates with long-range protein–water dynamics. *Journal of the American Chemical Society* 132: 12210–12211.

Ebbinghaus, S., K. Meister, M. B. Prigozhin et al. 2012. Functional importance of short-range binding and long-range solvent interactions in helical antifreeze peptides. *Biophysical Journal* 103: L20–L22.

Elber, R. and M. Karplus. 1986. Low-frequency modes in proteins: Use of the effective-medium approximation to interpret the fractal dimension observed in electron-spin relaxation measurements. *Physical Review Letters* 56: 394.

Ernst, J. A., R. T. Clubb, H.-X. Zhou, A. M. Gronenborn, and G. Clore. 1995. Demonstration of positionally disordered water within a protein hydrophobic cavity by NMR. *Science* 267: 1813–1817.

Evans, S. V. and G. D. Brayer. 1990. High-resolution study of the three-dimensional structure of horse heart metmyoglobin. *Journal of Molecular Biology* 213: 885–897.

Falconer, R. J. and A. G. Markelz. 2012. Terahertz spectroscopic analysis of peptides and proteins. *Journal of Infrared, Millimeter, and Terahertz Waves* 33: 973–988.

Fenimore, P. W., H. Frauenfelder, B. H. McMahon, and F. G. Parak. 2002. Slaving: Solvent fluctuations dominate protein dynamics and functions. *Proceedings of the National Academy of Sciences* 99: 16047–16051.

Fischer, H., I. Polikarpov, and A. F. Craievich. 2004. Average protein density is a molecular-weight-dependent function. *Protein Science* 13: 2825–2828.

Frölich, A., F. Gabel, M. Jasnin et al. 2009. From shell to cell: Neutron scattering studies of biological water dynamics and coupling to activity. *Faraday Discussions* 141: 117–130.

Fujiyoshi, Y., K. Mitsuoka, B. L. de Groot et al. 2002. Structure and function of water channels. *Current Opinion in Structural Biology* 12: 509–515.

Gallagher, T., P. Alexander, P. Bryan, and G. L. Gilliland. 1994. Two crystal structures of the B1 immunoglobulin-binding domain of streptococcal protein G and comparison with NMR. *Biochemistry* 33: 4721–4729.

Grantcharova, V. P., D. S. Riddle, J. V. Santiago, and D. Baker. 1998. Important role of hydrogen bonds in the structurally polarized transition state for folding of the src SH3 domain. *Nature Structural & Molecular Biology* 5: 714–720.

Grossman, M., B. Born, M. Heyden et al. 2011. Correlated structural kinetics and retarded solvent dynamics at the metalloprotease active site. *Nature Structural & Molecular Biology* 18: 1102–1108.

Halle, B. 2004. Protein hydration dynamics in solution: A critical survey. *Philosophical Transactions of the Royal Society of London. Series B: Biological Sciences* 359: 1207–1224.

Head-Gordon, T. 1995. Is water structure around hydrophobic groups clathrate-like? *Proceedings of the National Academy of Sciences* 92: 8308–8312.

Heugen, U., G. Schwaab, E. Bründermann et al. 2006. Solute-induced retardation of water dynamics probed directly by terahertz spectroscopy. *Proceedings of the National Academy of Sciences* 103: 12301–12306.

Heyden, M., E. Bründermann, U. Heugen et al. 2008. Long-range influence of carbohydrates on the solvation dynamics of water-answers from terahertz absorption measurements and molecular modeling simulations. *Journal of the American Chemical Society* 130: 5773–5779.

Heyden, M. and M. Havenith. 2010. Combining THz spectroscopy and MD simulations to study protein-hydration coupling. *Methods* 52: 74–83.

Heyden, M., J. Sun, S. Funkner et al. 2010. Dissecting the THz spectrum of liquid water from first principles via correlations in time and space. *Proceedings of the National Academy of Sciences* 107: 12068–12073.

Heyden, M., D. J. Tobias, and D. V. Matyushov. 2012. Terahertz absorption of dilute aqueous solutions. *The Journal of Chemical Physics* 137: 235103.

Hyman, A. A. and K. Simons. 2012. Beyond oil and water-phase transitions in cells. *Science* 337: 1047–1049.

Jepsen, P. U., D. G. Cooke, and M. Koch. 2011. Terahertz spectroscopy and imaging—Modern techniques and applications. *Laser & Photonics Reviews* 5: 124–166.

Jha, S. K., M. Ji, K. J. Gaffney, and S. G. Boxer. 2012. Site-specific measurement of water dynamics in the substrate pocket of ketosteroid Isomerase using time-resolved vibrational spectroscopy. *The Journal of Physical Chemistry B* 116: 11414–11421.

Kim, S. J., B. Born, M. Havenith, and M. Gruebele. 2008. Real-time detection of protein–water dynamics upon protein folding by terahertz absorption spectroscopy. *Angewandte Chemie International Edition* 47: 6486–6489.

Knab, J., J.-Y. Chen, and A. Markelz. 2006. Hydration dependence of conformational dielectric relaxation of lysozyme. *Biophysical Journal* 90: 2576–2581.

Knab, J. R., J.-Y. Chen, Y. He, and A. G. Markelz. 2007. Terahertz measurements of protein relaxational dynamics. *Proceedings of the IEEE* 95: 1605–1610.

Leitner, D. M., M. Gruebele, and M. Havenith. 2008. Solvation dynamics of biomolecules: Modeling and terahertz experiments. *HFSP Journal* 2: 314–323.

Leitner, D. M., M. Havenith, and M. Gruebele. 2006. Biomolecule large-amplitude motion and solvation dynamics: Modelling and probes from THz to X-rays. *International Reviews in Physical Chemistry* 25: 553–582.

Luong, T. Q., P. K. Verma, R. K. Mitra, and M. Havenith. 2011. Do hydration dynamics follow the structural perturbation during thermal denaturation of a protein: A terahertz absorption study. *Biophysical Journal* 101: 925–933.

Markelz, A. G. 2008. Terahertz dielectric sensitivity to biomolecular structure and function. *IEEE Journal of Selected Topics in Quantum Electronics* 14: 180–190.

Markelz, A. G., J. R. Knab, J. Y. Chen, and Y. He. 2007. Protein dynamical transition in terahertz dielectric response. *Chemical Physics Letters* 442: 413–417.

Matyushov, D. V. 2010. Terahertz response of dipolar impurities in polar liquids: On anomalous dielectric absorption of protein solutions. *Physical Review E* 81: 021914.

Meister, K., S. Ebbinghaus, Y. Xu et al. 2013. Long-range protein–water dynamics in hyperactive insect antifreeze proteins. *Proceedings of the National Academy of Sciences* 110: 1617–1622.

Murarka, R. K. and T. Head-Gordon. 2008. Dielectric relaxation of aqueous solutions of hydrophilic versus amphiphilic peptides. *The Journal of Physical Chemistry B* 112: 179–186.

Nandi, N., K. Bhattacharyya, and B. Bagchi. 2000. Dielectric relaxation and solvation dynamics of water in complex chemical and biological systems. *Chemical Reviews* 100: 2013–2046.

Niehues, G., M. Heyden, D. A. Schmidt, and M. Havenith. 2011. Exploring hydrophobicity by THz absorption spectroscopy of solvated amino acids. *Faraday Discussions* 150: 193–207.

Plusquellic, D. F., K. Siegrist, E. J. Heilweil, and O. Esenturk. 2007. Applications of terahertz spectroscopy in biosystems. *ChemPhysChem* 8: 2412–2431.

Ramírez, R., P. Kumar, and D. Marx. 2004. Quantum corrections to classical time-correlation functions: Hydrogen bonding and anharmonic floppy modes. *The Journal of Chemical Physics* 121: 3973.

Ruscio, J. Z., J. E. Kohn, K. A. Ball, and T. Head-Gordon. 2009. The influence of protein dynamics on the success of computational enzyme design. *Journal of the American Chemical Society* 131: 14111–14115.

Schmuttenmaer, C. A. 2004. Exploring dynamics in the far-infrared with terahertz spectroscopy. *Chemical Reviews* 104: 1759–1780.

Schröder, C., T. Rudas, S. Boresch, and O. Steinhauser. 2006. Simulation studies of the protein–water interface. I. Properties at the molecular resolution. *The Journal of Chemical Physics* 124: 234907.

Siegel, P. H. 2002. Terahertz technology. *IEEE Transactions on Microwave Theory and Techniques* 50: 910–928.

Siegel, P. H. 2004. Terahertz technology in biology and medicine. *IEEE Transactions on Microwave Theory and Techniques* 52: 2438–2447.

Siegrist, K., C. R. Bucher, I. Mandelbaum et al. 2006. High-resolution terahertz spectroscopy of crystalline trialanine: Extreme sensitivity to β-sheet structure and cocrystallized water. *Journal of the American Chemical Society* 128: 5764–5775.

Silvestrelli, P. L., M. Bernasconi, and M. Parrinello. 1997. Ab initio infrared spectrum of liquid water. *Chemical Physics Letters* 277: 478–482.

Szent-Gyorgyi, A. 1979. In *Cell-Associated Water*, eds. Clegg, J.S. and W. Drost-Hansen, Cambridge, MA: Academic Press.

Torii, H. 2011. Intermolecular electron density modulations in water and their effects on the far-infrared spectral profiles at 6 THz. *The Journal of Physical Chemistry B* 115: 6636–6643.

Whitmire, S. E., D. Wolpert, A. G. Markelz et al. 2003. Protein flexibility and conformational state: A comparison of collective vibrational modes of wild-type and D96N bacteriorhodopsin. *Biophysical Journal* 85: 1269–1277.

Winther, L. R., J. Qvist, and B. Halle. 2012. Hydration and mobility of trehalose in aqueous solution. *The Journal of Physical Chemistry B* 116: 9196–9207.

Wood, K., A. Frölich, A. Paciaroni et al. 2008. Coincidence of dynamical transitions in a soluble protein and its hydration water: Direct measurements by neutron scattering and MD simulations. *Journal of the American Chemical Society* 130: 4586–4587.

Wood, K., F. X. Gallat, R. Otten et al. 2013. Protein surface and core dynamics show concerted hydration-dependent activation. *Angewandte Chemie International Edition* 52: 665–668.

Xu, J., K. W. Plaxco, and S. J. Allen. 2006a. Collective dynamics of lysozyme in water: Terahertz absorption spectroscopy and comparison with theory. *The Journal of Physical Chemistry B* 110: 24255–24259.

Xu, J., K. W. Plaxco, and S. J. Allen. 2006b. Probing the collective vibrational dynamics of a protein in liquid water by terahertz absorption spectroscopy. *Protein Science* 15: 1175–1181.

Zhang, C. and S. M. Durbin. 2006. Hydration-induced far-infrared absorption increase in myoglobin. *The Journal of Physical Chemistry B* 110: 23607–23613.

Zhang, H., K. Siegrist, D. F. Plusquellic, and S. K. Gregurick. 2008. Terahertz spectra and normal mode analysis of the crystalline VA class dipeptide nanotubes. *Journal of the American Chemical Society* 130: 17846–17857.

Zhang, L., L. Wang, Y.-T. Kao et al. 2007. Mapping hydration dynamics around a protein surface. *Proceedings of the National Academy of Sciences* 104: 18461–18466.

Zhang, L., Y. Yang, Y.-T. Kao, L. Wang, and D. Zhong. 2009. Protein hydration dynamics and molecular mechanism of coupled water–protein fluctuations. *Journal of the American Chemical Society* 131: 10677–10691.

Zhong, D. 2009. Hydration dynamics and coupled water-protein fluctuations probed by intrinsic tryptophan. *Journal of Chemical Physics* 143: 83–149.

Zhong, D., S. K. Pal, D. Zhang, S. I. Chan, and A. H. Zewail. 2002. Femtosecond dynamics of rubredoxin: Tryptophan solvation and resonance energy transfer in the protein. *Proceedings of the National Academy of Sciences* 99: 13–18.

9

Terahertz Spectroscopy of Biological Molecules

Seongsin
Margaret Kim
University of Alabama

Mohammad
P. Hokmabadi
University of Alabama

9.1 Introduction

There has been a great research interest in applying terahertz (THz) spectroscopy to probe and characterize various biomolecules during the last decade. Biomolecules are closely related to a living organism. Well-known biomolecules are rather large macromolecules such as proteins, polysaccharides, lipids, deoxyribonucleic acids (DNAs), ribonucleic acids (RNAs), and nucleic acids. Small biomolecules, usually the weights are less than 900 Da (daltons), bind to biopolymers, act as an effector, and alter the function of the biopolymers. Small biomolecules can function through different cell types and species and are mostly used as therapeutic agents.

The low-frequency collective vibrational modes represent various information about biomolecules' conformational state. Many biological functions and their profound dynamic mechanisms can be revealed through the low-frequency collective motion or resonance in protein and DNA molecules, such as cooperative effects, allosteric transition, and intercalation of drugs into DNA. The vibrational modes of biological molecules have been studied in many different biomolecules with different emphasis (Dexheimer 2007; Laman et al. 2008). The status of the confirmation of molecules has been widely studied, such as in the proteins myoglobin, lysozyme, and bacteriorhodopsin and in the retinal chromophore (Markelz et al. 2002; Walther et al. 2000; Whitmire et al. 2003). The excited vibrational mode linked to chemical reactions in proteins and biological reaction in hemoglobin has been reported (Austin et al. 1989; Klug et al. 2002). The modes have also been shown to be affected by hydration oxidation and ligand binding (Balog et al. 2004; Chen et al. 2005; Kistner et al. 2007; Knab et al. 2006; Liu and Zhang 2006). Such low-frequency vibrational modes lie in the frequency range between 10 and 200 cm^{-1} (or 0.3–6.0 THz), and therefore, THz spectroscopy becomes a critical tool to probe such dynamics. Table 9.1 summarizes categories of the biomolecules conferred here and divided in three different groups in terms of the size and functionality.

9.2 Biomolecules in Terahertz Frequency

Due to their delocalized nature, low-frequency vibrational modes are strongly affected by the size and long-range order of the molecule. Small biological molecules such as nucleosides, amino acids, and sugars tend to have distinct, relatively isolated features (Bailey et al. 1997; Bandekar et al. 1983; Fischer et al. 2002;

TABLE 9.1   List of Biomolecules in Three Different Categories in Terms of the Size and Functionality

Biomonomers	Bio-Oligomers	Biopolymers
Amino acids	Oligopeptides	Polypeptides, proteins (hemoglobin)

(*continued*)

Monosaccharides

D-Glucose

L-Glucose

Oligosaccharides

Polysaccharides (cellulose)

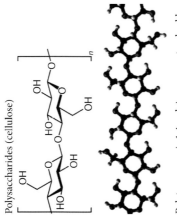

Isoprene

Terpenes

Polyterpenes: *cis*-1,4-polyisoprene natural rubber and
trans-1,4-polyisoprene gutta-percha

Nucleotides

Oligonucleotides

Polynucleotides, nucleic acids (DNA, RNA)

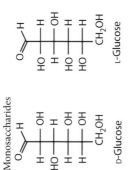

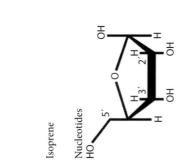

TABLE 9.1   (continued) List of Biomolecules in Three Different Categories in Terms of the Size and Functionality

Biomonomers	Bio-Oligomers	Biopolymers

Lee et al. 2000, 2001, 2004; Li et al. 2003; Nishizawa et al. 2003, 2005; Shen et al. 1981, 2004, 2007; Shi and Wang 2005; Upadhya et al. 2003; Walther et al. 2003; Yamaguchi et al. 2005; Yamamoto et al. 2005; Yu et al. 2004). Small polymers of these molecules such as oligopeptides have a larger number of modes that tend to increase in number and overlap with increasing polymer length (Kutteruf et al. 2003; Shotts and Sievers 1974). Large tandem repeats of peptides or nucleosides have some observable features, whereas nonperiodic biological molecules such as DNA and proteins have relatively featureless spectra (Austin et al. 1989; Dexheimer 2007; Klug et al. 2002; Markelz et al. 2002; Powell et al. 1987; Shotts and Sievers 1974; Xie et al. 1999; Yamamoto et al. 2002). In addition, the vibrational modes of large molecules tend to be internal (i.e., intramolecular) in nature, with large groups of atoms moving within the molecule. In contrast, small molecules tend to also have external (i.e., intermolecular) modes, where the entire molecule moves in concert within a crystalline lattice. It is not surprising that these modes are strongly dependent on the crystallinity of the material.

Many biological vibrational modes have been observed, particularly at the low frequencies. For example, the modes of biological interest have frequencies of 3.45 THz (115 cm^{-1}) for the conformational change of bacteriorhodopsin (Markelz et al. 2002), 1.515 THz (50.5 cm^{-1}) for chemical reactions in myoglobin (Whitmire et al. 2003), 1.17 THz (39 cm^{-1}) for the oxygen acceptance of hemoglobin (Walther et al. 2000), and 1.8 THz (60 cm^{-1}) for the primary event of vision (Austin et al. 1989). Historically, absorption lines with frequencies less than ~6 THz (200 cm^{-1}) were difficult to measure with conventional Fourier transform infrared (FTIR) spectroscopy due to the relatively poor performance of thermal sources and detectors in this frequency range. However, since the early 1990s, terahertz time-domain spectroscopy (THz-TDS) has been developed, which can exhibit a high signal/noise ratio (S/N) over a frequency range of ~100 GHz to 5 THz. The standard technique for sample preparation in both conventional FTIR spectroscopy and THz-TDS of biological molecules, and organic molecules in general, is to make a mixture of the material in powder form with a transparent host, such as polyethylene (PET), and press the mixture into a pellet, on the order of 1 mm thick with a 1 cm diameter. The transmission spectrum of the pellet is compared to that of a pure host pellet to obtain the absorption spectrum of the sample. This technique is quite effective and can be used for a very wide range of materials. Previous works have studied pellet samples of a number of biological molecules, including nucleobases and nucleosides, amino acids, proteins, retinal isomers, polypeptides, sugars, and benzoic and acetylsalicylic acids. These cited works have examined a large number of vibrational modes to a high degree. However, for many molecules, the resolution of the absorption spectra is limited not by the instrumental resolution but by the significant inhomogeneous broadening resulting from the disordered polycrystalline sample even at cryogenic temperatures.

9.3 Terahertz Spectroscopy of DNAs and RNAs

DNA, together with RNA and proteins, is one of the three important macromolecules that are essential for all known forms of life. It is composed of a chain of four different types of nucleotides, guanine (G), adenine (A), thymine (T), and cytosine (C), such that the sequence of these nucleotides determines the genetic information that organisms inherit. Most DNA molecules are double-stranded helices consisting of two long polymers of nucleotides. Every helix consists of a backbone made of alternating sugars and phosphate groups, with the nucleobases (G, A, T, C) attached to the sugars. By deep study of DNA, it is possible to fulfill major applications of genetics such as the early diagnoses of diseases through detection of changes in the DNA chain in patients' blood, controlling epidemics and plagues by the early detection of special specifications of viruses and bacteria, or controlling metabolic process via messenger RNA (mRNA) analysis and manipulation.

Recently, there has been an increasing interest in experimentally studying the low-frequency response (0.1–10 THz or 3–300 cm^{-1}) of biomolecules. The motivation is the theoretical results that predict resonances in absorption spectra of biological molecules in the THz frequency due to the vibrational and rotational intermolecular modes. These intermolecular bonding modes include the weak hydrogen bonds of the DNA base pairs or nonbonded interactions (Duong and Zakrzewska 1997; Feng and Prohofsky 1990; Lin et al. 1997; Mei et al. 1981; Sarkar et al. 1996; Saxena et al. 1991; van Zandt and Saxena 1994; Young et al. 1989). Hydrogen bonding plays a crucial role in determining the 3D structures of DNA and its conformations. Hydrogen bonding is also responsible for the distortion and destabilization of biological molecules such that they can control the properties of biomolecules and their interaction with enzymes, drugs, and so on.

One of the well-developed and widely used methods for biomolecular studies is the spectroscopic technique that utilizes the interaction of an applied electromagnetic (EM) field with the phonon (lattice vibration) field of the material. Among various spectroscopy methods in the far-infrared (FIR) and THz range, traditional Raman and Fourier transform spectroscopy (FTIR) methods have been extensively utilized to analyze biomolecules (Tominaga et al. 1985; Weidlich et al. 1990). Despite the fact that Raman spectroscopy is able to determine the phonon modes of DNA, Raman scattering is really a complicated process, and on the other hand, matching between theory and measurement is extremely challenging for DNA polymers by this method. FIR transmission studies are also limited due to the difficulty in accessing submillimeter ranges, especially below 50 cm^{-1}. Another spectroscopic method that recently has been used is THz-TDS. This technique is able to operate at so-called THz gap frequencies that are inaccessible by conventional FTIR methods. In addition, it provides rich information directly in time domain, which can be used for time-domain analysis and extracting time-dependent parameters of substances in Drude and Debye models.

The earliest study of DNA FIR absorption spectrum was done by A. Wittlin et al. (1986). They used four different spectroscopic methods including oversized-cavity technique, polarized-double-beam microwave interferometer, and two different Fourier transform spectrometers to cover the whole range between 3 and 450 cm^{-1}. They measured two highly oriented films of Li-DNA and Na-DNA in the temperature range of 5–300 K. Five low-frequency infrared-active vibrational modes were observed in which the lowest infrared-active mode at 45 cm^{-1} for Li-DNA and 41 cm^{-1} for Na-DNA were reported to soften upon sample hydration. They attributed the hydration-induced absorption at 10 cm^{-1} to relaxation processes due to the appearance of water at higher hydration levels. Their theoretical studies based on a simple lattice dynamics model showed that the mass loading of the molecular subunits of DNA by bound water molecules and the change in lattice constant associated with the conformational transition could be the main reason of the softening of the modes upon hydration. Woolard et al. (2002) have also done a comprehensive study of herring and salmon dry DNA samples in a large submillimeter wavelength spanning from 10 to 2000 cm^{-1} frequency range. They applied FTIR spectroscopy for their samples with different thicknesses of dried films in order to eliminate the etalon effect and distinguish the resonant modes of the DNAs. Through this study, they showed that DNA molecules exhibit intrinsic and specific phonon modes that could be useful for label-free detection of DNA molecules by using submillimeter spectroscopy.

The first pulsed THz spectroscopy measurement of lyophilized powder samples of calf thymus DNA was reported by Markle et al. (2000) in the range of 0.06–2 THz corresponding to the 2–67 cm^{-1} frequency range. In order to eliminate the multiple reflection etalon effect, they mixed lyophilized biomolecular powder samples with 200 mg of PET powder such that the final pellet thickness was obtained about 7.5 mm. This resulted in 0.4 cm^{-1} etalon spacing due to the multiple reflections, which was slightly less than 0.5 cm^{-1} spectral resolution of their measurement. Figure 9.1a shows the measured absorbance, for several DNA samples. The first two samples are just different in molar concentration of the mixed DNA, while the third one is a pure DNA without any mixture. From these measurements, it is observable that absorbance increases almost lineally versus frequency, confirming Beer's law. In order to compare the pure pellet data to the PET mixed pellet data, they normalized the absorbance of the pure pellets, by using $A_{norm} = A_p \times c_m l_m / c_n l_n$, where A_p is the absorbance of pure pellet and c_m (c_n), l_m (l_n) are the concentration and length of the mixed (pure) pellets, respectively. This normalization depends only on the measured pellet parameters c and l and is independent of the actual spectrum. Overlapping the data for samples #1 and #3 suggests that these assumptions are at least in part valid. One readily observes the similarity of the spectral shapes for sample #3 compared to those pressed in transparent PET. They have also measured sample #3 in hydration condition. Figure 9.1b shows the spectrum for hydrated DNA of sample #3. As it is demonstrated, the low-frequency absorption below 15 cm^{-1} diminishes from that in Figure 9.1a.

Figure 9.2 demonstrates the real part of refracted index at 25 cm^{-1} versus relative humidity for sample #3 calculated from the transmission spectra, and the inset shows the real part and absorption coefficient

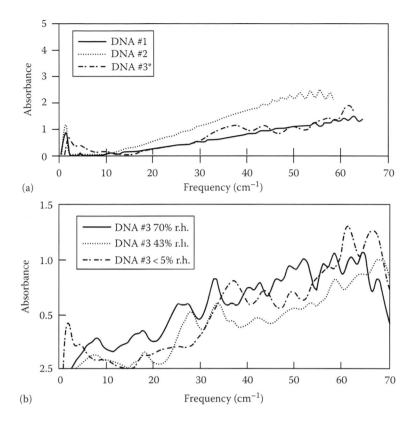

FIGURE 9.1 (a) Measured absorbance spectrum for three different samples of DNA at <5 r.h.; (b) absorbance spectrum for sample 3# at different hydration levels reported in Markelz et al. (2000).

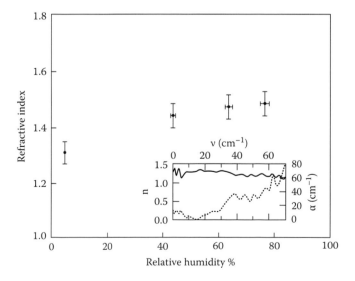

FIGURE 9.2 Real part of refractive index versus r.h. at 25 cm⁻¹. Density of DNA sample is 0.48 g cm⁻¹; inset shows the real part of refractive index and absorption coefficient for sample #3 reported in Markelz et al. (2000).

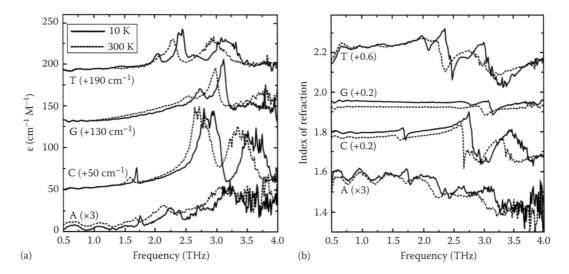

(a)

(b)

FIGURE 9.3   (a) Molar absorption coefficient and (b) index of refraction for the nucleobases A, G, C, and T, at temperatures of 10 K (solid curves) and 300 K (dashed curves) reported in Fischer et al. (2002).

of the same sample at <5% r.h. versus frequency. As it is observed from Figure 9.2, the real part of refractive index increases with hydration and finally reaches to a constant value. This behavior can be explained by assuming that the net measured dielectric response is an average from both the DNA and water of hydration. Because the refractive indices of dry DNA and water are 1.3 and 2.3 at 25 cm^{-1} respectively, therefore, one can expect increase in refractive index by raising the humidity.

Fischer et al. (2002) have studied and reported the first measurement of the FIR dielectric function (DF) for four nitrogenous nucleobases, adenine (A), guanine (G), cytosine (C), and thymine (T), in the frequency range of 0.5–4.0 THz and the corresponding nucleosides dA, dG, dC, and dT in the frequency range of 0.5–3.5 THz at 10 and 300 K temperatures. In Figure 9.3, the molar absorption coefficient and index of refraction are illustrated for the nucleobases A, G, C, and T, at temperatures of 10 K (solid curves) and 300 K (dashed curves). The curves for each sample have been offset vertically or multiplied by a factor for better representation as depicted in Figure 9.3. It is observed that at a temperature of 10 K, broad room temperature resonances split up into several narrowbands. Also, due to the decrease in bond lengths at low temperatures, the position of the bands generally moves to higher frequencies.

The same measurement of absorption coefficient and index of refraction is shown in Figure 9.4 for the nucleosides dA, dC, dG, and dT, at 10 K (solid curves) and 300 K (dashed curves) in the range of 0.5–3.5 THz. Like nucleobases, there are a series of resonances that upon cooling split up into several narrow resonances. But there exist two groups of resonances for nucleosides compared to nucleobases. One group of them is similar to those observed in the spectrum of the nucleobases, which can be characterized by relatively broad and intense line profiles above 1.5 THz. The other group of resonances, which is not seen in nucleobases, is located between 1 and 2 THz, with narrow, asymmetric line shapes. The author interpreted these additional lines as the vibrational signatures associated with the sugar groups attached to the nucleobases.

Furthermore, they have analyzed vibrational modes of the samples by using density functional theory (DFT) and have calculated the index of refraction of thymine in the range between 0.2 and 3.5 THz. For refractive index calculations, they assumed that each resonance can be described as a damped oscillator, resulting in a characteristic absorption and index profile described by

$$(n+ik)^2 = \varepsilon_\infty + \sum_j \frac{S_j \vartheta_j^2}{\vartheta_j^2 - \vartheta^2 - i\vartheta\Gamma_j},$$ (9.1)

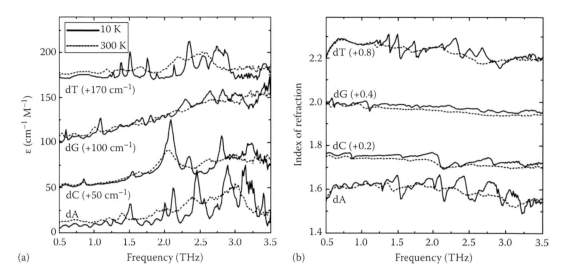

FIGURE 9.4 (a) Molar absorption coefficient and (b) index of refraction for the nucleosides dA, dG, dC, and dT, at temperatures of 10 K (solid curves) and 300 K (dashed curves) reported in Fischer et al. (2002).

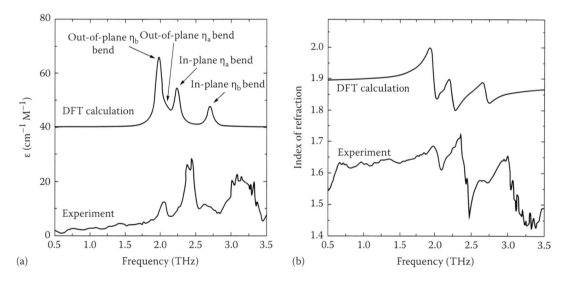

FIGURE 9.5 DFT calculated and measured (a) molar absorption coefficient and (b) index of refraction of thymine reported in Fischer et al. (2002).

where the sum is taken over the different oscillators with resonance frequencies ν_j, oscillator strengths S_j, and line widths Γ_j. As seen in Figure 9.5, their calculation is in a reasonable agreement with the experimentally observed positions based on the number of resonances and their vibrational frequencies. Vibrational analysis based on the DFT of thymine shows that the four lowest-frequency, infrared-active modes arise from intermolecular motion in the form of out-of-plane and in-plane vibrations of the hydrogen bond systems depicted by η_a and η_b, in Figure 9.5.

THz-TDS also has made it possible for label-free gen hybridization detection. Nagel et al. (2002a,b) designed an integrated THz sensor for this purpose. Most of the genetic sequence determination schemes are based on identifying the hybridization of unknown target DNA molecules to known single-stranded oligo- or polynucleotide probe DNA molecules. Most hybridization detection is recently based on

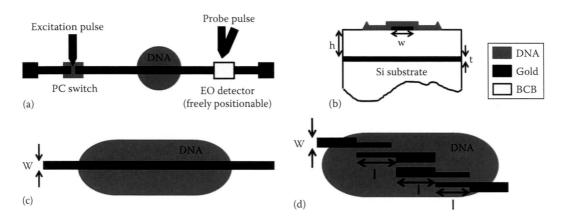

FIGURE 9.6 (a) Schematic illustration of integrated THz sensor for detecting hybridized and denatured DNAs including PC switch as the THz source, microstrip waveguide with DNA spot on it, and electrooptic THz detector, (b) cross-sectional view of the microstrip waveguide, (c) the basic structure of the sensor using microstrip waveguide, and (d) modified and enhanced structure of the sensor using THz resonator (band-pass filter) reported in Nagel et al. (2002a).

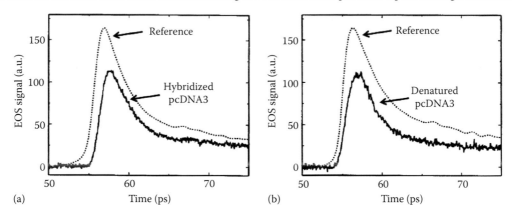

FIGURE 9.7 Measured detected THz signal of the sensor for (a) hybridized and (b) denatured DNA samples reported in Nagel et al. (2002a).

fluorescent labeling of DNA molecules. Although fluorescent labeling and photoluminescence detection have paved a route to very efficient diagnostic systems, alternative label-free detection methods are necessarily required; labeling with fluorescent chromospheres not only is an additional preparation step that may complicate genetic analysis, but it also eventually introduces modifications in the DNA strand conformation, which can lower the accuracy of gene detection (Ozaki and McLaughlin 1992). Labeling also ruins the quantifiability in comparative studies, since fluorescence efficiency is strongly site dependent and the additionally required processing steps induces labeling-dependent fluctuations (Larramendy et al. 1998; Zhu et al. 1994). The design has been demonstrated in Figure 9.6. It is based on femtosecond laser technology that uses integrated ultrafast photoconductive (PC) switches of low-temperature GaAs as THz sources with a bandwidth of 20 GHz to 2 THz with an electrooptic detection scheme. The generated THz pulse from the PC switch is coupled to the integrated THz microstrip waveguide. The microstrip lines are made from gold with benzocyclobutene (BCB) as dielectric material. The THz signal passes through the DNA sample deposited on the gold strip depicted in Figure 9.6b while guiding in the waveguide. They used two kinds of DNA samples, one was hybridized and the other was denatured DNA.

The sensor works based on change in effective permittivity of the microstrip waveguide. Therefore, the THz wave passing through this sensor will face different effective permittivities according to the DNA deposited on top of the sensor, and this difference will be reflected in the phase of the detected THz signal. Figure 9.7 shows the detected signal versus time that was obtained directly from the THz-TDS measurement. Figure 9.7a is the measurement illustration for hybridized DNA and Figure 9.7b is for denatured DNA. The hybridized DNA film exhibits a clear temporal shift of the THz signal induced by the DNA. The shift of the denatured sample is evidently much lower. The difference of the temporal shifts demonstrates that THz waveguide measurements are potentially attractive for the label-free analysis of the binding state of DNA molecules by sensitive monitoring of variations in the complex refractive index.

In their measurement, the denatured sample was thicker than hybridized DNA by approximately a factor of two, owing to the hybridization-dependent viscosity of the pipette DNA solution. This implies that the phase shift by hybridization probing could be even much larger for an identical sample thickness than observed in these experiments indicating high sensitivity of this sensor.

As it is seen in Figures 9.6d and 9.8a, they have modified their design by incorporating a band-pass filter as the THz resonator in the microstrip waveguide to enhance the interaction of THz radiation with the DNA sample. They could reach to femtomolar sensitivity. The resonator includes three coupled microstrip line resonators, in which the crucial length l is a quarter wavelength at the first passband center frequency in the medium of propagation. The resonator was designed to provide a passband around a center frequency of 610 GHz. The corresponding results of the modified design have been brought in Figure 9.8b. The dotted line shows the result of the simulation that matches well to experimental measurement result without any sample deposited on the sensor. As seen from Figure 9.8b, the frequency

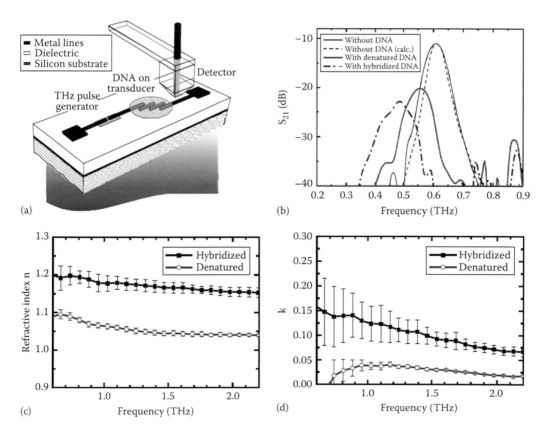

FIGURE 9.8 (a) Schematic of modified DNA label-free sensor; (b) detected transmitted signal including calculated (dashed), measured without DNA (solid thin), measured with denatured DNA (solid thick), and measured with hybridized DNA(dash-dotted); and (c) calculated real and (d) imaginary part of refractive indices from the transmission spectra for hybridized (black) and denatured (gray) DNAs reported in Bolivar et al. (2002).

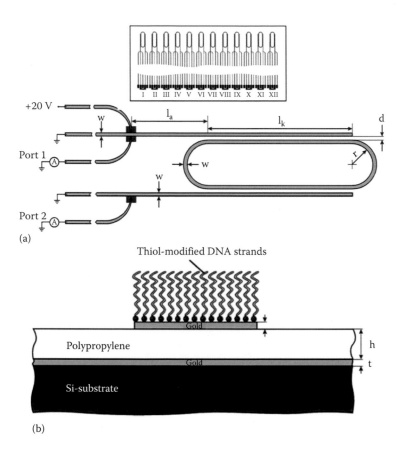

(a)

(b)

FIGURE 9.9 (a) Schematic of the fabricated chip consisting of 12 DNA sensors (top) and scheme of each individual sensors consisting of PC switches for generation and detection microstrip waveguides and ring resonator between them (below); (b) cross-sectional view of the thin-film microstrip with thiol-modified probe DNA strands on it reported in Nagel et al. (2003).

has been shifted toward lower ranges. This shift for hybridized DNA is bigger than that of denatured DNA. Actually, the shift in resonance frequency comes from the change in effective permittivity of the resonator that lowers the group velocity of the THz radiation and hence the resonance frequency. This indicates that the refractive index of hybridized DNA might be bigger than that of denatured one. In another report, the same group has calculated the corresponding refractive indices for both DNA samples by using detected signal spectrum, which have been demonstrated in Figure 9.8c and d (Bolivar et al. 2002). They attributed the large difference between these two refractive indices to the binding states of hybridized and denatured DNAs.

In another attempt, the same group has designed another label-free and more sensitive on-chip (OCH) sensor that is capable of sensing multiple DNA samples at the same time (Nagel et al. 2003). The sensitivity of the new design has been reported less than 40 fmol of 20-mer single-stranded DNA (ssDNA) molecules. The design has been depicted in Figure 9.9. They have used polypropylene (PP) instead of BCB as the dielectric between two gold strip lines that induces lower waveguide loss and dispersion at the functional frequency range of the sensor. The resonator between two microstrip lines has been designed such that it exhibits multiple resonance frequencies between 25 and 600 GHz. Binding and hybridization experiments were performed with complementary ssDNA 20-base oligonucleotides. Thiolated molecules with the sequence 5′-HS-(CH2)6-ACA CTG TGC CCA TCT ACG AG-3′ (probe) served as capture

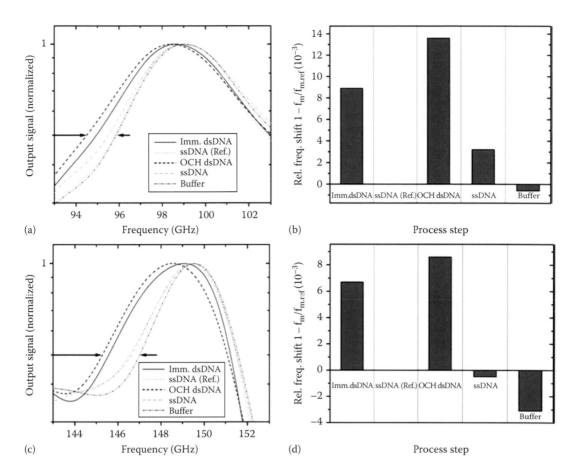

FIGURE 9.10 Reported normalized output signals versus frequency in Nagel et al. (2003) for the DNA sensor in the range of the resonances at (a) 99 GHz and (c) 149.5 GHz including respective sensing steps as shown in (b) and (d), which are, (1) immobilization of dsDNA (Imm. dsDNA), (2) denaturation step for sensor initialization (ssDNA Ref.), (3) OCH hybridization (OCH dsDNA), (4) redenaturation (ssDNA), and (5) control process with pure buffer solution without any DNA (buffer) and the corresponding relative frequency shift, $r_{shift} = 1-f_m/f_{m,ref}$, for each resonance measurement.

probes. Target molecules consisting of complementary ssDNA oligonucleotides with the sequence 5′-TGT GAC ACG GGT AGA TGC TC-(6-FAM)-3′ (target) hybridized to these capture probes.

The application of the sensor for the detection of 20-mer oligomolecule is as follows. The first step is the immobilization of the double-stranded DNA (dsDNA) molecules at the sensor (Imm. dsDNA) and the subsequent denaturation of them (ssDNA Ref.). This denaturation step only leaves immobilized ssDNA probe molecules at the sensor surface. This step is indeed the initialization of the sensor. The denaturation gives rise to a notable increase of resonance frequency as illustrated in Figure 9.10. This figure shows the output transmitted signal of the sensor at two resonance frequencies $f_r = 99$ and 149 GHz. The resonance frequency of ssDNA (Ref.) is used as the reference for measuring the resonance shift at different steps of sample processing. Correspondingly to compare the resonance shifts, relative frequency shifts are defined as $r_{shift} = 1-f_m/f_{m,ref}$ with f_m as the frequency at half the maximum amplitude of the output signal on the rising signal edge and $f_{m,ref}$ as the corresponding reference value of the initially functionalized sensor after ssDNA (Ref.) process. dsDNA resonance frequency is lower than that of the reference ssDNA, which proves the functionality of the device and target and probe samples. After the initializing step, the sensor is ready to actually measure the OCH dsDNA and redenatured sample.

The device is immersed into the solution with complementary ssDNA molecules, so that OCH rehybridization of the immobilized probe molecules occurs (OCH dsDNA).

After this process step, the output signal demonstrates a decrease of f_r as expected. However, demonstrated in Figure 9.10, the resonance frequency shift is slightly bigger than the first immobilized. dsDNA molecule. They attributed this difference to the existence of the ssDNA molecules in immobilized DNA sample before the first denaturation process. Afterwards, the DNA molecules at the resonator are denatured again to create ssDNA. Finally, the resonator is exposed to pure TE–NaCl buffer solution without DNA to exclude the effect of the different salts used for hybridization. The result for this measurement shows a slight deviation from the reference.

Advances in THz technology have recently brought about more research in DNA THz spectroscopic identification to find out the specific fingerprints of these molecules. Very recently, similar reports have been represented for DNA molecules with different structures and conformations by using THz-TDS including transmittance measurement of nucleobases by using difference frequency generation THz source (Nishizawa et al. 2005), investigation of polarization-dependent hydration dynamics of DNA films (Kistner et al. 2007), noncovalent interaction investigation in paired DNA nucleobases (King et al. 2011), and DNA spectroscopic detection in real aqueous solutions (Arora et al. 2012; Globus et al. 2006). Meanwhile, different novel THz sensors for DNA detection and analysis have been reported such as a photonic crystal sensor for determining the refractive index of DNA samples (Kurt and Citrin 2005), a membrane device for holding biomolecular samples (Yoneyama et al. 2008), and a metallic mesh-based THz biosensor for ssDNA and dsDNA detection (Hasebe et al. 2012). These achievements make a pavement for THz radiation to be used for noninvasive and marker-free DNA imaging and detection. However, there exist a few reports on harmful effects of THz radiation on human tissues thus far. Currently, Titova et al. has reported their investigation of the harmful biological effects of pulsed THz radiation on artificial human skin tissues. They have observed that exposure to intense THz pulses for 10 min leads to a significant induction of H2AX phosphorylation, which may cause DNA damage in exposed skin tissue. However, they also found a THz-pulse-induced increase in the levels of several proteins that is responsible for cell cycle regulation and tumor suppression, indicating that DNA damage repair mechanisms can quickly be activated (Titova et al. 2013).

9.4 Terahertz Spectroscopy of d-Glucose

THz-TDS has the potential for groundbreaking applications to biological systems because the vibrational modes of many organic macromolecules—proteins and glucose, for instance—lie within this spectral range. One of the major barriers to the use of THz remains water's high absorption in the range, so initial medical uses will probably be superficial, given the water content of our bodies. Skin, blood, and superficial soft tissue may be these first areas of application.

According to the National Institute of Diabetes and Digestive and Kidney Diseases (NIDDK), diabetes in some form affects 25.8 million people or 8.3% of the US population (2009). Therefore, checking blood sugar levels is an everyday activity for a large number of individuals. While the latest monitors can give an accurate reading with an extremely small amount of blood, the pricking is a source of discomfort. For this reason, a noninvasive monitor would be an extremely useful device for diabetics.

Work in the context of blood sugar is specifically interested in D-glucose because one of its hydroxyl groups is stereochemically flipped when compared to the L isomer. To some extent, sugars represent prototype systems of hydrogen-bonded networks both in the crystalline and in the amorphous state. In the solid state, saccharides are linked by a rigid network of hydrogen bonds, which in crystals are of a long-range order, resulting in highly regular lattice vibrations, or phonon modes, of the entire crystal structures (Fischer et al. 2007). Several researchers reported THz frequency absorption characteristics of the various state of D-glucose.

The early work is focused on applying the THz-TDS on powder (solid) of glucoses. Figure 9.11 shows the THz absorption spectra in comparison between L-glucose and D-glucose, the two stereoisomers of

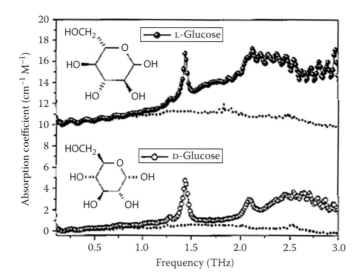

FIGURE 9.11 FIR absorption of pure D- and L-glucose at room temperature in pellets of 0.5 mm thickness. Dotted curves show the absorption for 1.7 M solutions of these samples. The curves for L-glucose have been vertically offset for clarity (Upadhya et al. 2003).

glucose (Upadhya et al. 2003). As shown in Figure 9.11, clear absorption peaks were distinguished for both D-glucose at 1.45 THz and 2.1 THz and a minor peak at 1.26 THz and L-glucose with a sharp peak at 1.45 THz and a broad feature at 2.12 THz.

Most of the case, the common method to prepare a sample is that each glucose powder was mixed with PET powder in a certain mass ratio. The crystals were crushed well before making the pellets in order that the particles were of submicron size, ensuring that the observed spectral features were not a result of Mie scattering. Since PET is nearly transparent in the THz frequency range, it is an ideal matrix material. Pellets were generally made with thicknesses between 1.2 and 1.4 mm. However, for room temperature measurements, the observed spectral features were less distinct and so samples were prepared as pure pellets with thicknesses between 0.3 and 0.7 mm.

Liu and Zhang reported dehydration kinetics of D-glucose monohydrate study using THz-TDS as shown in Figure 9.12. It shows the THz absorption spectra of anhydrous D-glucose and D-glucose

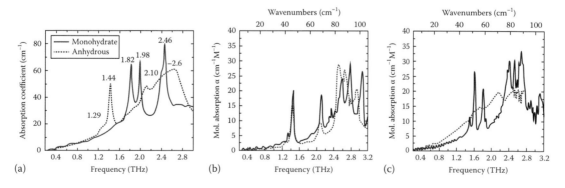

FIGURE 9.12 THz frequency absorption spectra of various D-glucose in different states. (a) Anhydrous D-glucose and D-glucose monohydrate at 25°C and molar absorption of (b) α-D-glucose and (c) β-D-glucose, recorded at room temperature (dashed lines) and cooled to 13 K (solid lines) (Fischer et al. 2007; Liu and Zhang 2006).

monohydrate at 25°C with the distinguishable characteristic peaks in absorption. For solid anhydrous D-glucose, peaks at 1.44, 2.10, and 2.6 THz are clearly resolved and well matched with the previous work. The two absorption spectra have evident differences, which have been assumed to result from the different intermolecular vibrational modes or phonon modes of the anhydrous and hydrated D-glucose and for D, respectively (Liu and Zhang 2006). Fisher et al. also demonstrated the individuality of the THz spectra of polycrystalline D-glucose in different phases, α-D-glucose and β-D-glucose as shown in Figures 9.12b and c. The high sensitivity of the spectra to small changes in the molecular structure enables even a differentiation between the spectra of the glucose anomers. While some similarities in the spectra are observed, for example, the first strong modes are only slightly shifted to higher frequencies in the β-anomer, clear differences appear in the range between 50 and 75 cm^{-1}.

While solid (usually polycrystalline) D-glucose has been relatively well profiled, no one has yet reported the ability to discern sharp, discrete spectra for aqueous D-glucose by THz spectroscopy; instead, spectra with very broad features were reported. For instance, using a mercury lamp and a THz attenuated total reflectance (ATR) spectroscopic technique, Suhandy et al. (2011) report the ability to predict aqueous glucose concentrations with a 95% confidence, but their results show only very broad features. Kim et al. proposed that the needed detailed spectra can, in fact, be probed using THz-TDS and postulated that there are frequencies that can distinguish glucose from the water solvent. Aqueous D-glucose solutions were prepared by diluting the anhydrous D-glucose powder in deionized water in concentrations that varied in range from 11% by mass to 32%. The solutions were then injected via a sterile syringe into a cavity created between two polymethylepentene (TPX) sheets by a silicone water sealant. The path length of this cavity varied from 0.3 to 0.55 mm. Figure 9.13 shows the absorption coefficients of three different concentrations studied by THz-TDS in transmission mode. Absorption coefficient was calculated using transmission ratios and with calculated complex refractive index of the solution (Bolus et al. 2013). Upon initial inspection, they seem to follow a spectral behavior of water; however, the glucose solutions deviate notably from water over specific frequency ranges. In this figure, it shows multiple peaks with distinct absorption coefficients compared to the ground signal. Specifically, 1.42 and 1.67 THz seem to be closely related to the D-glucose. Considering the features are not coming from the pure D-glucose, the absorption peaks are more likely related to the D-glucose molecules bounded in the water matrix, but nevertheless demonstrate deviation from a strong water background when it is referenced by. The hydrogen bonding of water molecules with other ions will significantly change in the absorption behavior as

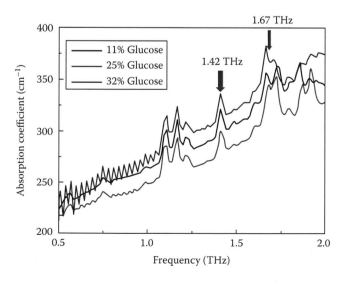

FIGURE 9.13   Absorption spectra of aqueous D-glucose in different concentrations.

we studied; the translation of the peaks has an important meaning. Furthermore, the absorption coefficient of D-glucose increases with increasing concentration, even though overabsorption of the solution is decreasing, since water molecules are replaced by the amount of glucose molecules, or it appears that the glucose concentration might affect intramolecular resonances.

Although the spectral features of the aqueous solutions are less sharp compared to those of the solid D-glucose pellets, the spectra are far sharper than those reported by Fischer et al. for such aqueous solutions. Furthermore, these absorption peaks also should not be chucked up to noise. Certainly, with the high absorption of water in the THz range, noise will play a role in some of the inconsistencies. It has been shown, therefore, that given an appropriate path length (~0.5 mm), transmission THz-TDS can, in fact, be used to gain meaningful spectral data from aqueous solutions. Moreover, given further effort, it may be possible to utilize THz spectroscopy not only by applying the Beer–Lambert law as seen in Suhandy et al.'s work but also by employing an ability to discern glucose signals in more complex water mixtures.

References

Arora, A., T. Q. Luong, M. Krüger et al. 2012. Terahertz-time domain spectroscopy for the detection of PCR amplified DNA in aqueous solution. *Analyst* 137: 575–579.

Austin, R. H., M. W. Roberson, and P. Mansky. 1989. Far-infrared perturbation of reaction rates in myoglobin at low temperatures. *Physical Review Letters* 62: 1912–1915.

Bailey, L. E., R. Navarro, and A. Hernanz. 1997. Normal coordinate analysis and vibrational spectra of adenosine. *Biospectroscopy* 3: 47–59.

Balog, E., T. Becker, M. Oettl et al. 2004. Direct determination of vibrational density of states change on ligand binding to a protein. *Physical Review Letters* 93: 028103.

Bandekar, J., L. Genzel, F. Kremer, and L. Santo. 1983. The temperature-dependence of the far-infrared spectra of L-alanine. *Spectrochimica Acta Part A: Molecular Spectroscopy* 39: 357–366.

Bolivar, P. H., M. Brucherseifer, M. Nagel et al. 2002. Label-free probing of genes by time-domain terahertz sensing. *Physics in Medicine and Biology* 47: 3815.

Bolus, M., S. Balci, D. S. Wilbert, P. Kung, and S. M. Kim. 2013. Effects of saline on terahertz absorption of aqueous glucose at physiological concentrations probed by THz spectroscopy. Presented at *Proceeding of 38th International Conference on Infrared, Millimeter and Terahertz Wave Conference,* Mainz, Germany.

Chen, J.-Y., J. R. Knab, J. Cerne, and A. G. Markelz. 2005. Large oxidation dependence observed in terahertz dielectric response for cytochrome c. *Physical Review E* 72: 040901.

Dexheimer, S. L. 2007. *Terahertz Spectroscopy: Principles and Applications.* London, U.K.: Taylor & Francis Group.

Duong, T. H. and K. Zakrzewska. 1997. Calculation and analysis of low frequency normal modes for DNA. *Journal of Computational Chemistry* 18: 796–811.

Feng, Y. and E. W. Prohofsky. 1990. Vibrational fluctuations of hydrogen bonds in a DNA double helix with nonuniform base pairs. *Biophysical Journal* 57: 547–553.

Fischer, B. M., H. Helm, and P. U. Jepsen. 2007. Chemical recognition with broadband THz spectroscopy. *Proceedings of the IEEE* 95: 1592–1604.

Fischer, B. M., M. Walther, and P. U. Jepsen. 2002. Far-infrared vibrational modes of DNA components studied by terahertz time-domain spectroscopy. *Physics in Medicine and Biology* 47: 3807.

Globus, T., D. Woolard, T. W. Crowe et al. 2006. Terahertz Fourier transform characterization of biological materials in a liquid phase. *Journal of Physics D: Applied Physics* 39: 3405.

Hasebe, T., S. Kawabe, H. Matsui, and H. Tabata. 2012. Metallic mesh-based terahertz biosensing of single-and double-stranded DNA. *Journal of Applied Physics* 112: 094702.

King, M. D., W. Ouellette, and T. M. Korter. 2011. Noncovalent interactions in paired DNA nucleobases investigated by terahertz spectroscopy and solid-state density functional theory. *The Journal of Physical Chemistry A* 115: 9467–9478.

Kistner, C., A. André, T. Fischer et al. 2007. Hydration dynamics of oriented DNA films investigated by time-domain terahertz spectroscopy. *Applied Physics Letters* 90: 233902.

Klug, D. D., M. Z. Zgierski, S. T. John et al. 2002. Doming modes and dynamics of model heme compounds. *Proceedings of the National Academy of Sciences* 99: 12526–12530.

Knab, J., J.-Y. Chen, and A. Markelz. 2006. Hydration dependence of conformational dielectric relaxation of lysozyme. *Biophysical Journal* 90: 2576–2581.

Kurt, H. and D. S. Citrin. 2005. Photonic crystals for biochemical sensing in the terahertz region. *Applied Physics Letters* 87: 041108.

Kutteruf, M. R., C. M. Brown, L. K. Iwaki et al. 2003. Terahertz spectroscopy of short-chain polypeptides. *Chemical Physics Letters* 375: 337–343.

Laman, N., S. S. Harsha, D. Grischkowsky, and J. S. Melinger. 2008. High-resolution waveguide THz spectroscopy of biological molecules. *Biophysical Journal* 94: 1010–1020.

Larramendy, M. L., W. e. El-Rifai, and S. Knuutila. 1998. Comparison of fluorescein isothiocyanate- and Texas red-conjugated nucleotides for direct labeling in comparative genomic hybridization. *Cytometry* 31: 174–179.

Lee, S. A., A. Anderson, W. Smith, R. H. Griffey, and V. Mohan. 2000. Temperature-dependent Raman and infrared spectra of nucleosides. Part I—Adenosine. *Journal of Raman Spectroscopy* 31: 891–896.

Lee, S. A., J. Li, A. Anderson et al. 2001. Temperature-dependent Raman and infrared spectra of nucleosides. II—Cytidine. *Journal of Raman Spectroscopy* 32: 795–802.

Lee, S. A., M. Schwenker, A. Anderson, and L. Lettress. 2004. Temperature-dependent Raman and infrared spectra of nucleosides. IV—Deoxyadenosine. *Journal of Raman Spectroscopy* 35: 324–331.

Li, J., S. Lee, A. Anderson et al. 2003. Temperature-dependent Raman and infrared spectra of nucleosides. III—Deoxycytidine. *Journal of Raman Spectroscopy* 34: 183–191.

Lin, D., A. Matsumoto, and N. Go. 1997. Normal mode analysis of a double-stranded DNA dodecamer d (CGCGAATTCGCG). *The Journal of Chemical Physics* 107: 3684–3690.

Liu, H.-B. and X.-C. Zhang. 2006. Dehydration kinetics of D-glucose monohydrate studied using THz time-domain spectroscopy. *Chemical Physics Letters* 429: 229–233.

Markelz, A., S. Whitmire, J. Hillebrecht, and R. Birge. 2002. THz time domain spectroscopy of biomolecular conformational modes. *Physics in Medicine and Biology* 47: 3797–3805.

Markelz, A. G., A. Roitberg, and E. J. Heilweil. 2000. Pulsed terahertz spectroscopy of DNA, bovine serum albumin and collagen between 0.1 and 2.0 THz. *Chemical Physics Letters* 320: 42–48.

Mei, W. N., M. Kohli, E. W. Prohofsky, and L. L. Van Zandt. 1981. Acoustic modes and nonbonded interactions of the double helix. *Biopolymers* 20: 833–852.

Nagel, M., P. Haring Bolivar, M. Brucherseifer et al. 2002a. Integrated planar terahertz resonators for femtomolar sensitivity label-free detection of DNA hybridization. *Applied Optics* 41: 2074–2078.

Nagel, M., P. Haring Bolivar, M. Brucherseifer et al. 2002b. Integrated THz technology for label-free genetic diagnostics. *Applied Physics Letters* 80: 154–156.

Nagel, M., F. Richter, P. Haring-Bolivar, and H. Kurz. 2003. A functionalized THz sensor for marker-free DNA analysis. *Physics in Medicine and Biology* 48: 3625.

Nishizawa, J.-i., T. Sasaki, K. Suto et al. 2005. THz transmittance measurements of nucleobases and related molecules in the 0.4-to 5.8-THz region using a GaP THz wave generator. *Optics Communications* 246: 229–239.

Nishizawa, J.-i., K. Suto, T. Sasaki, T. Tanabe, and T. Kimura. 2003. Spectral measurement of terahertz vibrations of biomolecules using a GaP terahertz-wave generator with automatic scanning control. *Journal of Physics D: Applied Physics* 36: 2958.

Ozaki, H. and L. W. McLaughlin. 1992. The estimation of distances between specific backbone-labeled sites in DNA using fluorescence resonance energy transfer. *Nucleic Acids Research* 20: 5205–5214.

Powell, J. W., G. S. Edwards, L. Genzel et al. 1987. Investigation of far-infrared vibrational modes in polynucleotides. *Physical Review A* 35: 3929.

Sarkar, M., S. Sigurdsson, S. Tomac et al. 1996. A synthetic model for triple-helical domains in self-splicing group I introns studied by ultraviolet and circular dichroism spectroscopy. *Biochemistry* 35: 4678–4688.

Saxena, V. K., B. H. Dorfman, and L. L. Van Zandt. 1991. Identifying and interpreting spectral features of dissolved poly (dA)-poly (dT) DNA polymer in the high-microwave range. *Physical Review A* 43: 4510.

Shen, S. C., L. Santo, and L. Genzel. 1981. Far infrared spectroscopy of amino acids, polypeptides and proteins. *Canadian Journal of Spectroscopy* 26: 126–133.

Shen, S. C., L. Santo, and L. Genzel. 2007. THz spectra for some bio-molecules. *International Journal of Infrared and Millimeter Waves* 28: 595–610.

Shen, Y. C., P. C. Upadhya, E. H. Linfield, and A. G. Davies. 2004. Vibrational spectra of nucleosides studied using terahertz time-domain spectroscopy. *Vibrational Spectroscopy* 35: 111–114.

Shi, Y. and L. Wang. 2005. Collective vibrational spectra of α-and γ-glycine studied by terahertz and Raman spectroscopy. *Journal of Physics D: Applied Physics* 38: 3741.

Shotts, W. J. and A. J. Sievers. 1974. The far-infrared properties of polyamino acids. *Biopolymers* 13: 2593–2614.

Suhandy, D., T. Suzuki, Y. Ogawa et al. 2011. A quantitative study for determination of sugar concentration using attenuated total reflectance terahertz (ATR-THz) spectroscopy. Presented at *Proceedings of SPIE*, p. 802705.

Titova, L. V., A. K. Ayesheshim, A. Golubov et al. 2013. Intense THz pulses cause H2AX phosphorylation and activate DNA damage response in human skin tissue. *Biomedical Optics Express* 4: 559.

Tominaga, Y., M. Shida, K. Kubota et al. 1985. Coupled dynamics between DNA double helix and hydrated water by low frequency Raman spectroscopy. *The Journal of Chemical Physics* 83: 5972.

Upadhya, P. C., Y. C. Shen, A. G. Davies, and E. H. Linfield. 2003. Terahertz time-domain spectroscopy of glucose and uric acid. *Journal of Biological Physics* 29: 117–121.

van Zandt, L. L. and V. K. Saxena. 1994. Vibrational local modes in DNA polymer. *Journal of Biomolecular Structure and Dynamics* 11: 1149–1159.

Walther, M., B. Fischer, M. Schall, H. Helm, and P. U. Jepsen. 2000. Far-infrared vibrational spectra of all-*trans*, 9-*cis* and 13-*cis* retinal measured by THz time-domain spectroscopy. *Chemical Physics Letters* 332: 389–395.

Walther, M., B. M. Fischer, and P. Uhd Jepsen. 2003. Noncovalent intermolecular forces in polycrystalline and amorphous saccharides in the far infrared. *Chemical Physics* 288: 261–268.

Weidlich, T., S. M. Lindsay, Q. Rui et al. 1990. A Raman study of low frequency intrahelical modes in A-, B-, and C-DNA. *Journal of Biomolecular Structure and Dynamics* 8: 139–171.

Whitmire, S. E., D. Wolpert, A. G. Markelz et al. 2003. Protein flexibility and conformational state: A comparison of collective vibrational modes of wild-type and D96N bacteriorhodopsin. *Biophysical Journal* 85: 1269–1277.

Wittlin, A., L. Genzel, F. Kremer et al. 1986. Far-infrared spectroscopy on oriented films of dry and hydrated DNA. *Physical Review A* 34: 493.

Woolard, D. L., T. R. Globus, B. L. Gelmont et al. 2002. Submillimeter-wave phonon modes in DNA macromolecules. *Physical Review E* 65: 051903.

Xie, A., Q. He, L. Miller, B. Sclavi, and M. R. Chance. 1999. Low frequency vibrations of amino acid homopolymers observed by synchrotron far-ir absorption spectroscopy: Excited state effects dominate the temperature dependence of the spectra. *Biopolymers* 49: 591–603.

Yamaguchi, M., F. Miyamaru, K. Yamamoto, M. Tani, and M. Hangyo. 2005. Terahertz absorption spectra of L-, D-, and DL-alanine and their application to determination of enantiometric composition. *Applied Physics Letters* 86: 053903.

Yamamoto, K., K. Tominaga, H. Sasakawa et al. 2002. Far-infrared absorption measurements of polypeptides and cytochrome c by THz radiation. *Bulletin of the Chemical Society of Japan* 75: 1083–1092.

Yamamoto, K., K. Tominaga, H. Sasakawa et al. 2005. Terahertz time-domain spectroscopy of amino acids and polypeptides. *Biophysical Journal* 89: L22–L4.

Yoneyama, H., M. Yamashita, S. Kasai et al. 2008. Membrane device for holding biomolecule samples for terahertz spectroscopy. *Optics Communications* 281: 1909–1913.

Young, L., V. V. Prabhu, and E. W. Prohofsky. 1989. Calculation of far-infrared absorption in polymer DNA. *Physical Review A* 39: 3173.

Yu, B., F. Zeng, Y. Yang et al. 2004. Torsional vibrational modes of tryptophan studied by terahertz time-domain spectroscopy. *Biophysical Journal* 86: 1649–1654.

Zhu, Z., J. Chao, H. Yu, and A. S. Waggoner. 1994. Directly labeled DNA probes using fluorescent nucleotides with different length linkers. *Nucleic Acids Research* 22: 3418–3422.

10

Terahertz Spectroscopic Insight into the Structure–Function Relation in Protein Molecules

Nikolay Brandt
Lomonosov Moscow State University

Andrey Chikishev
Lomonosov Moscow State University

Alexey Kargovsky
Lomonosov Moscow State University

Alexander Shkurinov
Lomonosov Moscow State University

10.1 Introduction

Recent interest in the relation of the enzyme structure and functioning has been driven mainly by the prospects for controlled enzyme functioning due to structural modifications. In this regard, it is expedient to employ new methods for the analysis of the protein structure: one of the methods is the THz absorption spectroscopy. One of the well-known examples of remarkable modifications of enzyme functioning owing to variations in the molecular environment is the inverse functioning of α-chymotrypsin (CT). This enzyme is known to catalyze hydrolysis of peptide bonds in aqueous medium, which is natural environment for many biomolecules (Northrop et al. 1948). However, the enzyme in the nonaqueous medium is involved in different reactions, and hence, different products can be formed (Ahern and Klibanov 1985; Klibanov 1989). The hydrolytic reactions catalyzed by proteases in water are transformed into transesterification or peptide synthesis in organic media under modified thermodynamic conditions (Ahern and Klibanov 1985; Debulis and Klibanov 1993; Klibanov 1989; Northrop et al. 1948).

Enzyme functioning in organic media yields numerous advantages (Ahern and Klibanov 1985; Debulis and Klibanov 1993; Dordick 1992; Klibanov 1989; Northrop et al. 1948), such as the following:

- An increase in the solubility of hydrophobic substrates and effectors
- A shift of the thermodynamic equilibrium of practically important reactions that involve the formation of amide or ether bonds toward the desired products
- Suppression of the side reactions that are possible in the presence of water
- An increase in the thermal stability of enzyme in comparison with the thermal stability in aqueous solutions
- Different substrate specificity and enantioselectivity of enzyme, which allow new possibilities in the synthesis of complex biomolecules and modifications at relatively high enantioselectivity

The serious disadvantage of nonaqueous systems is a dramatic loss of the enzyme activity compared to aqueous solutions, encouraging the search for the ways to control the catalytic activity in organic media. Adding sugars, amino acids, polyethylene glycols, polyelectrolyte salts, and crown ethers allows one to enhance the enzymatic activity in nonaqueous (nonconventional) media (Khmelnitsky et al. 1994; Reinhoudt et al. 1989; Triantafyllou et al. 1995; Volkin et al. 1991). Crown ethers are known to increase the activity of serine proteases, tyrosinases (Broos et al. 1995a,b; Unen et al. 1998a,b), and lipases (Itoh et al. 1997a,b) suspended in organic solvents. It was demonstrated that the enzymatic activity of CT in organic solvents (cyclohexane, acetonitrile, etc.) increases by several orders of magnitude due to interaction with crown ethers (Broos et al. 1995b; Unen et al. 1998b).

Several methods are used for the structural analysis of biomolecules. In particular, CT is characterized by x-ray diffraction (Blevins and Tulinsky 1985; Tsukada and Blow 1985), FTIR absorption spectroscopy (Byler and Susi 1986; Susi et al. 1985), Raman spectroscopy (Brandt et al. 2001; Lord and Yu 1970), CD spectroscopy (Jibson et al. 1981; Volini and Tobias 1969), polarization-sensitive CARS (Brandt et al. 2000a; Chikishev et al. 1992), ROA, VCD, etc. X-ray analysis is the most informative method for the study of molecular structure, which can be employed only in crystals. The methods of vibrational spectroscopy make it possible to characterize molecular structure in various phase states. Several vibrational bands in Raman spectra of proteins are known to be conformation sensitive (Carey 1982). Amide I (1640–1660 cm^{-1}) and amide III (1200–1240 cm^{-1}) bands are sensitive to the secondary structure. Tyrosine doublet (830 and 850 cm^{-1}) and tryptophan marker (1361 cm^{-1}) are sensitive to conformational state of these residues and the adjacent groups. The relative intensities of the bands at 510, 525, and 540 cm^{-1} are indicative of the conformational state of the disulfide bridges. Alternative methods and modifications of the vibrational spectroscopic techniques employ Fourier-transform spectroscopy and time- and frequency-domain methods. However, all of them characterize the vibrational structure of the molecule under study. Note the complementary character of the infrared (IR) absorption and Raman spectroscopic techniques related to the physical principles of the methods. In this regard, the THz absorption spectroscopy provides the data that supplement the results of the low-frequency (0–400 cm^{-1}) Raman measurements.

The study of the low-frequency oscillations of biomolecules is a topical problem of modern biophysics. In most works on the IR and Raman spectroscopy, the measurements are performed in the *fingerprint* interval (500–2000 cm^{-1}), and the low-frequency interval appears to be less studied due to several technical problems. The topicality of the low-frequency spectroscopy of biological macromolecules is due to the fact that their functioning may lead to the oscillations at such frequencies (Ebeling et al. 2002). For the first time, the low-frequency bands at 25–30 cm^{-1} were detected in the Raman spectra of CT and pepsin in solid state in Brown et al. (1972). Then the bands in the same range were detected in the Raman spectrum of lysozyme in Genzel et al. (1976). However, the observations of the low-frequency Raman bands of proteins in the native (aqueous) medium are missing. The Raman spectra of lysozyme in the samples with various contents of water were measured in Caliskan et al. (2002). At room temperature, an increase in the water content in the sample leads to an increase in the intensity of the quasi-elastic scattering and the band at ca. 20 cm^{-1} vanishes. The absence of developed low-frequency peaks in the Raman spectra of aqueous solutions of proteins can be related to overdamping of the corresponding oscillations in water (McCammon and Wolynes 1977). The damping can be due to H-bonding of the surface amino acids of protein and water molecules. The effect of H-bonding on the low-frequency oscillations of simple organic molecules is studied in Brandt et al. (2007). Apparently, the application of the THz absorption spectroscopy can be helpful in the study of the low-frequency oscillations.

A physical problem may require the study of model systems. In the vibrational spectroscopy of proteins, model systems can be helpful in the study of amino groups, whose vibrational bands are peaked at frequencies of about 1600 cm^{-1} and, hence, are overlapped with relatively strong broad amide I band. In addition, deformation oscillations of water molecules are manifested in the same frequency interval. Thus, the analysis of the interaction of the protein amino groups with surrounding molecules can be solved using the methods of vibrational spectroscopy with the aid of model molecules with amino groups that do not exhibit additional bands in the interval 1550–1700 cm^{-1}. With regard to the THz measurements, the model systems can be interesting objects and the analysis of the corresponding spectra may reveal new regularities. By way of example, we consider the interaction of CT with crown ethers. Since crown ethers are known to interact strongly with amino groups, the influence of crown ethers on CT may be, in particular, related to the complexation of the surface amino groups of the protein (Brandt et al. 2000b, 2001, 2003; Izatt et al. 1978; Mankova et al. 2013). Tris(hydroxymethyl)aminomethane or 2-amino-2-hydroxymethyl-propane-1,3-diol $(HOCH_2)_3CNH_2$ (tris) and its complexes with crown ether can be used as the chemical models (Brandt et al. 2012; Costantino et al. 1997). In addition to the amino groups of proteins, we study the disulfide bridges, which are important elements that stabilize the protein structure. The structural features and modifications of the bridges due to various interactions can also be studied using the chemical models (Brandt et al. 2008). However, note that the spectroscopic study of proteins is also possible, since the bands of the disulfide bridges in protein molecules are not overlapped with different bands (see, e.g., Brandt et al. 2005).

10.2 Technique for the THz Studies of Protein Molecules

10.2.1 FTIR Spectrometer

For the FTIR measurements in the transmission configuration, we used a Nicolet 6700 (Thermo Electron Corporation) spectrometer. The measurements are performed in the spectral interval 50–600 cm^{-1} with a spectral resolution of 2 cm^{-1}.

10.2.2 THz-TDS Apparatus I

The time-domain spectroscopy (Sakai 2005) apparatus has been reported previously (Nazarov et al. 2007). In brief, a mode-locked Ti:sapphire laser (Tsunami, Spectra-Physics) that is tunable from 720 to 995 nm, with a typical output power of 1.5 W and a minimum pulse duration of 60 fs, was used to generate THz radiation from GaAs wafer or from ZnTe crystal. After passing the sample, the THz wave was collected to a plate of ZnTe crystal and measured by electrooptical sampling. The dielectric properties of the samples were obtained by the Fourier analysis of the temporal waveform of the measured THz electric field.

10.2.3 Broadband Terahertz Time-Domain Spectroscopy Apparatus

Several experimental techniques including photoconductive antennas, optical rectification, and electrooptical sampling are widely used for the generation and detection of THz-TDS pulses (Zhang and Xu 2009). These techniques have disadvantages related to a very narrow spectrum of the THz pulses. In this work, we employ an alternative approach based on the light-induced breakdown of gases using high-intensity femtosecond pulses (Balakin et al. 2010; Dai et al. 2009) and the air-biased coherent detection (ABCD) (Dai et al. 2006; Frolov et al. 2012).

In the experiments, we use a Zomega ZAP-ABCD THz-TDS spectrometer equipped with a Newport Spectra-Physics MAI TAI femtosecond laser with a pulse duration of 40 fs, a repetition rate of 1 kHz, and a central wavelength of 800 nm. The THz radiation in the frequency interval 0.1–10 THz can be generated. The spectrometer is based on the THz emission resulting from the photoionization of gases with two-color (ω and 2ω) laser field.

The ABCD technique is used for the wideband THz-TDS detection (Karpowicz et al. 2008). The setup is placed in a purged gas (N_2) chamber to eliminate the THz absorption of water vapor. To determine the validity of the experimental data, we estimate the dynamic range of the spectrometer using the method from Jepsen and Fischer (2005). In particular, we estimate the maximum absorption coefficient (α_{max}) that can be measured on the spectrometer at the given thickness of the sample. This quantity is related to the dynamic range of the spectrometer in the measurements of the following given sample:

$$\alpha_{max} = \frac{2}{d}\log[DR]. \tag{10.1}$$

Here, d is thickness of the sample and DR is the dynamic range of spectrometer that results from the normalization of the spectrum of reference signal by the spectrum of noise signal (Jepsen and Fischer 2005). For the estimation of DR, we use the averaged spectra. We perform the calculation of α_{max} for pure tris (pH 10), which exhibits the maximum absorbance (Figure 10.1). In this work, relatively narrow spectral windows in all the figures in this chapter are due to a crucial decrease in the signals having passed through the samples under study at frequencies of greater than 3.5–8.0 THz (depending on the sample) and the corresponding decrease in the signal-to-noise ratio.

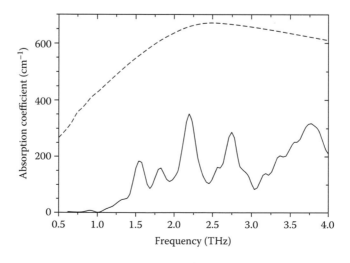

FIGURE 10.1 (Dashed line) Maximum absorption curve and (solid line) the absorption spectrum of unprotonated tris.

10.2.4 THz-TDS Data Processing and Fitting Procedures

After measuring the temporal profiles with and without the sample, the absorption coefficient of the sample is roughly calculated as

$$\alpha(\omega) = -\frac{2}{d}\log\left|\frac{E_{\text{sample}}(\omega)}{E_{\text{ref}}(\omega)}\right|, \qquad (10.2)$$

where

$E_{\text{sample}}(\omega)$ is the amplitude spectrum of temporal profile with sample
$E_{\text{ref}}(\omega)$ is the spectrum of temporal profile without sample

The THz spectra are processed using the computer codes from Nazarov et al. (2008). Note that the signal-to-noise ratio is relatively low and the Savitzky–Golay smoothing procedure with a 30-point smoothing window was used (the total number of points in each THz spectrum is about 200).

10.2.5 Computational Method

IR vibrational spectra of molecules or crystals can be analyzed using quantum theory. Here, we briefly describe a general procedure of calculation without a deep dive into the quantum chemistry. The starting point is the Schrödinger equation for the system of nuclei and electrons, in which motions can be decoupled due to their huge mass difference (Born–Oppenheimer approximation). Then, electronic relaxation with respect to nuclear motion is almost instantaneous. As such, it is convenient to compute electronic energies for fixed nuclear positions in ground state. Then, we find a geometric configuration of the system corresponding to the minimum of energy. The final step is obtaining vibrational states and probabilities of transitions between them due to the external field (for instance, in harmonic approximation). The theoretical evaluation of harmonic vibrational frequencies is efficiently done by the evaluation of second derivatives of the total energy with respect to Cartesian (or internal) coordinates.

There are many methods of solving of the Schrödinger equation for electronic subsystem.

Density functional theory (DFT) is one of the most popular and successful quantum mechanical approaches to matter. The premise behind DFT is that the energy of a nonrelativistic Coulomb system can be determined from the electron density. This theory originated with a theorem by Hohenberg and Kohn (1964) that stated this was possible for the ground state. A practical application of DFT was developed by Kohn and Sham (1965). This approach forms the basis of the majority of electronic-structure calculations in physics and chemistry. The electric, magnetic, and structural properties of materials can be calculated using DFT, and the extent to which DFT has contributed to the science of molecules is reflected by the 1998 Nobel Prize in Chemistry, which was awarded to Walter Kohn (and John A. Pople). For further details, we refer the reader to monographs (Cramer 2005; Jensen 2007; Koch et al. 2001).

Solid-state DFT calculations for protonated and unprotonated tris and 18-crown-6 (CE) are performed using the DMol³ (version 5.5) software package (Delley 1990, 2000) with *fine* grid sizes (corresponding to a k-point separation of 0.04 Å^{-1}) and geometry optimization convergence criteria (program option corresponding to an energy convergence of 10^{-6} Ha, max energy gradient of 10^{-3} Ha Å^{-1}, max displacement of 5×10^{-3}Å), the threshold for density convergence during the SCF minimization of 10^{-6}, the DNP (double numerical with d and p polarization) basis set that is comparable to a 6–31G(d, p) Gaussian-type basis set, and the generalized-gradient approximation (GGA) density functional Perdew–Burke–Ernzerhof (PBE) (Perdew et al. 1996). The Tkatchenko–Scheffler dispersion correction scheme is used (Tkatchenko and Scheffler 2009). Atomic coordinates within the unit cell are optimized within the cell parameters specified by the x-ray diffraction studies of tris (Golovina et al. 2002), protonated tris (Rudman et al. 1983), and CE (Maverick et al. 1980) at room temperature.

Normal-mode analysis is performed in the harmonic approximation. The solid-state IR intensities are calculated from the square of the change in dipole moments for the unit cell that results from atomic displacements along each normal-mode coordinate $(\partial\mu/\partial Q_k)^2$ using Hirshfeld's (1977) charge analysis. Cell dipole moment is given by

$$\mu = \sum_i q_i^H r_i + \mu_i^H, \tag{10.3}$$

where

   i is the number of the atom that is located at coordinates r_i

   q_i^H and μ_i^H are Hirshfeld's effective atomic charge and dipole moment

For dipole calculations, a more tight threshold for density convergence during the SCF minimization (10^{-8}) is used.

The calculated spectra are in qualitative agreement with the measured spectra. However, not all of the calculated spectral amplitudes are in agreement with the experimental ones. Thus, we only present the calculated frequencies and the calculated amplitudes can be found in supplementary table.

Harmonic vibrational frequencies for peptides and isolated complex of protonated tris and 18-crown-6 are calculated using Firefly quantum-chemical package (Granovsky 2009). The computations employ DFT with the Becke hybrid exchange and the Lee–Yang–Parr correlation functionals (B3LYP) (Becke 1993) and the Dunning's augmented correlation-consistent polarized valence double-zeta basis set aug-cc-pVDZ (Dunning Jr. 1989). Geometry optimization is carried out using direct inversion in iterative subspace with the following parameters: the largest component and RMS value of the energy gradient are 10^{-6} and 3×10^{-7} Ha/Bohr, respectively. Numerical integration is performed using a grid with 99 radial shells (974 angular points per shell) per each atom (Lebedev and Skorokhodov 1992). The Hessian is calculated semianalytically: the first derivatives of the energy calculated analytically are numerically differentiated.

The calculations were performed using the SKIF MSU *Chebyshev* and *Lomonosov* supercomputers of Lomonosov Moscow State University Supercomputing Center.

10.3 Spectroscopy of Disulfide Bridges as Elements That Stabilize the Protein Structure

We chose sulfur-containing amino acids for the studies, because the sulfur atoms being involved in the creation of disulfide bond link two cysteine residues that play an important role in the stabilization of the spatial structure of a protein molecule. The disulfide bridges represent the elements of protein primary structure that link the fragments of a single polypeptide chain or different chains. Sulfur atom is significantly heavier than oxygen, carbon, or nitrogen atoms, predominantly contained in the protein molecules. Therefore, the vibrational modes of the disulfide bridges should potentially give rise to both Raman and THz absorption lines in the low-frequency spectral range (Matei et al. 2005). The low-frequency marker bands of the disulfide bridges can be used in the study of the protein structure and in the interpretation of the results obtained with the methods of molecular dynamics. The marker bands are widely used in the study of the protein conformational states (Aoki et al. 1982; Chen and Lord 1976; Kudryavtsev et al. 1998; Podstawka et al. 2004). Both theoretical and experimental works are devoted to the disulfide bridges in proteins and model compounds (Nakamura et al. 1997; Thamann 1999; Weiss-Lopez et al. 1986; Yoshida and Matsuura 1998). The potential energy profiles are calculated in terms of the valence angles in the bridges (Görbitz 1994; Yoshida et al. 1992). The distributions of the protein molecules with respect to the corresponding angles are presented in Thornton (1981).

TABLE 10.1 Substances under Study

	Substance Name	Chemical Structure	Mol. Weight, Da
1st pair	Cysteine (H–Cys–OH)	HN–CH–COOH	121
		H$_2$C–SH	240
	Cystine	H–Cys–OH	
		H–Cys–OH	
2nd pair	LC	H–Leu–Cys–OH	234
	(LC)$_2$	H–Leu–Cys–OH	466
		H–Leu–Cys–OH	
3rd pair	IC	H–Ile–Cys–OH	234
	(IC)$_2$	H–Ile–Cys–OH	466
		H–Ile–Cys–OH	

To observe and assign the low-frequency vibrational lines of disulfide bridges, we study THz spectra of three pairs of sulfur-containing di- and tetrapeptides. Each pair of substances can be considered as a monomer–dimer pair, and each of the dimers is formed owing to the disulfide bridge. Apparently, the THz absorption spectra of the dimers should contain marker bands of disulfide bridges.

Table 10.1 shows chemical structures of the substances under study. Cysteine and the corresponding dimer (cystine) are the substances of the first pair. The monomer of the second pair is the LC dipeptide consisting of leucine and cysteine. The monomer of the third pair is the IC dipeptide consisting of isoleucine and cysteine. All of the peptides were characterized by HPLC analysis, the melting points, and the ESI mass and ^{1}H NMR spectra. Based on the reversed-phase HPLC analysis, the purity of the peptides under study was higher than 96%.

The stretching vibrations of the disulfide bridges are manifested in the protein fingerprint range Raman spectra at 480–550 cm^{-1} (Thamann 1999). It is commonly accepted that there exist three basic conformations for the disulfide bridge: *gauche–gauche–gauche* (*g-g-g*), *gauche–gauche–trans* (*g-g-t*), and *trans–gauche–trans* (*t-g-t*). In the protein spectrum, the Raman lines assigned to these conformations are normally peaked at 510, 525, and 540 cm^{-1}, respectively, and serve as the markers of the disulfide bridges (Kitagawa et al. 1979; Sugeta et al. 1973; Tu 1986). It is demonstrated in VanWart et al. (1973) that the frequency of the S–S stretching vibration linearly depends on the dihedral angle C–S–S–C (ξ). In accordance with these data, at $\xi < 40°$, the vibrational frequency ranges from 480 to 505 cm^{-1}, and at $50° \leq \xi \leq 80°$, the frequency of the stretching mode is about 510 cm^{-1}. Note that the preceding dependence is obtained for the gauche conformation of the CS–SC bond. Devlin et al. (1990) observed the vibrational bands of the disulfide bond at 487 and 490 cm^{-1} in allyl methyl trisulfide and diallyl disulfide, respectively. In these substances, the angle is $\xi \approx 25°$, and the authors classify the corresponding conformation as near-*cis*. The Raman band peaked at 499 cm^{-1} in the spectrum of L-cystine is assigned to the disulfide bond with $\xi \approx 74°$ (Pearson et al. 1959). This assignment contradicts the aforementioned linear dependence, and an alternative interpretation of the experimental result is proposed in VanWart et al. (1973). The frequency of the S–S stretching vibration may substantially be affected by the minor variations in the CS–SC dihedral angle. Therefore, the straightforward assignment based only on the linear dependence of the vibrational frequency on the C–S–S–C dihedral angle may yield ambiguous results. A combined effect of the two dihedral angles on the frequency of the S–S stretching vibration can be a reason for the result obtained for L-cystine.

All of THz absorption spectra show several absorption peaks and continuous absorption, which increases with frequency. In addition to the thermal background, several sources contribute to this continuous absorption. First, this signal can originate from the scattering, since, at the highest frequencies (about 4 THz), the particle size in the sample becomes comparable with the wavelength. Then, another contribution to this background absorption may come from amorphous phase of the sample. We present the THz absorption spectra without background subtraction.

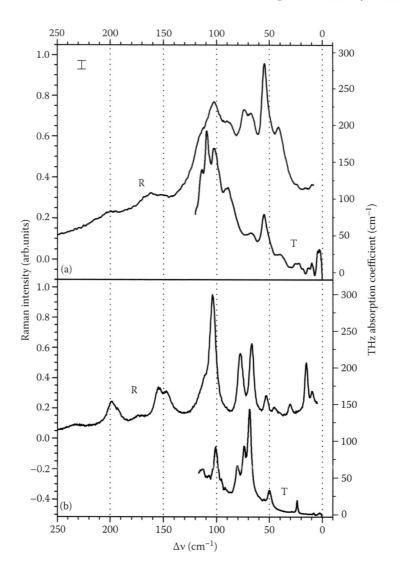

FIGURE 10.2 Low-frequency Raman (R) and THz absorption (T) spectra of (a) L-cysteine and (b) L-cystine.

Figure 10.2 demonstrates the Raman and THz absorption spectra of cysteine and cystine. Table 10.2 compares the measured absorption frequencies with the literature data. With regard to different backgrounds, the Raman and THz absorption spectra of the same sample are similar in general (Figure 10.2a and b). Low-frequency Raman and THz absorption spectra of cysteine and IC are similar (Figure 10.3).

Let us consider the spectral components that can be assigned to the disulfide bridge. The Raman spectra of cystine, $(LC)_2$, and $(IC)_2$ exhibit lines peaked in the range 480–550 cm^{-1}. In the spectrum of cystine, a strong Raman line is peaked at 497 (DFT value is 491) and a weaker line is peaked at 542 cm^{-1}. In the preceding range, the spectrum of $(LC)_2$ can reasonably be fitted with three Gaussian curves peaked at 499, 514, and 528 cm^{-1}. Note that the position of the last spectral component is determined with an error of about 10 cm^{-1}. The Raman lines of $(IC)_2$ are peaked at 506 and 531 cm^{-1}. Based on the Raman data, we conclude that all of the samples exhibit the conformational heterogeneity of the disulfide bridge (contain *g-g-g*, *g-g-t*, and *t-g-t* conformations).

In the polycrystalline samples under study, we can expect inter- and intramolecular vibrations in the low-frequency range (Walther et al. 2000) owing to the H-bond networks (Korter et al. 2006;

TABLE 10.2   THz Absorption Lines of L-Cysteine, L-Cystine, IC and LC

L-Cysteine		L-Cystine		IC	LC
Expt.	Ref. data	Expt.	Ref. data	Expt.	Expt.
		8			
		23	23.74[b]		
		28			
43	46[a], 45.05[b]	50	49.74[b]		
55	56[a], 54.67[b]	55		56	
66		69	68.87[b]		
	71[a]	74			
	80[a]	81		80	80
90		90			
102	97[a]	101		95	95

All frequencies are in cm^{-1}.

[a] Data taken from Korter et al. (2006).

[b] Data taken from Yamamoto et al. (2005).

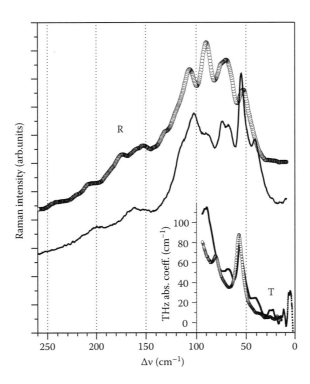

FIGURE 10.3   Comparison of the Raman (R) and THz absorption spectra (T) of (solid line) L-cysteine and (circles) IC.

Plazanet et al. 2002). The solid-state effects can yield the splitting of normal modes of isolated molecules in the crystal into several modes of different symmetries. Some of these modes are allowed in the IR absorption, and some of them are Raman active (Hochstrasser 1966). In THz range, the effect of splitting was studied in all-*trans*-retinal molecular crystals (Gervasio et al. 1998) and in dibromobenzophenone crystal (Volovšek et al. 2002).

Figures 10.2a and 10.3 show the THz absorption spectrum of L-cysteine. The five major absorption peaks at 43, 55, 66, 90, and 102 cm^{-1} are observed. Under room temperature, the L-cysteine molecule crystallizes in the space group $P2_12_12_1$ with four molecules in the unit cell (Kerr and Ashmore 1973), and the molecule exists as a zwitterion with a 3D network of H-bonds. Each normal mode of the isolated L-cysteine molecule splits in the crystal into four components, two Raman and two IR active. Thus, instead of two vibrational modes predicted by the calculations (DFT using exchange-correlation functional B3LYP) peaked at 43 and 88.5 cm^{-1} that are assigned to the COO and C^α–C^β torsions, respectively, we observe four picks at 43, 55, 66, and 90 cm^{-1}. THz absorption spectra of dipeptides IC and LC are shown in Figures 10.3 and 10.4, and the positions of the vibrational modes are shown in Table 10.2. For IC, the three absorption peaks at 56, 80, and 95 cm^{-1} and, for LC, two absorption peaks at 80 and 95 cm^{-1} are observed. We note that these lines are also observed in L-cysteine.

L-Cystine is a dimer of L-cysteine and it crystallizes in two polymorphic forms: a tetragonal phase ($P4_1$) (Chaney and Steinrauf 1974) and a hexagonal phase ($P6_122$) (Dahaoui et al. 1999). Both the tetragonal and hexagonal phases of L-cystine crystallize with the molecule in its zwitterionic form. The S–S bond distances in the two polymorphs do not differ significantly from each other. The same comment applies to the C–S–S–C torsional angle, which is positive in both forms (69° and 75°, respectively). The principal intermolecular interactions take the form of S···S contacts and NH···O hydrogen bonds, formed between the ammonium and carboxylate moieties. The NH···O hydrogen bonds lead to the formation of layers. The hydrogen-bonded layers are linked to each other on one side by covalent S–S bridges and on the other side by NH···O hydrogen bonds (Moggach et al. 2005).

Calculated normal modes of the isolated L-cystine are 20.37, 27.62, 29.19 (corresponds to relative intermolecular torsions of monomers), 41.34, 50.47 (relative torsions of COO+$C^{\alpha 0}$–$C^{\beta 0}$ and C^0OO+C^α–C^β), 55.64, and 75.86 cm^{-1} (deformation modes are relative motions of monomers). The normal modes of the crystal will be composed of different molecular modes. Because of the crystalline symmetry, some of vibrational modes of the isolated molecule may split in the crystal, and it should be a property of the deformation and relative motions of L-cysteine monomers. The IR and Raman frequencies may also differ by few wave numbers due to the group splitting (Hochstrasser 1966).

Figure 10.2b shows the Raman and THz absorption spectra of L-cystine. The THz spectra exhibit ten bands peaked at 8, 23, 28, 50, 55, 69, 74, 81, 90, and 101 cm^{-1}. The complex crystal structure of the sample does not allow us to assign the observed lines to the calculated normal modes, but we would like to summarize our observations in general. THz absorption and Raman spectra contain clear similarities, but the number of THz absorption bands in the frequency range below 120 cm^{-1} differs from the number of Raman lines. Note that, in the spectra of L-cysteine, the Raman doublet at 67 and 78 cm^{-1} becomes a triplet at about 69, 74, and 80 cm^{-1} in the THz absorption spectrum. We should also mention the existence of group of THz (23 and 28 cm^{-1}) and Raman (15 and 29 cm^{-1}) spectral features that are not observed in the L-cysteine. Based on our DFT calculations, we assign these spectral features to the relative motions of the hydrogen-bonded layers of the crystal that involve the S–S cysteine bond linkage.

Hexagonal symmetry of the crystal permits the splitting of equations of motion into two groups: librational and translational. Note that zero frequencies are assigned to translational motions. This property of the crystalline system allows us to interpret the low-frequency bands lying below 10 cm^{-1} as the translational motions of hydrogen-bonded layers. These motions also involve the S–S bridge and thus should be reflected in the low-frequency Raman and THz absorption spectra.

Such vibrations are also manifested in Raman and THz absorption spectra of (IL)$_2$ and (LC)$_2$. However, without proper identification of the crystal structure of the sample, their precise identification is not reasonable. The most low-frequency Raman lines are peaked at 10 and 15 cm^{-1} (cystine) and 9 and 18 cm^{-1} ((IC)$_2$). However, the lines with similar frequencies are not observed in the spectrum of (LC)$_2$. The featureless THz spectra are obtained for (IL)$_2$ and (LC)$_2$ (Figure 10.4). This behavior may take place due to the lack of the long-range order in the LC sample, which is typical of amorphous samples (Walther et al. 2003). However, we cannot conclude that the (IL)$_2$ and (LC)$_2$ samples are amorphous, since their refractive indices do not differ from the refractive indices of the other samples under study.

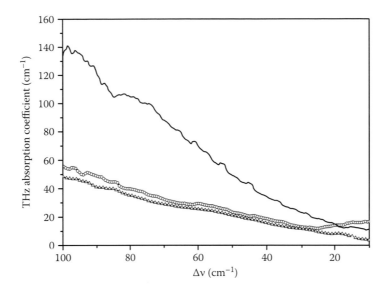

FIGURE 10.4 Comparison of the THz absorption spectra of (solid line) LC, (circles) (IC)$_2$, and (triangles) (LC)$_2$.

10.4 Inverse Functioning of Enzymes in Inorganic Solvents

The analysis of biological molecules is among the central problems in physics, chemistry, and biology. Enzymes are important biological molecules that control functioning of living organisms. Enzymatic activity is normally related to their conformational dynamics, so that the structural features of enzymes play an important role in functioning.

Water is natural environment for a significant part of enzymes. When the environment is modified, the free energy of protein globule changes and a new equilibrium is reached due to variations in spatial structure. The analysis of the function-related structural modifications of enzymes in various media is necessary when the control of enzymatic activity is considered.

In water, serine proteases provide the hydrolysis of peptide bonds, and, for example, for CT, the maximum activity is reached at pH 7.8 (Fersht 1999). When the lyophilized proteases are placed in organic solvents, the functioning changes to transesterification and the activity decreases by a few orders of magnitude (Debulis and Klibanov 1993; Khmelnitsky et al. 1994). However, the activity in organic solvents substantially increases if the serine protease is lyophilized from the aqueous solution containing crown ethers. The maximum activities are different in different solvents and are reached at different relative molar concentrations of enzyme and crown ether (e.g., 1:250 and 1:50 for CT in cyclohexane and acetonitrile, respectively) (Broos et al. 1995b; Unen et al. 1998a). One of the possible reasons for such an increase in the enzymatic activity is the interaction of crown ether molecules with the surface amino groups of protein (Brandt et al. 2000b,c, 2001, 2003; Engbersen et al. 1996; Tsukube et al. 2001). To simulate such interaction, we employ a chemical model based on tris(hydroxymethyl)aminomethane. The application of more close models based on amino acids and their derivatives is impeded by the spectral overlapping of the bands of amino groups and carboxyl group. Tris is a primary amine that exhibits typical properties (e.g., condensation with aldehydes). Therefore, it can be used as a chemical model of the protein amino groups in the absence of the strong amide bands. It was demonstrated (Brandt et al. 2012; Mankova et al. 2012) that unprotonated tris weakly interacted with crown ether, whereas a relatively stable complex was possible for the tris with protonated amino group (protonated tris). The first results on the THz spectroscopy of tris can be found in Harsha and Grischkowsky (2010).

Crown ether molecules are relatively large macrocycles with oxygen atoms in ethylene bridges –CH$_2$–O–CH$_2$–. Crown ethers exhibit complexation due to unshared electron pairs of oxygen that are oriented inside the cycle. In the experiments, we use 18-crown-6 (1,4,7,10,13,16-hexaoxacyclooctadecane $C_{12}H_{24}O_6$). It is demonstrated in Izatt et al. (1978) and Trueblood et al. (1982) that the interaction of CEs with amines leads to the formation of complexes in which the amino group is located inside the cycle (similar complexes are formed with alkali metals). It is commonly accepted that the complex formation is due to the H-bonding of amino group and CE oxygens.

THz-TDS and FTIR spectroscopic techniques are applied in the study of interaction of CT (serine protease) and protonated tris with CE. Both spectroscopic methods make it possible to study low-frequency vibrational modes.

10.4.1 Spectroscopy of the Tris–Crown System (Model for the Interaction of the Surface Amino Groups of Protein)

The tris–CE mixtures are lyophilized from aqueous solutions at several relative molar concentrations. First, tris is dissolved in bidistilled water and pH 3 is reached using hydrochloric acid. The solution predominantly contains the protonated tris, since $pK_{tris} = 8.06$. Then, CE is added to the solution and the resulting tris-to-CE relative molar concentrations are 1:1, 1:2, 1:5, and 1:10. The relative concentration of CE in the samples increases, since the maximum enzymatic activity in nonaqueous solvents is reached when the number of the CE molecules per one protein molecule is noticeably greater than the number of the alleged binding sites of the protein molecule. The lyophilization yields powders consisting of microcrystals with sizes of 40–100 μm. For THz-TDS measurements, the powders are pressed in tablets. In the FTIR measurements, relatively high absorption coefficients of the samples under study necessitate a decrease in the thickness of tablets to several tens of microns. Such thin tablets are mechanically unstable and we pressed the powders on parafilm circles with a thickness of 120 μm.

Figure 10.5 demonstrates the THz-TDS and FTIR spectra of the protonated tris and the tris–CE mixtures. The CE spectrum (Figure 10.5f and f′) exhibits the bands peaked at 2.0, 3.2, 4.3, 4.7, 5.6, 6.8, and 8.0 THz (75, 108, 143, 169, 190, 228, and 266 cm^{-1}). The vertical bars in Figure 10.5f′ show the calculated frequencies, which are in agreement with the experimental data. The spectra of tris (Figure 10.5a and a′) exhibit the bands peaked at 3.0, 4.1, 5.6, 7.2, and 8.1 THz (100, 138, 186, 240, and 275 cm^{-1}).

Relatively strong interaction with the possible formation of a stable complex for the 1:1 sample, which was demonstrated in Brandt et al. (2000b,c, 2001, 2003), Engbersen et al. (1996), and Tsukube et al. (2001), also follows from Figure 10.5b and b′. The spectra of the 1:1 complex crucially differ from the spectra of pure substances and cannot be represented as linear combinations of the spectra of components. Note also that the absolute values of the absorption coefficient of the 1:1 sample are not additive with respect to absorption coefficients of the components (Figure 10.5b and b′). We assume that the bands of tris peaked at 3.0 and 4.1 THz are broadened and red-shifted and the background absorption increases in the high-frequency part of the spectrum. Several bands from the spectrum of the 1:1 sample (3.4, 4.1, 6.4, and 7.5 THz [115, 135, 214, and 249 cm^{-1}]) are not observed in the CE and tris spectra.

The spectra of the nonequimolar samples (1:2, 1:5, and 1:10) cannot be represented as the sums of the spectra of equimolar sample and CE with the corresponding coefficients depending on the relative concentrations. The spectra of the nonequimolar samples are virtually free of the bands of pure tris, but an increase in the relative concentration of CE gives rise to the bands of CE (3.2, 4.5, 5.6, and 6.8 THz [108, 150, 190, and 228 cm^{-1}]). The maximum absorbance of the 1:2 sample (about 400 cm^{-1}) is significantly greater than the maximum absorbances of CE (about 240 cm^{-1}) and equimolar sample (about 330 cm^{-1}). The maximum absorbance of the 1:5 sample is even higher (about 430 cm^{-1}). The spectrum of the 1:2 sample is free of the developed band of the equimolar sample peaked at 250 cm^{-1}.

Thus, the spectra of the samples are gradually modified with an increase in the relative concentration of CE. The bands that are peaked at 3.4 and 4.1 THz (115 and 135 cm^{-1}) in the spectrum of the

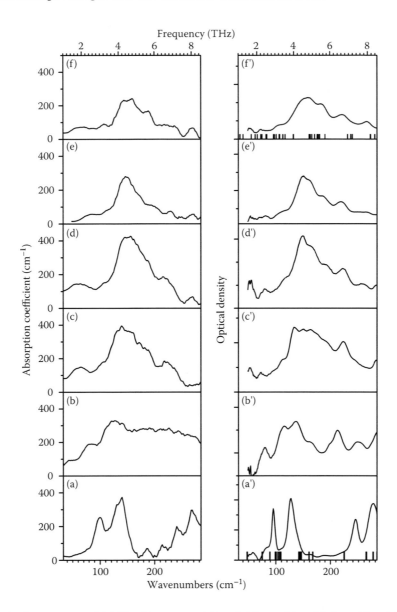

FIGURE 10.5 (a–f) THz-TDS and (a′–f′) FTIR spectra of (a) and (a′) tris; (b) and (b′) 1:1, (c) and (c′) 1:2, (d) and (d′) 1:5, and (e) and (e′) 1:10 tris–CE samples; and (f) and (f′) CE. The bars in panels (a′) and (f′) demonstrate the calculated frequencies.

equimolar sample gradually decrease with an increase in the CE content. The band that is peaked at 193 cm^{-1} in the spectrum of the 1:5 sample is shifted to 186 cm^{-1} in the spectrum of the 1:10 sample and to 182 cm^{-1} in the spectrum of CE. We conclude that the spectra of tris–CE samples cannot be represented as superpositions of the spectra of components.

Based on the preceding results, we assume that the tris–CE samples represent at least partially ordered structures with cooperative vibrational modes. A hypothetical unit cell of the 1:2 (1:5) sample may contain 1/3 (1/5) of tris molecules and 2/3 (5/6) of CE molecules. Even when the relative molar concentration of CE is significantly greater than the relative molar concentration of tris, the spectrum of the mixture may differ from the spectrum of pure CE.

10.4.2 Spectroscopic Characterization of the Enzyme Interaction with Crown Ether

The CT–CE mixtures are lyophilized from aqueous solutions at relative molar concentrations of 1:100 and 1:250. The tablets for THz-TDS and FTIR measurements are prepared as in the previous section. Figure 10.6 shows the THz-TDS (panels a through d) and FTIR (panels a′ through d′) spectra of CT–CE samples. As in the aforementioned experiments, the spectral data obtained using the THz spectroscopy are in agreement with the FTIR results. General modifications of the spectra with an increase in the relative concentration of CE are similar to the spectral modifications of the tris–CE samples. Evidently, the minimum number of the possible binding sites on the protein surface is equal to the number of protonated surface amino groups of protein. Note also the possibility of binding with the surface OH groups (Brandt et al. 2000c, 2001). The number of the binding sites can be estimated using the x-ray data from Northrop et al. (1948). At pH < 8, the number of the protonated surface amino groups of CT is no greater than 20. Formal estimation of the surface area of the protein molecule allows about 50 CE molecules to be placed on the surface. Thus, the samples under study contain excess amounts of CE molecules with respect to the number of the possible binding sites on the protein surface.

The CT spectrum represents a relatively smooth curve with weakly developed spectral features peaked at 131, 154, and 234 cm^{-1}. The spectrum of the 1:100 sample substantially differs from the spectrum of protein, and the spectrum of the 1:250 sample differs from the spectrum of protein and appears to be

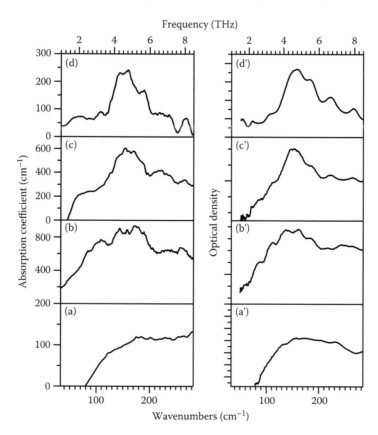

FIGURE 10.6 (a–d) THz-TDS and (a′–d′) FTIR spectra of (a) and (a′) CT; (b) and (b′) 1:100 (c) and (c′) 1:250 CT–CE samples; and (d) and (d′) CE.

generally similar to the spectrum of CE. Note significantly different maximum absorption coefficients and nonmonotonic variations in the absorbance with a decrease (increase) in the relative concentrations of components.

The general similarity of the spectral shapes and the presence of the spectral features from the CE spectrum in the spectra of the 1:100 and 1:250 samples make it possible to assume that the spectra of the CT–CE samples can be represented as superpositions of the spectra of pure substances. Under such an assumption, we can calculate the spectra of the CT–CE samples using the measured spectra of the pure substances (CT and CE) and the relative concentrations of components. We calculate the optical densities of the 1:100 and 1:250 samples assuming the absence of interaction of components (additivity of optical densities) with allowance for the mass ratios of the components (28 wt% of protein and 72 wt% of CE in the 1:250 sample and 49 wt% of protein and 51 wt% of CE in the 1:100 sample).

Figure 10.7 shows that the spectra of the 1:100 and 1:250 samples (solid lines) cannot be represented as linear combinations of the spectra of components with the weighting coefficients determined by the relative concentrations (dashed lines). The comparison makes it possible to assume the interaction of protein and CE, which possibly leads to CE binding. An alternative interpretation may involve the formation of heterogeneous substance in which the CT molecules are embedded in the matrix of CE molecules or vice versa. The 1:100 solid sample can be considered as a lattice of closely packed CT molecules in which the free space between the protein molecules is filled with CE. Then, in 1:250 sample, the distance between the protein molecules can be greater than the mean diameter of protein globule by approximately 20%.

Assume that the difference of the spectra of 1:100 and 1:250 samples is only due to additional amount of CE in the last sample. In other words, assume that 1:250 sample is an additive mixture of 1:100 sample and CE. Then, the spectrum of 1:250 sample must be identical to the linear combination of the spectrum of 1:100 sample and the spectrum of CE with the corresponding weighting coefficients (we assume the absence of free protein molecules, so that the concentration of 1:100 species in 1:250 sample is identical to the concentration of protein). Figure 10.8 compares the measured and calculated spectra. It is seen that the spectral shapes are different and the calculated optical density is generally lower than experimental optical density. Thus, the additional CE molecules in 1:250 sample must be at least partially involved in the interaction with the protein molecule and/or cause transformation of the heterogeneous CT–CE system.

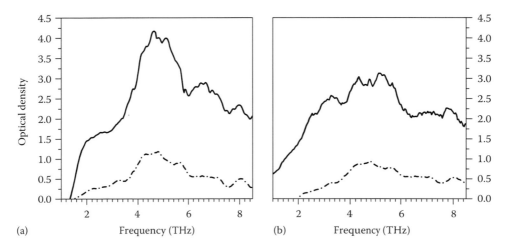

FIGURE 10.7 (Solid lines) THz-TDS spectra of (a) 1:250 and (b) 1:100 CT–CE samples and (dashed-and-dotted lines) calculated spectra (see text for details).

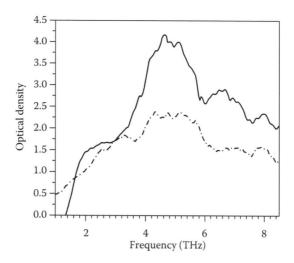

FIGURE 10.8 (Solid line) THz-TDS spectrum of 1:250 CT–CE sample and (dashed-and-dotted line) calculated spectrum (see text for details).

10.5 Conclusions

Using the existing spectroscopic data and the results of additional Raman measurements, we assume that the lines in THz absorption spectra correspond to various vibrations of disulfide bond (e.g., SSC bending mode and SS torsional mode). Such lines may serve as the low-frequency markers of the disulfide bridge. The differences between Raman and THz absorption frequencies can be due to different intermolecular interactions. Thus, the lowest vibrational frequencies may correspond to lattice vibrations. We conclude that the low-frequency Raman and THz absorption lines reflect both inter- and intramolecular motions. With regard to the problem of CT activation in nonaqueous media in the presence of crown ethers possibly due to the interaction with CT surface amino groups, we studied the tris–CE interaction. It is demonstrated that protonated tris actively interacts with CE: the structures that are formed at different relative molar concentrations of components exhibit spectral features that are absent in the spectra of pure substances. Note that protonated amino groups of tris interact with CE more actively than unprotonated groups do. Experimental results are in qualitative agreement with the calculated vibrational modes of tris and CE unit cells. The CT spectrum demonstrates the broadband IR absorption with only three weak spectral lines centered at 131, 154, and 234 cm^{-1}. The spectra of CT–CE mixtures show several developed spectral features. The spectra of the mixtures significantly differ from the spectra of components. The estimations show that the spectra of the 1:100 and 1:250 CT–CE samples cannot be represented as superpositions of the spectra of components, and the spectrum of the 1:250 sample cannot be represented as the superposition of the spectrum of the 1:100 sample and the spectrum of CE.

The changes in the spectra of the CT–CE mixtures with an increase in the relative molar concentration of CE are similar to the corresponding changes in the spectra of the tris–CE samples. Thus, we assume that an increase in the functional activity of CT in nonaqueous solvents is related to the interaction of the protonated amino groups of the protein with the CE molecules.

References

Ahern, T. J. and A. M. Klibanov. 1985. The mechanisms of irreversible enzyme inactivation at 100C. *Science* 228: 1280–1284.

Aoki, K., H. Okabayashi, S. Maezawa et al. 1982. Raman studies of bovine serum albumin-ionic detergent complexes and conformational change of albumin molecule induced by detergent binding. *Biochimica et Biophysica Acta (BBA)-Protein Structure and Molecular Enzymology* 703: 11–16.

Balakin, A. V., A. V. Borodin, I. A. Kotelnikov, and A. P. Shkurinov. 2010. Terahertz emission from a fem-tosecond laser focus in a two-color scheme. *Journal of the Optical Society of America B* 27: 16–26.

Becke, A. D. 1993. Density-functional thermochemistry. III. The role of exact exchange. *The Journal of Chemical Physics* 98: 5648–5652.

Blevins, R. A. and A. Tulinsky. 1985. The refinement and the structure of the dimer of alpha-chymotrypsin at 1.67-A resolution. *Journal of Biological Chemistry* 260: 4264–4275.

Brandt, N. N., A. Y. Chikishev, V. I. Dolgovskii, and S. I. Lebedenko. 2007. Laser Raman spectroscopy of the effect of solvent on the low-frequency oscillations of organic molecules. *Laser Physics* 17: 1133–1137.

Brandt, N. N., A. Y. Chikishev, J. Greve et al. 2000a. CARS and Raman spectroscopy of function-related conformational changes of chymotrypsin. *Journal of Raman Spectroscopy* 31: 731–737.

Brandt, N. N., A. Y. Chikishev, A. V. Kargovsky et al. 2008. Terahertz time-domain and Raman spec-troscopy of the sulfur-containing peptide dimers: Low-frequency markers of disulfide bridges. *Vibrational Spectroscopy* 47: 53–58.

Brandt, N. N., A. Y. Chikishev, A. A. Mankova et al. 2012. THz and IR spectroscopy of molecular systems that simulate function-related structural changes of proteins. *Spectroscopy: An International Journal* 27: 429–432.

Brandt, N. N., A. Y. Chikishev, and I. K. Sakodinskaya. 2003. Raman spectroscopy of tris-(hydroxymethyl) aminomethane as a model system for the studies of α-chymotrypsin activation by crown ether in organic solvents. *Journal of Molecular Structure* 648: 177–182.

Brandt, N. N., A. Y. Chikishev, A. I. Sotnikov et al. 2005. Ricin, ricin agglutinin, and the ricin bind-ing subunit structural comparison by Raman spectroscopy. *Journal of Molecular Structure* 735: 293–298.

Brandt, N. N., V. V. Molodozhenya, I. K. Sakodynskaya, and A. Y. Chikishev. 2000b. The Raman spectra of a crown complex of tris (hydroxymethyl) aminomethane. *Russian Journal of Physical Chemistry A* 74: 1883–1887.

Brandt, N. N., I. K. Sakodinskaya, and A. Y. Chikishev. 2000c. Raman spectroscopy of conformational changes of alpha-chymotrypsin in a reaction with 18-crown-6: The effect of enzyme activation in organic solvents. *Doklady Physical Chemistry* 375: 235–239.

Brandt, N. N., I. K. Sakodynskaya, and A. Y. Chikishev. 2001. A study of interaction between alpha-chymotrypsin and 18-crown-6 in organic solvents by Raman spectroscopy. *Russian Journal of Physical Chemistry* 75: 928–932.

Broos, J., J. F. J. Engbersen, I. K. Sakodinskaya, W. Verboom, and D. N. Reinhoudt. 1995a. Activity and enantioselectivity of serine proteases in transesterification reactions in organic media. *Journal of the Chemical Society, Perkin Transactions 1* 2899–12905.

Broos, J., I. K. Sakodinskaya, J. F. J. Engbersen, W. Verboom, and D. N. Reinhoudt. 1995b. Large activation of serine proteases by pretreatment with crown ethers. *Journal of the Chemical Society, Chemical Communications* 255–256.

Brown, K. G., S. C. Erfurth, E. W. Small, and W. L. Peticolas. 1972. Conformationally dependent low-frequency motions of proteins by laser Raman spectroscopy. *Proceedings of the National Academy of Sciences* 69: 1467–1469.

Byler, D. M. and H. Susi. 1986. Examination of the secondary structure of proteins by deconvolved FTIR spectra. *Biopolymers* 25: 469–487.

Caliskan, G., A. Kisliuk, A. M. Tsai, C. L. Soles, and A. P. Sokolov. 2002. Influence of solvent on dynamics and stability of a protein. *Journal of Non-Crystalline Solids* 307: 887–893.

Carey, P. 1982. *Biochemical Applications of Raman and Resonance Raman Spectroscopes*. New York: Access Online via Elsevier.

Chaney, M. O. and L. K. Steinrauf. 1974. The crystal and molecular structure of tetragonal L-cystine. *Acta Crystallographica Section B: Structural Crystallography and Crystal Chemistry* 30: 711–716.

Chen, M. C. and R. C. Lord. 1976. Laser-excited Raman spectroscopy of biomolecules. VIII. Conformational study of bovine serum albumin. *Journal of the American Chemical Society* 98: 990–992.

Chikishev, A. Y., G. W. Lucassen, N. I. Koroteev, C. Otto, and J. Greve. 1992. Polarization sensitive coherent anti-Stokes Raman scattering spectroscopy of the amide I band of proteins in solutions. *Biophysical Journal* 63: 976–985.

Costantino, H. R., K. Griebenow, R. Langer, and A. M. Klibanov. 1997. On the pH memory of lyophilized compounds containing protein functional groups. *Biotechnology and Bioengineering* 53: 345–348.

Cramer, C. J. 2005. *Essentials of Computational Chemistry: Theories and Models.* Chichester, U.K.: John Wiley & Sons, Inc.

Dahaoui, S., V. Pichon-Pesme, J. A. K. Howard, and C. Lecomte. 1999. CCD charge density study on crystals with large unit cell parameters: The case of hexagonal L-cystine. *The Journal of Physical Chemistry A* 103: 6240–6250.

Dai, J., N. Karpowicz, and X.-C. Zhang. 2009. Coherent polarization control of terahertz waves generated from two-color laser-induced gas plasma. *Physical Review Letters* 103: 023001.

Dai, J., X. Xie, and X.-C. Zhang. 2006. Detection of broadband terahertz waves with a laser-induced plasma in gases. *Physical Review Letters* 97: 103903.

Debulis, K. and A. M. Klibanov. 1993. Dramatic enhancement of enzymatic activity in organic solvents by lyoprotectants. *Biotechnology and Bioengineering* 41: 566–571.

Delley, B. 1990. An all-electron numerical method for solving the local density functional for polyatomic molecules. *The Journal of Chemical Physics* 92: 508–517.

Delley, B. 2000. From molecules to solids with the DMol approach. *The Journal of Chemical Physics* 113: 7756–7764.

Devlin, M. T., G. Barany, and I. W. Levin. 1990. Conformational properties of asymmetrically substituted mono-, di-and trisulfides: Solid and liquid phase Raman spectra. *Journal of Molecular Structure* 238: 119–137.

Dordick, J. S. 1992. Designing enzymes for use in organic solvents. *Biotechnology Progress* 8: 259–267.

Dunning Jr., T. H. 1989. Gaussian basis sets for use in correlated molecular calculations. I. The atoms boron through neon and hydrogen. *The Journal of Chemical Physics* 90: 1007–1123.

Ebeling, W., L. Schimansky-Geier, and Y. M. Romanovsky. 2002. *Stochastic Dynamics of Reacting Biomolecules.* Singapore, Singapore: World Scientific.

Engbersen, J. F. J., J. Broos, W. Verboom, and D. N. Reinhoudt. 1996. Effects of crown ethers and small amounts of cosolvent on the activity and enantioselectivity of α-chymotrypsin in organic solvents. *Pure and Applied Chemistry* 68: 2171–2178.

Fersht, A. 1999. *Structure and Mechanism in Protein Science: A Guide to Enzyme Catalysis and Protein Folding.* New York: W.H. Freeman.

Frolov, A. A., A. V. Borodin, M. N. Esaulkov, I. I. Kuritsyn, and A. P. Shkurinov. 2012. Theory of a laser-plasma method for detecting terahertz radiation. *Journal of Experimental and Theoretical Physics* 114: 893–905.

Genzel, L., F. Keilmann, T. P. Martin et al. 1976. Low-frequency Raman spectra of lysozyme. *Biopolymers* 15: 219–225.

Gervasio, F. L., G. Cardini, P. R. Salvi, and V. Schettino. 1998. Low-frequency vibrations of all-trans-retinal: Far-infrared and Raman spectra and density functional calculations. *The Journal of Physical Chemistry A* 102: 2131–2136.

Golovina, N. I., A. V. Raevskii, B. S. Fedorov et al. 2002. Temperature-dependent structure–energy changes in crystals of compounds with poly (hydroxymethyl) grouping. *Journal of Solid State Chemistry* 164: 301–312.

Görbitz, C. H. 1994. Conformational properties of disulphide bridges. 2. Rotational potentials of diethyl disulphide. *Journal of Physical Organic Chemistry* 7: 259–267.

Granovsky, A. A. 2009. *Firefly version 7.1.G.* http://classic.chem.msu.su/gran/firefly/index.html.

Harsha, S. S. and D. Grischkowsky. 2010. Terahertz (far-infrared) characterization of tris (hydroxymethyl) aminomethane using high-resolution waveguide THz-TDS. *The Journal of Physical Chemistry A* 114: 3489–3494.

Hirshfeld, F. L. 1977. Bonded-atom fragments for describing molecular charge densities. *Theoretica Chimica Acta* 44: 129–138.

Hochstrasser, R. M. 1966. *Molecular Aspects of Symmetry*. New York: W. A. Benjamin.

Hohenberg, P. and W. Kohn. 1964. Inhomogeneous electron gas. *Physical Review* 136: B864–B871.

Itoh, T., K. Mitsukura, K. Kaihatsu et al. 1997a. Remarkable acceleration of a lipase-catalyzed reaction by a thiacrownether additive: Buffer-free highly regioselective partial hydrolysis of4-acetoxy-2-methylbut-2-enyl acetate. *Journal of the Chemical Society, Perkin Transactions 1* 2275–12278.

Itoh, T., K. Mitsukura, W. Kanphai et al. 1997b. Thiacrown ether technology in lipase-catalyzed reaction: Scope and limitation for preparing optically active 3-hydroxyalkanenitriles and application to insect pheromone synthesis. *The Journal of Organic Chemistry* 62: 9165–9172.

Izatt, R. M., N. E. Izatt, B. E. Rossiter, J. J. Christensen, and B. L. Haymore. 1978. Cyclic polyether-protonated organic amine binding: Significance in enzymatic and ion transport processes. *Science* 199: 994–996.

Jensen, F. 2007. *Introduction to Computational Chemistry*. Chichester, U.K.: John Wiley & Sons, Inc.

Jepsen, P. U. and B. M. Fischer. 2005. Dynamic range in terahertz time-domain transmission and reflection spectroscopy. *Optics Letters* 30: 29–31.

Jibson, M. D., Y. Birk, and T. A. Bewley. 1981. Circular dichroism spectra of trypsin and chymotrypsin complexes with bowman-birk or chickpea trypsin inhibitor. *International Journal of Peptide and Protein Research* 18: 26–32.

Karpowicz, N., J. Dai, X. Lu et al. 2008. Coherent heterodyne time-domain spectrometry covering the entire "terahertz gap". *Applied Physics Letters* 92: 011131.

Kerr, K. A. and J. P. Ashmore. 1973. Structure and conformation of orthorhombic L-cysteine. *Acta Crystallographica Section B: Structural Crystallography and Crystal Chemistry* 29: 2124–2127.

Khmelnitsky, Y. L., S. H. Welch, D. S. Clark, and J. S. Dordick. 1994. Salts dramatically enhance activity of enzymes suspended in organic solvents. *Journal of the American Chemical Society* 116: 2647–2648.

Kitagawa, T., T. Azuma, and K. Hamaguchi. 1979. The Raman spectra of Bence-Jones proteins. Disulfide stretching frequencies and dependence of Raman intensity of tryptophan residues on their environments. *Biopolymers* 18: 451–465.

Klibanov, A. M. 1989. Enzymatic catalysis in anhydrous organic solvents. *Trends in Biochemical Sciences* 14: 141–144.

Koch, W., M. C. Holthausen, and E. J. Baerends. 2001. *A Chemist's Guide to Density Functional Theory*. Weinheim, Germany: FVA-Frankfurter Verlagsanstalt GmbH.

Kohn, W. and L. J. Sham. 1965. Self-consistent equations including exchange and correlation effects. *Physical Review* 140: A1133–A1138.

Korter, T. M., R. Balu, M. B. Campbell et al. 2006. Terahertz spectroscopy of solid serine and cysteine. *Chemical Physics Letters* 418: 65–70.

Kudryavtsev, A. B., S. B. Mirov, L. J. DeLucas et al. 1998. Polarized Raman spectroscopic studies of tetragonal lysozyme single crystals. *Acta Crystallographica Section D: Biological Crystallography* 54: 1216–1229.

Lebedev, V. I. and A. L. Skorokhodov. 1992. Quadrature formulas of orders 41, 47, and 53 for the sphere. *Russian Academy of Sciences Doklady Mathematics* 45: 587–592.

Lord, R. C. and N.-T. Yu. 1970. Laser-excited Raman spectroscopy of biomolecules: II. Native ribonuclease and α-chymotrypsin. *Journal of Molecular Biology* 51: 203–213.

Mankova, A. A., A. V. Borodin, A. V. Kargovsky et al. 2012. Terahertz time-domain and FTIR spectroscopy of tris-crown interaction. *Chemical Physics Letters* 554: 201–207.

Mankova, A. A., A. V. Borodin, A. V. Kargovsky et al. 2013. Terahertz time-domain and FTIR spectroscopic study of interaction of α-chymotrypsin and protonated tris with 18-crown-6. *Chemical Physics Letters* 560: 55–59.

Matei, A., N. Drichko, B. Gompf, and M. Dressel. 2005. Far-infrared spectra of amino acids. *Chemical Physics* 316: 61–71.

Maverick, E., P. Seiler, W. B. Schweizer, and J. Dunitz. 1980. 1, 4, 7, 10, 13, 16-Hexaoxacyclooctadecane: Crystal structure at 100 K. *Acta Crystallographica Section B: Structural Crystallography and Crystal Chemistry* 36: 615–620.

McCammon, J. A. and P. G. Wolynes. 1977. Nonsteady hydrodynamics of biopolymer motions. *The Journal of Chemical Physics* 66: 1452–1456.

Moggach, S. A., D. R. Allan, S. Parsons, L. Sawyer, and J. E. Warren. 2005. The effect of pressure on the crystal structure of hexagonal L-cystine. *Journal of Synchrotron Radiation* 12: 598–607.

Nakamura, K., S. Era, Y. Ozaki et al. 1997. Conformational changes in seventeen cystine disulfide bridges of bovine serum albumin proved by Raman spectroscopy. *FEBS Letters* 417: 375–378.

Nazarov, M. M., A. P. Shkurinov, E. A. Kuleshov, and V. V. Tuchin. 2008. Terahertz time-domain spectroscopy of biological tissues. *Quantum Electronics* 38: 647.

Nazarov, M. M., A. P. Shkurinov, V. V. Tuchin, and O. S. Zhernovaya. 2007. Modification of terahertz pulsed spectrometer to study biological samples. Presented at *Optical Technologies in Biophysics and Medicine VIII*, vol. 6535, pp. 65351–65357.

Northrop, J. H., M. Kunitz, and R. M. Herriott. 1948. *Crystalline Enzymes*. New York: Columbia University Press.

Pearson, W. B., L. D. Caivert, J. M. Bijvoet, and J. D. Dunitz. 1959. *Structure Reports: Metals and Inorganic Compounds. For 1959*: International Union of Crystallography.

Perdew, J. P., K. Burke, and M. Ernzerhof. 1996. Generalized gradient approximation made simple. *Physical Review Letters* 77: 3865–3868.

Plazanet, M., N. Fukushima, and M. R. Johnson. 2002. Modelling molecular vibrations in extended hydrogen-bonded networks–crystalline bases of RNA and DNA and the nucleosides. *Chemical Physics* 280: 53–70.

Podstawka, E., Y. Ozaki, and L. M. Proniewicz. 2004. Adsorption of S–S containing proteins on a colloidal silver surface studied by surface-enhanced Raman spectroscopy. *Applied Spectroscopy* 58: 1147–1156.

Reinhoudt, D. N., A. M. Eendebak, W. F. Nijenhuis et al. 1989. The effect of crown ethers on enzyme-catalysed reactions in organic solvents. *Journal of the Chemical Society, Chemical Communications* 399–400.

Rudman, R., R. Lippman, D. S. Sake Gowda, and D. Eilerman. 1983. Polymorphism of crystalline poly (hydroxymethyl) compounds. VIII. Structures of the tris (hydroxymethyl) aminomethane hydrogenhalides,(HOH2C) 3CNH3+. X-(X = F, Cl, Br, I). *Acta Crystallographica Section C: Crystal Structure Communications* 39: 1267–1271.

Sakai, K. 2005. *Terahertz Optoelectronics*. Berlin, Germany: Springer.

Sugeta, H., A. Go, and T. Miyazawa. 1973. Vibrational spectra and molecular conformations of dialkyl disulfides. *Bulletin of the Chemical Society of Japan* 46: 3407–3411.

Susi, H., D. M. Byler, and J. M. Purcell. 1985. Estimation of β-structure content of proteins by means of deconvoled FTIR spectra. *Journal of Biochemical and Biophysical Methods* 11: 235–240.

Thamann, T. J. 1999. A vibrational spectroscopic assignment of the disulfide bridges in recombinant bovine growth hormone and growth hormone analogs. *Spectrochimica Acta Part A: Molecular and Biomolecular Spectroscopy* 55: 1661–1666.

Thornton, J. M. 1981. Disulphide bridges in globular proteins. *Journal of Molecular Biology* 151: 261–287.

Tkatchenko, A. and M. Scheffler. 2009. Accurate molecular van der Waals interactions from ground-state electron density and free-atom reference data. *Physical Review Letters* 102: 073005.

Triantafyllou, A. Ö., E. Wehtje, P. Adlercreutz, and B. Mattiasson. 1995. Effects of sorbitol addition on the action of free and immobilized hydrolytic enzymes in organic media. *Biotechnology and Bioengineering* 45: 406–414.

Trueblood, K. N., C. B. Knobler, D. S. Lawrence, and R. V. Stevens. 1982. Structures of the 1:1 complexes of 18-crown-6 with hydrazinium perchlorate, hydroxylammonium perchlorate, and methylammonium perchlorate. *Journal of the American Chemical Society* 104: 1355–1362.

Tsukada, H. and D. Blow. 1985. Structure of α-chymotrypsin refined at 1.68 Å resolution. *Journal of Molecular Biology* 184: 703–711.

Tsukube, H., T. Yamada, and S. Shinoda. 2001. Crown ether strategy toward chemical activation of biological protein functions. *Journal of Heterocyclic Chemistry* 38: 1401–1408.

Tu, A. T. 1986. Peptide backbone conformation and microenvironment of protein side chains. In *Spectroscopy of Biological Systems*, ed. Clark, R. J. H. and R. E. Hester, pp. 47–112. Chichester, U.K.: John Wiley & Sons, Inc.

Unen, D.-J., J. F. J. Engbersen, and D. N. Reinhoudt. 1998a. Large acceleration of α-chymotrypsin-catalyzed dipeptide formation by 18-crown-6 in organic solvents. *Biotechnology and Bioengineering* 59: 553–556.

Unen, D.-J., I. Sakodinskaya, J. J. Engbersen, and D. Reinhoudt. 1998b. Crown ether activation of cross-linked subtilisin Carlsberg crystals in organic solvents. *Journal of the Chemical Society, Perkin Transactions 1* 3341–3344.

Van Wart, H. E., A. Lewis, H. A. Scheraga, and F. D. Saeva. 1973. Disulfide bond dihedral angles from Raman spectroscopy. *Proceedings of the National Academy of Sciences* 70: 2619–2623.

Volini, M. and P. Tobias. 1969. Circular dichroism studies of chymotrypsin and Its derivatives: Correlation of changes in dichroic bands with deacylation. *Journal of Biological Chemistry* 244: 5105–5109.

Volkin, D. B., A. Staubli, R. Langer, and A. M. Klibanov. 1991. Enzyme thermoinactivation in anhydrous organic solvents. *Biotechnology and Bioengineering* 37: 843–853.

Volovšek, V., D. Kirin, L. Bistričič, and G. Baranović. 2002. Low-wavenumber lattice vibrations and dynamics of 4,4′-dibromobenzophenone. *Journal of Raman Spectroscopy* 33: 761–768.

Walther, M., B. Fischer, M. Schall, H. Helm, and P. U. Jepsen. 2000. Far-infrared vibrational spectra of all-*trans*, 9-*cis* and 13-*cis* retinal measured by THz time-domain spectroscopy. *Chemical Physics Letters* 332: 389–395.

Walther, M., B. M. Fischer, and P. Uhd Jepsen. 2003. Noncovalent intermolecular forces in polycrystalline and amorphous saccharides in the far infrared. *Chemical Physics* 288: 261–268.

Weiss-Lopez, B. E., M. H. Goodrow, W. K. Musker, and C. P. Nash. 1986. Conformational dependence of the disulfide stretching frequency in cyclic model compounds. *Journal of the American Chemical Society* 108: 1271–1274.

Yamamoto, K., M. H. Kabir, and K. Tominaga. 2005. Terahertz time-domain spectroscopy of sulfur-containing biomolecules. *Journal of the Optical Society of America B* 22: 2417–2426.

Yoshida, H., I. Kaneko, H. Matsuura, Y. Ogawa, and M. Tasumi. 1992. Importance of an intramolecular 1, 5-CH... O interaction and intermolecular interactions as factors determining conformational equilibria in 1, 2-dimethoxyethane: a matrix-isolation infrared spectroscopic study. *Chemical Physics Letters* 196: 601–606.

Yoshida, H. and H. Matsuura. 1998. Density functional study of the conformations and vibrations of 1, 2-dimethoxyethane. *The Journal of Physical Chemistry A* 102: 2691–2699.

Zhang, X. C. and J. Xu. 2009. *Introduction to THz Wave Photonics*. New York: Springer.

11

Protein Dielectric Response at Terahertz Frequencies: Correlated and Diffusive Contributions

Deepu K. George
University at Buffalo

Andrea G. Markelz
University at Buffalo

11.1 Introduction: Hydrated Protein System and General Dielectric Response

The picosecond timescale would seem ideal for studies of protein dynamics. This is because a number of modeling approaches have found long-range structural vibrations for a variety of proteins and poly-nucleotides at terahertz frequencies. Since it is well understood that the structure of biomacromolecules plays an operative role in their biological function, the understanding of long-range motions of the structure through vibrational spectroscopy holds significant promise for understanding both biological networks and engineering of biomolecules for both biomedical and technological outcomes. Initial calculations performed by solid-state physicists used simplified lattice models in particular for DNA to determine the vibrational spectrum. These calculations did not include hydration. However, biology does not occur in dehydrated environments, and the biochemistry associated with proteins and polynucleotides requires a minimal hydration. While water is necessary in biological systems, it is highly inconvenient for far-infrared optical studies. Because of the large absorption coefficient, measurements using the new method of terahertz time domain spectroscopy (THz TDS) initially used freeze-dried powders of proteins and polynucleotides. The measurements found smooth absorption spectra increasing rapidly with frequency with a so-called *glassy* response. That is, the response looked like that of a glass, with a standard Debye dielectric relaxation response. It was naively hoped that this glassy response arose from the disorder that necessarily occurs as a result of freeze-drying of the protein or DNA. It is well known that the protein structure changes with lyophilization (Griebenow and Klibanov 1995). Thus, a typical lyophilized sample consists of a heterogeneous sampling of conformations, some of which may be partially denatured. Thus, even if distinct spectra arise from particular protein structures, the averaging of the absorbance spectra for all the different conformations present in a typical lyophilized sample would completely wash out the so-called fingerprinting features. In addition to the nonuniformity of the sample due to lyophilization effects, often lyophilized powders are crystalline powders, with particle size ~1 μm. It was also hoped that the rapid increase in the apparent absorption with frequency was in part caused by Rayleigh scattering from these particles.

One can achieve a more uniform native-state structure and remove scattering concerns by using solution phase samples. To avoid strong background from bulk solvent, one can form thin (~100 μm) films of solution and dry these in a controlled way, using, for example, a closed cell with humidity control. The hydration can be set high enough to achieve protein structural homogeneity in the film, but sufficiently low so that bulk water is not present and cannot contribute to absorption. Terahertz measurement of such samples, however, still had glasslike absorption. While closer examination of the complex permittivity of the hydrated films revealed that the response could not be entirely accounted for with just relaxational response, but also required at least one damped harmonic oscillator (Knab et al. 2006), it is clear that for fully hydrated proteins, we must consider multiple contributions to the dielectric response. By proper decomposition of the data, enhanced by systematic variation of temperature, solvent content, or protein functional state, one can isolate the component arising from the protein structural vibrations. We can write down an approximate decomposition for solution phase proteins using the following formula:

$$\varepsilon = \varepsilon_W + \varepsilon_{bW} + \varepsilon_{p,relax} + \varepsilon_{p,vib}, \tag{11.1}$$

where the subscripts W, bW, p, $relax$, and p, vib refer to permittivity contributions from the bulk water, bound water, protein relaxational response, and protein long-range structural vibrational response, respectively. This decomposition is a vast simplification, ignoring any interaction between different components, but it expresses the various sources of dielectric response for the complex sample. The first three components are modeled very well using dielectric relaxation, and the last component is a sum of damped harmonic oscillators. In this chapter, we will discuss the different relaxational contributions to the terahertz response for hydrated proteins and methods to remove these so that one can extract the contribution from protein structural motions. We will discuss general aspects of relaxational response, the permittivity of water, in all its various forms, present in biological samples, and the relaxational response from amino acid side chains. Finally, we will discuss, in light of all the various models for relaxational effects, how the current measurements for protein samples suggest that the long-range vibrations of proteins do contribute to the terahertz response.

11.2 Dielectric Relaxation

The polarization of a material due to an external electric field E is given by the relation

$$\vec{P} = \varepsilon_0(\hat{\varepsilon} - 1)\vec{E}, \tag{11.2}$$

where
   ε_0 is the dielectric permittivity of free space
   $\hat{\varepsilon}$ is the relative permittivity of the material

The total polarization $P(t)$ has an instantaneous contribution P_1 and a time-dependent contribution $P_2(t)$ (Harrop 1972). The instantaneous polarization refers to all processes with characteristic response time $\tau \ll 2\pi / \omega$. For processes with $\tau >\approx 2\pi / \omega$, we have a time-dependent polarization $P_2(t)$ giving a net polarization

$$P(t) = P_1 + P_2(t). \tag{11.3}$$

The time-dependent polarization $P_2(t)$ increases from zero to a saturation value $P_2(\infty)$. Considering the boundary conditions on dielectric permittivity

$$\hat{\varepsilon} \to \varepsilon_\infty \quad \text{for } \omega \to \infty \; (t = 0)$$

$$\hat{\varepsilon} \to \varepsilon_s \quad \text{for } \omega \to 0 \; (t = \infty),$$

where
  ε_s is the static permittivity
  ε_∞ is the permittivity in the high-frequency limit, we arrive at equations for polarizations at these
    limits:

$$P(0) = P_1 = \varepsilon_0 (\varepsilon_\infty - 1) E \tag{11.4}$$

$$P(\infty) = P_1 + P_2(\infty) = \varepsilon_0 (\varepsilon_s - 1) E. \tag{11.5}$$

From Equations 11.3 and 11.4, we get

$$P_2(\infty) = \epsilon_0 (\varepsilon_s - \varepsilon_\infty) E. \tag{11.6}$$

When an external voltage is applied to a material, it takes a finite time for the dipoles to align to the electric field. Similarly, when the electric field is turned off, it takes a finite time for the dipoles to be randomly oriented. This phenomenon is called dipole relaxation.

Debye (1960) developed a model to describe the dipole relaxation of noninteracting molecules. Suppose a dc voltage is applied to a polar dielectric and then is turned off, the time-dependent polarization $P_2(t)$ decays exponentially as

$$P_2(t) = P_2(\infty)[1 - e^{t/\tau}] \tag{11.7}$$

where τ is the relaxation time. The time derivative of the polarization can be written as

$$\frac{dP_2(t)}{dt} = -P_2(\infty)\left(-\frac{1}{\tau}\right)e^{-\frac{t}{\tau}} = \frac{P_2(\infty)e^{-\frac{t}{\tau}}}{\tau} \tag{11.8}$$

Substituting Equation 11.7 into 11.8, we get

$$\frac{dP_2(t)}{dt} = \frac{P_2(\infty) - P_2(t)}{\tau} \tag{11.9}$$

If a time-dependent electric field $E = E_0 \exp(i\omega t)$ is applied to the dielectric field, we can write Equation 11.9 with the help of Equation 11.6 as

$$\frac{dP_2(t)}{dt} = \frac{\int_0 (\varepsilon_s - \varepsilon_\infty) E_0 e^{i\omega t} - P_2(t)}{\tau}. \tag{11.10}$$

The solution of the preceding first-order differential equation can be written as

$$P_2(t) = Ce^{-t/\tau} + \frac{\varepsilon_0 (\varepsilon_s - \varepsilon_\infty) E(t)}{1 + i\omega\tau}. \tag{11.11}$$

At time $t \gg \tau$, the first term can be neglected. Following Equations 11.3, 11.4, and 11.11, the total polarization is given by

$$P(t) = \varepsilon_0 (\varepsilon_\infty - 1) E(t) + \frac{\varepsilon_0 (\varepsilon_s - \varepsilon_\infty) E(t)}{1 + i\omega\tau} \tag{11.12}$$

$$\hat{\varepsilon}(\omega) = \varepsilon_\infty + \frac{\varepsilon_s - \varepsilon_\infty}{1 + i\omega\tau}. \tag{11.13}$$

Equation 11.13 is known as the Debye equation for permittivity. The Debye equation can be divided into real and imaginary parts:

$$\varepsilon'(\omega) = \varepsilon_\infty + \frac{\varepsilon_s - \varepsilon_\infty}{1 + (\omega\tau)^2}, \quad \varepsilon''(\omega) = \frac{(\varepsilon_s - \varepsilon_\infty)\omega\tau}{1 + (\omega\tau)^2}. \tag{11.14}$$

where $\hat{\varepsilon} = \varepsilon' - i\varepsilon''$. Most materials deviate from this ideal behavior. Therefore, Debye model has been modified in several ways by later empirical models. The physical meaning of τ is the effective time for the dipole to rotate to follow the applied field. When this dipole realignment requires large movement, that is, over several angstroms, this structural rearrangement time is less, on the order of microsecond to second, and is typically referred to as α relaxation. There is a distinct difference between α and β relaxations, and it is important to understand this difference. The relaxation types are distinguished by timescale (microsecond versus picosecond), spatial scale (>2 Å cooperative motion versus <1 Å local motion), and temperature dependence of the relaxation time. The classification is particularly clear through the temperature dependence of the relaxation time, which is determined by the energetics associated with the dipole alignment motion. Bond breaking and rearrangement associated with α relaxation give rise to the Vogel–Tammann–Fulcher–Hesse temperature dependence with $\tau = \tau_o \exp(DT_o/(T - T_o))$, where D, τ_o, and T_o are fitting coefficients (Angell et al. 2000). The somewhat less complex bond changing associated with β relaxation gives rise to an Arrhenius temperature dependence with $\tau = \tau_0 e^{E_A/k_b T}$, where E_A is dependent on the energetics of the bond breaking associated with the dipole motion. As we are focused on motions in the picosecond timescale, all relaxations we consider are β relaxations.

11.2.1 Empirical Models

11.2.1.1 Cole–Cole Relaxation

Cole and Cole showed (Cole and Cole 1941) that a plot of ε'' against ε' yields a semicircle for materials exhibiting Debye relaxation. Such a plot is known as a complex plane plot of $\hat{\varepsilon}$. Polar molecules that have more than one relaxation time do not satisfy Debye equations. Cole–Cole showed that, in this case, the semicircle will have its center displaced below the ε' axis and suggested an empirical equation for the complex dielectric constant as

$$\hat{\varepsilon}^{*} = \varepsilon_{\infty} + \frac{\varepsilon_s - \varepsilon_{\infty}}{1 + (i\omega\tau_{c-c})^{1-\alpha}}; \quad 0 \leq \alpha \leq 1, \tag{11.15}$$

where τ_{c-c} is the mean relaxation time. α varies between 0 and 1. $\alpha = 0$ results in Debye relaxation.

11.2.1.2 Cole–Davidson Relaxation

Cole–Davidson have suggested (Davidson and Cole 1951) an empirical equation for the complex dielectric constant

$$\varepsilon^{*} = \varepsilon_{\infty} + \frac{\varepsilon_s - \varepsilon_{\infty}}{(1 + i\omega\tau_{d-c})^{\beta}}; \quad 0 \leq \beta \leq 1, \tag{11.16}$$

where β is a characteristic constant of the material. The real and imaginary parts of the equation can be written as (Davidson and Cole 1951)

$$\varepsilon' - \varepsilon_{\infty} = (\varepsilon_s - \varepsilon_{\infty})(\cos(\phi))^{\beta} \cos(\phi\beta) \tag{11.17}$$

$$\varepsilon'' = (\varepsilon_s - \varepsilon_{\infty})(\cos(\phi))^{\beta} \sin(\phi\beta), \tag{11.18}$$

where $\tan(\phi) = \omega\tau_0$.

In the Cole–Davidson equation, setting $\beta = 1$ will yield Debye relation. For materials holding Davidson–Cole equation, values of ε_s, ε_{∞}, and β may be determined directly from a plot of RHS of Equation 11.16 against ω. The complex plot of ε'' against ε' yields a skewed semicircle for the Davidson–Cole relation.

11.2.1.3 Havriliak and Negami Relaxation

For complex molecules like polymers, both Cole–Cole and Cole–Davidson relations prove to be inadequate. Havriliak and Negami have measured the dielectric properties of several molecules as a function of temperature. They found the complex plane plot to be linear at high frequencies (Cole–Davidson) and circular at low frequencies (Cole–Cole). Combining Cole–Cole and Cole–Davidson relations, Havriliak and Negami suggested a function for complex dielectric response of polymers as

$$\frac{\varepsilon^{*} - \varepsilon_{\infty}}{\varepsilon_s - \varepsilon_{\infty}} = \left[1 + (i\omega\tau_{H-N})^{1-\alpha}\right]^{-\beta}, \tag{11.19}$$

where $\beta = 1$ produces the circular arc of Cole–Cole relation and $\alpha = 0$ results in the skewed circle by Cole–Davidson relation. Setting $\alpha = 0$ and $\beta = 1$ reduces the function to Debye relation (Havriliak and Negami 1967).

Previously, we had attempted to fit the dielectric response of protein films exclusively with the relaxational response. Different empirical models give a great deal of flexibility in fitting the complex permittivity. Nevertheless, even with the most accommodating Havriliak and Negami model, we were not able to reproduce the lysozyme data and found that we could best reproduce data by using the combination of simple type I relaxation and a single damped harmonic oscillator. We want to emphasize to the reader however that there is a great deal of literature discussing the relaxational response of polymers, and proteins in polynucleotides constitute polymers, albeit highly structured polymers. For the rest of this chapter, we will use Debye relaxation exclusively to attempt to describe the relaxational response of the various components of hydration water and the protein.

11.3 Water

The first and foremost relaxational contribution that needs to be considered for analysis of the terahertz protein spectroscopy is that from water. Typical minimal hydration necessary for biochemical function for many proteins is on the order of 0.2–1.0 g water/grams protein (Rupley and Careri 1991). As will be discussed, this water does not have a uniform dielectric response. For sufficiently high hydration, there are possibly three different contributions to the permittivity from water: bound water, biological water, and bulk water. We will now discuss these three types of water as currently understood.

11.3.1 Bulk Water

Bulk water in the gigahertz and terahertz range is generally described as a sum of Debye relaxational terms and damped harmonic oscillators (Nandi et al. 2000):

$$\varepsilon_{Bulk} = \varepsilon_{\infty} + \frac{\Delta\varepsilon_1}{1+i\omega\tau_1} + \frac{\Delta\varepsilon_2}{1+i\omega\tau_2} + \frac{A_1}{\omega_1^2 - \omega^2 - i\omega\gamma_1} + \frac{A_2}{\omega_2^2 - \omega^2 - i\omega\gamma_2}. \tag{11.20}$$

In the 0.2–3.0 THz range, it is typically sufficient to have only two relaxation components and one damped harmonic oscillator. Several values determined by a variety of measurements for these quantities are shown in Table 11.1. We note that there have not been extensive measurements on the temperature dependence of these relaxation times; however, Ronne et al. (1997) were able to find Arrhenius behavior with $\Delta E_1 \sim 170$ meV and $\Delta E_2 \sim 160$ meV, suggesting somewhat similar activation energies for the two processes. We will return to this when we discuss the different contributions to the dielectric response at low temperatures.

11.3.2 Bound Water

Adding solute to solvent changes the organization of the solvent and the dynamics of the solvent in proximity to the solute. In addition, some of the solvent can form hydrogen bonds with the solute. There is consensus with respect to the number of strongly bonded waters to the surfaces of amino acids and proteins, typically 0.2–0.3 h (g water/g protein) (Nandi and Bagchi 1997; Nandi et al. 2000; Pethig 1995; Rupley and Careri 1991). However, the response of these strongly bonded waters is in some dispute. Some authors suggest that these do not contribute to the dielectric response at all (Vinh et al. 2011), and some suggest that the bound water has a Debye relaxation response, but with retarded relaxation time approximately seven to eight times bulk (Comez et al. 2013; Rupley and Careri 1991). There is strong correspondence between hydration numbers for the strongly bonded water and the unfrozen water found when proteins are cooled below 273 K. Neutron scattering measurements on deuterated samples as a function of temperature have allowed for selective detection of only this hydration water and its relaxation (Capaccioli et al. 2011). This combination of measurements strongly suggests that the retarded relaxation model for the strongly bonded water is most likely correct.

TABLE 11.1 Dielectric Relaxation Times and Oscillator Strengths of Bulk Water as Determined by Various Studies

Work	ε_{∞}	$\Delta\varepsilon_1$	τ_1 (ps)	$\Delta\varepsilon_2$	τ_2 (ps)	$\omega_1/2\pi$ (THz)	$\gamma_1/2\pi$ (THz)
Kindt and Schmuttenmaer (1996)	3.48	73.43	8.24	4.93	0.18	—	—
Møller et al. (2009)	2.68	72.3	8.34	2.12	0.36	5.01	7.06
Yada et al. (2008)	2.50	74.9	9.43	1.63	0.25	5.30	5.35
Ronne et al. (1997)	3.3	72.1	8.5	1.9	0.17	—	—
Sato et al. (2005)	3.96	72.2	8.32	2.14	0.39	—	—

11.3.3 Biological Water

There is considerable debate on how to properly treat the water beyond the strongly bound layer. In Rupley and Careri's review paper, they show correlation time calculations for water as one moves away from the protein surface, plotted as a function of h (Rupley and Careri 1991). Up to 0.25 h, the correlation time is small and flat with increasing h, then between 0.25 and 0.38 h (where 0.38 h corresponds to <2 Å thickness of the hydration shell), the correlation time drops by an order of magnitude, rapidly approaching the bulk correlation times. This somewhat older result is probably a fairly accurate description of the water dynamics beyond the protein; however, the debate continues on how to best treat the effective permittivity for the region beyond the strongly bound water where the permittivity is rapidly dropping toward bulk. While some investigators assume that the solution phase protein consists of protein, bound water, and bulk water, other investigators suggest that, in addition, there is a well-defined region beyond the solute surface with a permittivity distinct from bulk water, the hydration shell. This construct with abrupt interfaces can be useful for calculating the net dielectric response; however, as shown by a variety of authors, the correlation time in the vicinity of the solute is changing smoothly without any sharp interfaces. We will now discuss this debate and notice that there are two chief problems in the comparison of the results from various groups: the concentration range and the decomposition method for determining the permittivity contribution from different constituents. Without addressing these two inconsistencies, this debate cannot be resolved. The strongly bound water can be considered somewhat distinct from this continuum, and presumably, this is why there is such consistency among different measurement techniques.

For systematic measurements to study the changes in water dynamics due to the presence of a biological solute, one needs a method that can avoid or overcome the large absorbance for water. This has been done using absorption measurements with high-power terahertz sources such as amplified Gunn diode systems (Vinh et al. 2011), free electron lasers (Xu et al. 2006a,b, 2007), and p-doped germanium lasers (Ebbinghaus et al. 2007, 2008; Heugen et al. 2006; Heyden et al. 2008). Another method that can avoid the large absorbance in the terahertz range is to use light scattering methods with the incident light frequency outside the water absorption band (Comez et al. 2013; Paolantoni et al. 2009). To achieve high precision in the dielectric response, variable cell path lengths (Kindt and Schmuttenmaer 1996; Koeberg et al. 2007; Tielrooij et al. 2010; Vinh et al. 2011) and modulation techniques have been used (Heyden et al. 2008). To extract the solute's effect on the water response, systematic solute concentration measurements have been made, and the resulting trends analyzed. This brings up one of the chief inconsistencies between different measurements, the concentration range. Measurements have focused on either protein concentrations greater than 4 mM or less than 1 mM. The low concentration measurements suggest that the biological water extends so far beyond the solute surface that at concentrations above 3 mM, one already has an overlap of these hydration shells so that one will sense only two components (Comez et al. 2013; Ebbinghaus et al. 2007, 2008; Heugen et al. 2006), which will be misinterpreted as bulk and bound water rather than the retarded hydration shell and bound water. If this is correct, then all the measurements at higher concentrations performed by other groups will not perceive this extended hydration shell since their minimal concentrations already have no *bulk* water. This lack of experimental overlap between groups reporting different results for the biological water extent is avoided in the case of the sugars glucose and trehalose (Comez et al. 2013; Heyden et al. 2008). In this case, there is still a difference in the water perturbed by the solute, but this difference likely arises from the different analysis required for the absorbance measurements versus light scattering.

We now turn to data analysis used by different groups to determine the response and the extent of the biological water. In one case, the method of fractional volumes is used:

$$\varepsilon = \sum_i f_i \varepsilon_i, \tag{11.21}$$

where f_i is the percent volume for component i with permittivity ε_i. The group compared the concentrations variation of the measured permittivity to the expected variation for a two-component system. The two-component system consisted of solute, which was assumed to have zero imaginary permittivity, that is, zero absorbance, and bulk water. For such a two-component system with a fractional volume permittivity model, one expects a linear decrease in absorbance due to the removal of the highly absorbing solvent by the excluded volume of the solute. The measured absorbance change compared to pure solvent did not follow this linear decrease, and the authors accounted for the discrepancy by using a three-component model with the third component arising from the biological water. The biological water region had a well-defined permittivity different from bulk water. This work did not express the permittivity of the extended shell in terms of a net shift in the relaxation time such as the retardation factor discussed for bound water. However, they found good agreement with their estimates of the permittivity variation with molecular dynamics simulations. Overall, they found a net slowing down of the dynamics of the extended hydration shell compared to the bulk. The estimates of the extension of the hydration shell from the surface of the solute were approximately 20 Å, somewhat larger than even the largest estimates determined by other methods. Using similar methods, the same group found that the extended shell for sugars such as trehalose was on the order of 6.5 Å (Heyden et al. 2008). This was in considerable disagreement with polarized light scattering measurements, which found that the hydration layers extended only 2.2 Å from the surface (Comez et al. 2013; Paolantoni et al. 2009), which is less than a one complete hydration layer, (estimated water diameter is 2.75 Å). In this particular case, the concentration range overlapped for the two techniques, thus removing the concentration range as being the source of the disagreement. We will return to the trehalose example to illustrate the differences in the approaches of data analysis, partial volumes method versus effective medium theory (EMT).

The partial volume approach to determine the net dielectric response for a mixture while straightforward to implement assumes distinct regions that do not interact electrostatically. This is a particularly reasonable method for layered media. For a homogeneous mixture of media, with component size much less than the wavelength, one needs to determine the net response averaged over the media. Students have encountered this through the Clausius–Mossotti derivation. This overall strategy is called EMT, and the most popular model is the Bruggeman formula. The Bruggeman formula for homogeneous spheroid inclusions in a uniform matrix is given by

$$\sum_i f_i \frac{\varepsilon_i - \varepsilon}{\varepsilon_i + 2\varepsilon} = 0, \tag{11.22}$$

where f_i is the volume percent of component i with permittivity ε_i, and the net permittivity is given by ε. For only two components, it is straightforward to solve for ε; however, for increasingly complex systems, it becomes more difficult to determine the roots. The effective medium model has been used successfully for THz measurements of solid-state systems (Baxter and Schmuttenmaer 2006; Hendry et al. 2006) and recently was applied to protein solutions (Vinh et al. 2011). In the case of applying the EMT to proteins, the authors assumed that the protein, hen egg white lysozyme (HEWL), was a pure dielectric with no net absorption and the matrix pure bulk water, with the relaxation parameters such as those given in Table 11.1. They were able to reproduce the measured permittivity if they removed ~165 waters/protein from the bulk water matrix. They labeled this *missing* water as the bound water. This estimate of bound water for HEWL is slightly lower than other estimates of strongly bound water of ~0.3 h = 242 waters (Rupley and Careri 1991). Their lower estimate may be due to the assumption of zero absorption from the protein itself. The net result of only 165 water molecules per protein removed from the bulk response is a considerable deviation from that of the extended hydration shell of 20 Å (Ebbinghaus et al. 2007). However, the authors suggested that this may arise from different concentration ranges measured. Here, we would also suggest that the different analysis technique also might make a considerable impact on determining the composition of the solvent.

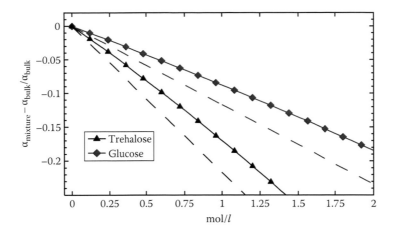

FIGURE 11.1 Calculations of the expected change in the absorption coefficient of solutions of glucose (gray) or trehalose (black) as a function of molar concentration. The dashed lines show the expected change assuming that the permittivities are additive, with the net change arising from excluded volume of the solute. The solid lines with markers show the expected change using the Bruggeman effective formula. We note that the similarity in the effective medium results and those measures in Heyden et al. (2008).

To illustrate the importance of using EMT, we consider the particular case of trehalose and glucose solutions. These particular sugars have been studied by both THz measurements with partial volumes analysis and polarization relaxation light scattering measurements (Comez et al. 2013; Heyden et al. 2008). There is a large disagreement between the two measurement techniques for the extent of the water perturbed by the solute. Here we simply compare the calculated response using the hydration number from light scattering and the two models: fractional volume and EMT. The fractional volume f_i is given in terms of the molar concentration c, molecular weight MW, and the density ρ of the molecule V_{mol} as $f_i = c_i * \text{MW}/\rho$. The values we used for MW and ρ for glucose (trehalose) were 180.16 (342.3) and 1.54 g/mL (1.58 g/mL), respectively. We assumed zero absorbance for glucose and trehalose, with real permittivity = 4 for both. The matrix was assumed to be pure water with relaxation times from row one in Table 11.1. The results are plotted in Figure 11.1. The dashed lines show the expected concentration variation assuming partial volume analysis (i.e., the linear decrease with increasing excluded volume). The solid lines show the expected concentration variation with the Bruggeman effective medium formula. Figure 11.1 looks very similar to Figure 2 in Heyden et al. (2008). In particular, the solid lines appear to be nearly identical to the measured result. It seems possible that the deviation that the authors ascribe to an extended hydration shell may arise from the averaging of the permittivity for a homogeneous mixture. The extended hydration shell controversy goes beyond the simple analysis presented here. It is clear that more measurements over a comprehensive concentration range with EMT analysis are required to resolve this.

11.4 Relaxational Response of Proteins: Amino Acid Side Chains

As discussed in Section 11.2, dielectric relaxation response can be understood as macroscopic dipoles rotating to align along the applied field direction. The realignment requires a rearrangement of bonds, typically van der Waals and hydrogen bonding. The relaxational response depends on both the effective dipole moment and the local bonding of the group. We have focused on the relaxation of water molecules and discussed how the bound water has relaxation times retarded compared to the bulk due to the bonding with the protein surface. We can reverse our view and note that the amino acid side chains at the protein surface have net dipole moments or can have induced dipoles with the applied field, and again these dipoles can follow the time-varying field with a relaxation time dependent on the bond energies associated with the molecular rearrangement for the dipole rotation. We focus on the surface residues as the internal residues for the native folded state have strong steric hindrance and hydrogen bonding that would prevent their realignment. The description of residue relaxation is complicated by the possible overlap with water relaxation as mentioned in the bound water section. High-frequency concentration-dependent dielectric relaxation measurements on amino acids need to be pursued to carefully consider this. At this time, measurements have found the relaxation times for amino acids to be somewhat longer than those for water. The measurements that have been done for amino acids by various groups are in good agreement. In particular, glycine has been found to have the relaxation time on the order of 35 ps and an effective dipole moment of approximately 20 Debye (Rodríguez-Arteche et al. 2012; Sato et al. 2005). The relaxation time for other amino acids has been measured between 50 and 200 ps (Rodríguez-Arteche et al. 2012). The strength of the glycine relaxation contribution to the permittivity is sizable considering that the dipole moment of water is only 1.85 Debye (Rodríguez-Arteche et al. 2012; Sato et al. 2005). Thus, it is clear that any decomposition of the net dielectric response from a protein solution sample must include the relaxational contribution from the amino acid side chains.

11.5 Decomposition Example: Frozen Protein Solutions

In recent efforts for the determination of the protein hydration shell, the protein has been treated as transparent. However, in this section, we will demonstrate that the protein in the terahertz range is certainly not transparent and its contribution cannot be described as solely arising from relaxational response. Our goal using THz TDS is to access information on the correlated motions of proteins, the long-range vibrational modes. However, the direct decomposition of protein solution permittivity to extract only the correlated protein motion contribution is particularly complex at room temperature as the character of the so-called hydration shell is still ambiguous. The question of the extended hydration shell can be avoided by measuring solutions at less than 273 K, where a variety of measurements have found that only ~0.32 h hydration water remains unfrozen (Rupley and Careri 1991). The frozen water, ice, has a much smaller absorbance compared to room temperature. Further, a variety of measurements have found that in the range of 210–273 K, many proteins are functional. The unfrozen water and protein response in the 80–273 K range has been extensively studied as part of an effort to understand the so-called protein dynamical transition (Angell et al. 2000; Capaccioli et al. 2011; Doster et al. 2010; He et al. 2008; Schirò et al. 2011).

The dynamical transition for proteins refers to a rapid increase in the averaged atomic mean squared displacement at approximately 210 K seen for a variety of proteins. Because several proteins also become functional at this same minimal hydration (0.032 h) and temperature range ($T > 210$ K), considerable effort has been made to determine if it is an outgrowth of fundamental dynamics of biomolecules. Recently, it has become clear that the minimal temperature requirement of 210 K is not universal, and a number of proteins remain functional down to lower temperature. The rapid increase in protein atomic displacements has been determined to be associated with thermally activated motions of the unfrozen solvent. Thus, one can readily achieve measurements of fully hydrated systems without a large bulk water background. We summarize the results as follows: (1) a thin layer of water immediately adjacent to the protein remains unfrozen; (2) this water has a dielectric relaxation response with a relaxation time that follows an Arrhenius temperature dependence; (3) the temperature range where the relaxation time of the solvent decreases to less than 10 ns corresponds to a large increase in picosecond protein motions. This temperature is typically approximately 210 K; and (4) a wide variety of proteins become functional for temperatures greater than 210 K, suggesting that at this temperature, the molecule can access the necessary structural motions for function. Given these results, we see that by measuring the terahertz permittivity at temperatures 210 K < T < 273 K, we avoid bulk water absorption and still access the permittivity contribution from the protein structural motions.

As an example, we will consider measurements of lysozyme solutions at 240 K. Standard THz TDS measurements were performed on solution phase 200 mg/mL HEWL solutions as a function of temperature as discussed in He et al. (2008). The measured complex index $N = n + ik$ was extracted from the data, and the real and imaginary parts of the permittivity were determined by $\varepsilon' = n^2 - \kappa^2$ and $\varepsilon'' = 2n\kappa$.

We have three different media that contribute to the permittivity: ice, unfrozen bound water, and protein. Ideally, we would use EMT to decompose ε; however, with three unique materials contributing versus only two for our trehalose and glucose examples, the extraction of ε is somewhat more difficult. Thus, we will use the simpler fractional volume approximation that gives

$$\varepsilon'' = f_p \varepsilon_p'' + f_{bw} \varepsilon_{bw}'' + f_{ice} \varepsilon_{ice}'', \tag{11.23}$$

where ε_p'', ε_{bw}'', and ε_{ice}'' are the imaginary dielectric constants for the protein, bound water, and ice, respectively. We note that using partial volumes considerably simplifies the analysis as it allows us to readily separate ε' and ε''. This is not the case for EMT, where there is no simple method for extracting ε'' in Equation 11.22.

To calculate the fractional volumes, we will start from how the typical solution is made: adding mass M_p of lyophilized purified protein powder to volume V_s of buffered solvent. The density of lysozyme has been found to be $\rho_p = 1.38$ g/mL (Svergun et al. 1998), so that the volume of the mass M_p of lyophilized purified protein powder is $V_p = M_p / \rho_p$. When the protein powder is added to the solvent, some fraction of the solvent binds to the protein surface, and the density of this bound water is higher than that for bulk water. For HEWL, we will assume that the bound water is all the unfrozen water 0.32 h (Rupley and Careri 1991). It has been found that for HEWL, the density of the bound water is $\rho_{bw} = 1.1$ g/mL (Svergun et al. 1998), so that the volume of the unfrozen water is $V_{bw} = 0.32 {*} M_p / \rho_{bw}$. The remaining volume of the bulk water is $V_{ice} = V_s - 0.32 {*} M_p$ (mL). The total volume is given by $V_{tot} = V_p + V_{bw} + V_{ice}$, and the fractional volumes are given by $f_p = V_p / V_{tot}$, $f_{bw} = V_{bw} / V_{tot}$, and $f_{ice} = V_{ice} / V_{tot}$. For a sample concentration of 200 mg/mL, these volume fractions were calculated as 0.13, 0.05, and 0.82 for protein, unfrozen water, and ice, respectively. We need to now determine ε_{bw}'' and ε_{ice}'' to extract ε_p'' from the data using Equation 11.23.

For unfrozen water, we start from a Debye relaxation form for ε_{bw}'':

$$\varepsilon_{bw} = \frac{\Delta \varepsilon_{bw} \omega \tau_{bw}}{1 + \omega^2 \tau_{bw}^2}. \tag{11.24}$$

We need to determine τ_{bw} and $\Delta \varepsilon_{bw}$. As discussed, extensive measurements have been made for the bound water within the context of trying to understand the dynamical transition. These measurements have found that for this unfrozen water adjacent to the solutes there are two relaxation processes. One, a slow relaxation process referred to as the α process, which typically has a Vogel–Fulcher–Tammann–Hesse temperature dependence. The timescale for this slow relaxation is 0.1–100 ms in the 200–270 K range. Thus, this α process will not contribute to the THz dielectric response. A second fast relaxation process has an Arrhenius temperature dependence in the 200–270 K range with activation energy of the order of 49 kJ/mol and has the timescale of 100 ns to 10 ps in the same temperature range, and this unfrozen solvent will contribute to the terahertz response. The fast relaxation process is assumed similar to that responsible for the 8.2 ps relaxation in bulk water, but is retarded due to the interaction with the solute to approximately 32 ps. Using the activation energy of 49 kJ/mol for the Arrhenius temperature dependence we find that at 240 K this fast relaxation time is 2.3 ns. We account for this unfrozen water contribution using the dielectric strength weight of the different relaxation processes $\Delta \varepsilon_{bw}(240\ \text{K}) = 25$ as measured previously by Capaccioli et al. for deoxyribose (Capaccioli et al. 2011).

To account for the various contributions to the THz response for a hydrated biomolecule at $220\ \text{K} < T < 273\ \text{K}$, one can therefore consider the following contributions:

For the frozen water, we have the imaginary part of the permittivity related to the real and imaginary parts of the index by $\varepsilon_{ice}'' = 2nk_{ice}$. If one rapidly freezes water to below 130 K, the result is an amorphous glass; however, for temperatures above 130 K, the glass becomes a highly viscous liquid and immediately can crystallize. Thus, for temperatures above 130 K, we assume the frozen water is crystalline ice.

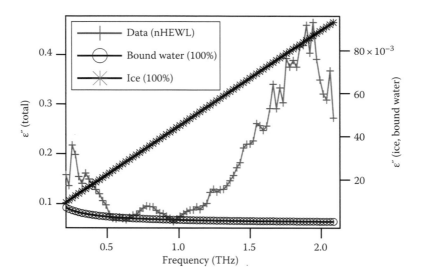

FIGURE 11.2 Plot of the total ε'' for native-state HEWL solution at 240 K. Plot with + (left axis) shows ε'' measured using THz TDS. The plot with * (right axis) shows the calculated dielectric response for pure ice at 240 K. The plot with O (right axis) shows the calculated dielectric response for pure bound water at 240 K.

Zhang and coworkers characterized crystalline ice with THz TDS and found that the real and imaginary parts of the index, n and k, respectively, can be fit by

$$n = 1.79 \quad \text{and} \quad \kappa \approx C_2 \nu \quad \text{with}$$

$$C_2 = \frac{1}{4\pi c} \frac{1.391 \times 10^5 \text{cm}^{-1}\text{K}}{T} \left\{ \frac{e^{hc\nu_o/k_B T}}{\left(e^{hc\nu_o/k_B T} - 1\right)^2} \frac{1}{\nu_o^2} \right\} \tag{11.25}$$

and $\nu_o = 233$ cm^{-1}. At $T = 240$ K, at which the discussion data discussed in the following section were measured, we estimated $C_2 = 1.25 \times 10^{-2}$ / THz.

In Figure 11.2, we show the measured ε'' and the calculated ε'' for pure ice at 240 K and pure bound water at 240 K.

From the calculations given earlier and the estimated volume fractions, the actual contribution from the proteins can be calculated by the following relation:

$$\varepsilon_p'' = \frac{\left(\varepsilon'' - f_{bw}\varepsilon_{bw}'' - f_{ice}\varepsilon_{ice}''\right)}{f_p}. \tag{11.26}$$

This is the imaginary part to the total protein permittivity. Previously, we discussed how the protein response includes relaxational response from the amino acid side chains and vibrational response from correlated motions of the protein structure. That is, we can approximate the total ε_p'' as

$$\varepsilon_p'' = \frac{\Delta\varepsilon_{1p}\omega\tau_p}{1 + \left(\omega\tau_p\right)^2} + \frac{\Delta\varepsilon_{2p}\omega\gamma}{\left[1 - \left(\dfrac{\omega}{\omega_0}\right)^2\right]^2 + \omega^2\gamma^2}, \tag{11.27}$$

where

τ_p is the relaxation time
ω_0 is the resonance frequency
γ is the damping constant
$\Delta\varepsilon_1 = \varepsilon_0 - \varepsilon_\infty$
$\Delta\varepsilon_2 = \varepsilon_1 - \varepsilon_\infty$.

The fit of the extracted ε_p'' to Equation 11.27 is shown in Figure 11.3. The values obtained from the fit for these coefficients were $\Delta\varepsilon_{1p} = 217$, $\tau_p = 400$ ps, $\omega_o/2\pi = 1.8$ THz, $\Delta\varepsilon_{2p} = 0.74$, and $\gamma = 0.02$. It was not possible to arrive at any reasonable and physically possible fit with just multiple relaxations and no resonant contributions.

At room temperature, the relaxation time for a number of amino acids was determined to be between ~30 and 200 ps (Rodríguez-Arteche et al. 2012; Sato et al. 2005). Since the relaxation time increases with decreasing temperature due to the Arrhenius temperature dependence, the value of 400 ps for the relaxational part of the protein contribution for the frozen solutions is in reasonable agreement with this response coming from the surface side chains. However, before leaving this particular fit result, we would like to discuss another possibility and that is the relaxation from the fast water. As discussed in the bulk water section, at room temperature, the bulk water response is generally modeled with two relaxation times, a slow relaxation time $\tau_1 \sim 8.2$ ps and a fast relaxation time $\tau_2 \sim 0.2$ ps. We have focused on the retardation and temperature dependence of the 8.2 ps relaxation and have ignored the fast relaxation. The slow relaxation has little contribution at 240 K; however, we can consider a similar analysis for the fast relaxation here. We will assume the same retardation factor due to interaction with the solute surface as was used for the slow relaxation, with 0.2 ps going to 0.8 ps. As for additional slowing down with temperature, we do not have the extensive temperature-dependent characterization of the fast relaxation as was done for the slow relaxation, simply because neutron facilities cannot easily access the subpicosecond timescale. However, THz TDS measurements of the temperature dependence above freezing for bulk water have been done, and the activation energies determined were $E_1 = 172$ meV for τ_1 and $E_2 = 163$ meV for τ_2 (Ronne et al. 1997). That is, the two activation energies were found to be somewhat similar for the bulk water result. For the sake of a first-order estimate, we will assume that the bound water τ_1 and τ_2 also have similar activation energies. Using the same 0.511 eV activation energy we used earlier for τ_1 for the bound water, we find that the fast water at 240 K τ_2 (240 K) ~ 57 ps, somewhat shorter than the fit to

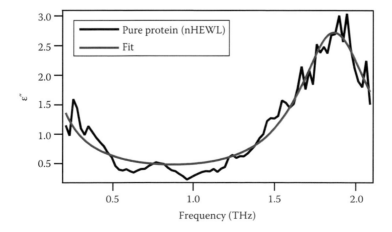

FIGURE 11.3 Plot of the extracted protein permittivity from the net permittivity plotted in Figure 11.2. Unfrozen water and ice contributions have been removed. The plot also shows a fit to the permittivity using a relaxational modes and single resonance.

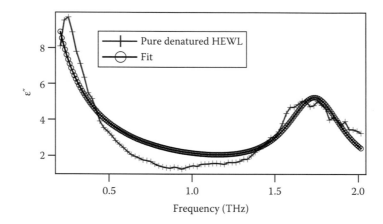

FIGURE 11.4 Plot of the extracted protein contribution to ε'' for denatured HEWL solution (red), and a fit to the data using a single relaxational term and single resonant term.

the data in Figure 11.3. In addition, this contribution is too long to be responsible for the peak at 1.8 THz. Thus, the decomposition of the native-state HEWL solution at 240 K reveals that there is a relaxational response at 400 ps consistent with side chain relaxation and a resonant response at 1.8 THz or 60 cm^{-1} that may be associated with long-range correlated motions of the protein structure.

We repeat the same procedure for measurements of solution phase denatured HEWL. Here we have that with the loss of the 3D structure, the bound water increases to 0.75 h (Rupley and Careri 1991). After removing the ice and bound water contributions, we fit the result with Equation 11.27. We show the result in Figure 11.4 with the measured ε''_p and the fit. Immediately, one can see a large increase in the low-frequency component for the denatured sample. The values of coefficients for the fit were $\Delta\varepsilon_{1p} = 1500$, $\tau_p = 130$ ps, $\omega_o/2\pi = 1.75$ THz, $\Delta\varepsilon_{2p} = 0.97$, and $\gamma = 0.02$. While the coefficients for the relaxational component vastly changed, surprisingly, a resonance was still observed, with a slightly redshifted frequency. As the resonance was associated with long-range correlations, it suggests that perhaps some structure remains for the denatured sample. The large increase in the relaxational component is consistent with the increase in exposure of the residues and removal of internal bonding that suppressed rotational relaxation.

11.6 Conclusion

In this chapter, we have attempted to give the reader an overview of the various contributions to solution phase protein terahertz spectroscopy. It is clearly a complex problem, which still has many unresolved issues, the most outstanding of which are the proper description of the hydration shell and the determination of the relaxational contribution of residue side chains. We note that while we have discussed non-Debye relaxation models, only Debye models were used to analyze the data. It is likely that this is an oversimplification. More careful and complete THz TDS measurements are needed to determine the correct description of, for example, the residue relaxation contribution.

We have found that the problem can be considerably simplified by minimizing the bulk water present. While one might make the argument that measurements of proteins at nonphysiological temperatures are not biologically relevant, we would suggest that a great deal of physical understanding of materials occurs for conditions far from ambient, such as neutron scattering and x-ray crystallography. New terahertz methods are currently being developed to address the removal of all the various relaxational contributions to the terahertz response, such as anisotropy measurements. As this major challenge is overcome, we suggest that THz may be the tool that will allow characterization of long-range protein motions, which will become increasingly important as a parameter in protein engineering.

References

Angell, C. A., K. L. Ngai, G. B. McKenna, P. F. McMillan, and S. W. Martin. 2000. Relaxation in glassforming liquids and amorphous solids. *Journal of Applied Physics* 88: 3113–3157.

Baxter, J. B. and C. A. Schmuttenmaer. 2006. Conductivity of ZnO nanowires, nanoparticles, and thin films using time-resolved terahertz spectroscopy. *Journal of Physical Chemistry B* 110: 25229–25239.

Capaccioli, S., K. L. Ngai, S. Ancherbak, P. A. Rolla, and N. Shinyashiki. 2011. The role of primitive relaxation in the dynamics of aqueous mixtures, nano-confined water and hydrated proteins. *Journal of Non-Crystalline Solids* 357: 641–654.

Cole, K. S. and R. H. Cole. 1941. Dispersion and absorption in dielectrics I. Alternating current characteristics. *The Journal of Chemical Physics* 9: 341–351.

Comez, L., L. Lupi, A. Morresi et al. 2013. More is different: Experimental results on the effect of biomolecules on the dynamics of hydration water. *The Journal of Physical Chemistry Letters* 4: 1188–1192.

Davidson, D. W. and R. H. Cole. 1951. Dielectric relaxation in glycerol, propylene glycol, and n-propanol. *The Journal of Chemical Physics* 19: 1484.

Debye, P. J. W. 1960. *Polar Molecules*. New York: Dover Publications.

Doster, W., S. Busch, A. M. Gaspar et al. 2010. Dynamical transition of protein-hydration water. *Physical Review Letters* 104: 098101.

Ebbinghaus, S., S. J. Kim, M. Heyden et al. 2007. An extended dynamical hydration shell around proteins. *Proceedings of the National Academy of Sciences* 104: 20749–20752.

Ebbinghaus, S., S. J. Kim, M. Heyden et al. 2008. Protein sequence- and pH-dependent hydration probed by terahertz spectroscopy. *Journal of the American Chemical Society* 130: 2374–2375.

Griebenow, K. and A. M. Klibanov. 1995. Lyophilization-induced reversible changes in the secondary structure of proteins. *Proceedings of the National Academy of Sciences* 92: 10969–10976.

Harrop, P. J. 1972. *Dielectrics*. New York: Wiley.

Havriliak, S. and S. Negami. 1967. A complex plane representation of dielectric and mechanical relaxation processes in some polymers. *Polymer* 8: 161–210.

He, Y., P. I. Ku, J. R. Knab, J. Y. Chen, and A. G. Markelz. 2008. Protein dynamical transition does not require protein structure. *Physical Review Letters* 101: 178103.

Hendry, E., M. Koeberg, B. O'Regan, and M. Bonn. 2006. Local field effects on electron transport in nanostructured TiO_2 revealed by terahertz spectroscopy. *Nano Letters* 6: 755–759.

Heugen, U., G. Schwaab, E. Bründermann et al. 2006. Solute-induced retardation of water dynamics probed directly by terahertz spectroscopy. *Proceedings of the National Academy of Sciences* 103: 12301–12306.

Heyden, M., E. Bründermann, U. Heugen et al. 2008. Long-range influence of carbohydrates on the solvation dynamics of water-answers from terahertz absorption measurements and molecular modeling simulations. *Journal of the American Chemical Society* 130: 5773–5779.

Kindt, J. T. and C. A. Schmuttenmaer. 1996. Far-infrared dielectric properties of polar liquids probed by femtosecond terahertz pulse spectroscopy. *The Journal of Physical Chemistry* 100: 10373–10379.

Knab, J., J.-Y. Chen, and A. Markelz. 2006. Hydration dependence of conformational dielectric relaxation of lysozyme. *Biophysical Journal* 90: 2576–2581.

Koeberg, M., C. C. Wu, D. Kim, and M. Bonn. 2007. THz dielectric relaxation of ionic liquid: Water mixtures. *Chemical Physics Letters* 439: 60–64.

Møller, U., D. G. Cooke, K. Tanaka, and P. U. Jepsen. 2009. Terahertz reflection spectroscopy of Debye relaxation in polar liquids [Invited]. *Journal of the Optical Society of America B* 26: A113–A125.

Nandi, N. and B. Bagchi. 1997. Dielectric relaxation of biological water. *The Journal of Physical Chemistry B* 101: 10954–10961.

Nandi, N., K. Bhattacharyya, and B. Bagchi. 2000. Dielectric relaxation and solvation dynamics of water in complex chemical and biological systems. *Chemical Reviews* 100: 2013–2046.

Paolantoni, M., L. Comez, M. E. Gallina et al. 2009. Light scattering spectra of water in trehalose aqueous solutions: Evidence for two different solvent relaxation processes. *The Journal of Physical Chemistry B* 113: 7874–7878.

Pethig, R. 1995. Dielectric studies of protein hydration, chapter 4. In *Protein-Solvent Interactions*, ed. Gregory, R., p. 265. New York: Taylor & Francis.

Rodríguez-Arteche, I., S. Cerveny, Á. Alegría, and J. Colmenero. 2012. Dielectric spectroscopy in the GHz region on fully hydrated zwitterionic amino acids. *Physical Chemistry Chemical Physics* 14: 11352–11362.

Ronne, C., L. Thrane, P. O. Astrand et al. 1997. Investigation of the temperature dependence of dielectric relaxation in liquid water by THz reflection spectroscopy and molecular dynamics simulation. *Journal of Chemical Physics* 107: 5319–5331.

Rupley, J. A. and G. Careri. 1991. Protein hydration and function. *Advances in Protein Chemistry* 41: 37–172.

Sato, T., R. Buchner, S. Fernandez, A. Chiba, and W. Kunz. 2005. Dielectric relaxation spectroscopy of aqueous amino acid solutions: Dynamics and interactions in aqueous glycine. *Journal of Molecular Liquids* 117: 93–98.

Schirò, G., C. Caronna, F. Natali, M. M. Koza, and A. Cupane. 2011. The "protein dynamical transition" does not require the protein polypeptide chain. *The Journal of Physical Chemistry Letters* 2: 2275–2279.

Svergun, D. I., S. Richard, M. H. J. Koch et al. 1998. Protein hydration in solution: Experimental observation by x-ray and neutron scattering. *Proceedings of the National Academy of Sciences* 95: 2267–2272.

Tielrooij, K. J., N. Garcia-Araez, M. Bonn, and H. J. Bakker. 2010. Cooperativity in ion hydration. *Science* 328: 1006–1009.

Vinh, N. Q., S. J. Allen, and K. W. Plaxco. 2011. Dielectric spectroscopy of proteins as a quantitative experimental test of computational models of their low-frequency harmonic motions. *Journal of the American Chemical Society* 133: 8942–8947.

Xu, J., K. W. Plaxco, and S. J. Allen. 2006a. Collective dynamics of lysozyme in water: Terahertz absorption spectroscopy and comparison with theory. *Journal of Physical Chemistry B* 110: 24255–24259.

Xu, J., K. W. Plaxco, and S. J. Allen. 2006b. Probing the collective vibrational dynamics of a protein in liquid water by terahertz absorption spectroscopy. *Protein Science* 15: 1175–1181.

Xu, J., K. W. Plaxco, S. J. Allen, J. E. Bjarnason, and E. R. Brown. 2007. 0.15–3.72 THz absorption of aqueous salts and saline solutions. *Applied Physics Letters* 90: 031908.

Yada, H., M. Nagai, and K. Tanaka. 2008. Origin of the fast relaxation component of water and heavy water revealed by terahertz time-domain attenuated total reflection spectroscopy. *Chemical Physics Letters* 464: 166–170.

12

Nonlinear Interaction of Amino Acids and Proteins with Terahertz Waves

Gurpreet Kaur
Intel Corporation

Xi-Cheng Zhang
*Huazhong University of
Science and Technology*

12.1 Introduction

Terahertz (THz) radiation is generally considered to be safe and noninvasive for biomedical applications due to its low photon energy as compared to gamma- and x-rays. The resonant coupling of electromagnetic radiation with biomolecules has been of interest since the early theoretical predictions by Fröhlich that biological systems exhibit longitudinal electric modes (Foster et al. 1987; Fröhlich 1975; Gabriel et al. 1987). This possibility of resonant coupling has had an impact on the regulations of microwave radiation exposure (Kuster et al. 2004; Maier et al. 2000; Repacholi 2001). Past studies have indicated that when biomolecules like nucleosides and carbohydrates are irradiated with mid- and long-infrared (IR) radiation from a free-electron laser for time ranges of 30–120 min, they undergo structural changes (Dlott and Fayer 1991; Yang et al. 2005, 2007). The structural changes are dependent on time as well as frequency of irradiation and can be detected using Fourier transform infrared (FTIR) spectroscopy. Biomolecules like proteins and amino acids also exhibit vibrational modes in the far-IR or THz region of the electromagnetic spectrum (Markelz et al. 2000; Plusquellic et al. 2007).

Most of the past THz radiation–biomolecule interaction studies have focused on linear THz spectroscopy to investigate the molecular structure using vibrational spectroscopy. Recent development of high-power THz sources (Bartel et al. 2005; Cook and Hochstrasser 2000; Xie et al. 2006) has led to a surge in studies investigating the nonlinear dynamics of semiconductors induced by THz radiation. However, there are only a handful of studies offering the detailed or in-depth investigations of the dynamics of biomolecules under intense THz radiation. With the development of high-power THz sources, it is imperative to understand the resonant effects of THz radiation with molecules and to investigate whether any molecular structural changes can be induced by THz radiation. The understanding of nonlinear effects of THz radiation is also crucial to address health/radiation concerns regarding impact of intense THz radiation on cells and tissues. This chapter focuses on the interaction of high-power THz radiation with amino acids and proteins. The technique used to investigate the nonlinear effects is THz-radiation-induced fluorescence (FL) modulation in proteins and amino acids.

FL is one of the well-established and widely used techniques in the field of biological sciences. Thereby, FL from the molecules acts as excellent probes for investigating the structural/molecular changes induced by intense THz radiation. By monitoring the THz-radiation-induced FL modulation, the short- and long-term dynamical changes leading to structural modifications of biomolecules are investigated.

12.2 THz-Induced Fluorescence Modulation in Amino Acids

12.2.1 Introduction to Fluorescence

FL is a subclass of luminescence with a lifetime of few nanoseconds. FL occurs when, after absorbing some type of electromagnetic energy, an electron relaxes from an excited state to a ground state and gives away its energy in the form of photons. Important characteristics of FL are Stokes shift, quantum yield, and FL lifetime. Stokes shift refers to the fact that the energy of emission is less than the energy of absorption and the emission spectrum is typically independent of the absorption wavelength. Quantum yield refers to the efficiency of the emission process and is defined as the ratio of emitted photons to absorbed photons. Quantum yield is always less than 1 because of competing nonradiative pathways by which an electron can relax to a ground state. The molecules that emit FL are called fluorophores and are divided into two main classes: intrinsic and extrinsic. Intrinsic fluorophores are naturally occurring, while extrinsic fluorophores are added to the sample to change the emission/absorption properties or to make nonfluorescent materials fluoresce. Few examples of intrinsic fluorophores are aromatic amino acids, flavins, and chlorophyll. Table 12.1 lists some examples of commonly available intrinsic fluorophores.

The intrinsic FL of proteins is due to the presence of aromatic amino acids: tryptophan (Trp), tyrosine, and phenylalanine (Demchenko 1988). The FL of Trp offers a very convenient way to probe molecular changes due to Trp's strong ultraviolet (UV) absorbance, large FL quantum yield, and high sensitivity to

TABLE 12.1   List of Commonly Available Intrinsic Fluorophores

Fluorophores	Excitation Wavelength (nm)	Emission Wavelength (nm)
Amino acids		
Tryptophan	280	320
Tyrosine	260	303
Phenylalanine	250	290
Proteins		
Whey	270	330
GFP	460	510
BSA	270	308
Tonic water	400	450
Spinach	267	365
Onion bulb	267	440
Fingernails	400	450
25% cotton paper	400	440, 678

the local environment (Callis and Burgess 1997). Trp metabolism within different tissues is associated with physiological functions and plays a role in the regulation of growth, mood, behavior, and immune responses in the brain (Le Floc'h et al. 2011). It is thus considered to be useful for understanding the chemical pathways for drug development for the treatment of depression (Waider et al. 2011).

FL spectroscopy is one of the most powerful and attractive imaging and detection modalities across the life sciences (Lakowicz 1999). With its high sensitivity and specificity, it can be used for investigating dynamical changes within proteins and live cells (Lakowicz 1999; Udenfriend 1971). The optical properties of fluorophores are extremely sensitive to changes in their local environment (Loew and Harris 2000). Fluctuations in temperature, pH, hydration, and protein conformation can all lead to changes in the FL quantum yield (Attallah and Lata 1968; Heyduk 2002; Martin and Lindqvist 1975; Ohmae et al. 1996; Wallach and Zahler 1966). Furthermore, shifts in the emission spectrum and changes in FL lifetime can be observed (Levitt et al. 2009). Thus, FL can be a resourceful tool for studying the transient and long-term effects of THz radiation.

Amino acids are molecules containing an amine group, a carboxylic acid group, and a side chain that varies for different amino acids. Amino acids serve as building blocks for proteins with different amino acids linking together in varying sequences to form a vast variety of proteins. Only 20 amino acids are naturally incorporated into polypeptides and are called standard or proteinogenic amino acids. Trp is one of the standard amino acids and is the only amino acid containing an indole functional group. Indole is an aromatic organic compound consisting of a six-membered benzene ring fused to a five-membered nitrogen containing a pyrrole ring.

The indole functional group acts as the chromophore of Trp and absorbs strongly in the near-UV part of the electromagnetic spectrum (Demchenko 1988). The strong UV absorption is the result of overlapping $\pi\pi^*$ transitions to two excited states. These states are denoted using Platt notation as 1_{L_a} and 1_{L_b} as shown in Figure 12.1.

These two states have similar energies and either of them can have the lowest energy depending on the environment. According to Kasha's rule (Kasha 1950), FL emission occurs from the lowest energy state. The directions of transition dipoles for 1_{L_a} and 1_{L_b} states are nearly perpendicular to each other (Albinsson and Norden 1992; Albinsson et al. 1989; Callis 1991; Song and Kurtin 1969; Yamamoto and Tanaka 1972). The fractional contribution of each state to the absorption transition determines the absorption/emission anisotropy. The maximal anisotropy is at 300 nm and is characteristic of absorption/emission from the 1_{L_a} state (Petrich et al. 1983). The absorption to the 1_{L_a} state is less structured compared to the 1_{L_b} state as shown in Figure 12.2.

The FL of Trp exhibits double exponential decay with lifetimes of 0.5 and 3.1 ns and is due to different rotational isomers of Trp (Creed 1984; Petrich et al. 1983). By the mirror image rule, the emission spectrum is generally the mirror image of the $S_0 \rightarrow S_1$ sorption due to the symmetric nature of transitions involved in both absorption and emission and the similarities of vibrational energy levels of S_0 and S_1 Trp

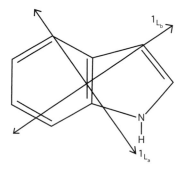

FIGURE 12.1 Electronic absorption transitions in Trp. (Reprinted with permission from Callis, P. R., *J. Chem. Phys.*, 95, 4230, 1991. © 1991, AIP Publishing LLC.)

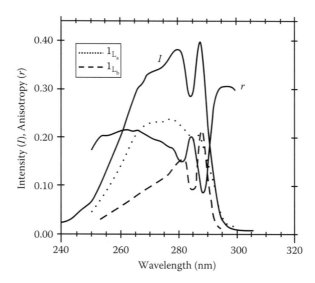

FIGURE 12.2 FL excitation spectra (solid lines) and anisotropy spectra (*r*) of Trp. The absorption spectra are resolved into 1_{L_a} (.....) and 1_{L_b} (- - -) states. (Adapted with permission from Eftink, M. R., Selvidge, L. A., Callis, P. R., and Rehms, A. A., *J. Phys. Chem.*, 94, 3469, 1990, © 1990, American Chemical Society.)

emits from the 1Lb state when the local environment of the chromophore is unpolar. The local electric field may also influence the emission spectrum of the indole (Callis 1991).

Thus, the FL spectrum and yield provide information about the local environment of the chromophore, changes in the vibrational energies, local electric field, etc. Thereby, monitoring the FL provides a convenient way to probe the effects of external THz radiation. In the following sections, the experimental setup used to probe the changes in the FL from Trp molecules and the observed results are discussed.

12.2.2 Experimental Setup

A schematic representation of the experimental setup for investigating THz-radiation-induced FL modulation is illustrated in Figure 12.3.

A UV beam with a wavelength of 270 nm was isolated from the broadband light generated by a xenon-enhanced arc lamp source (150 W lamp, model no. 6254, Newport) using interference filters. A tunable far-IR gas laser (SIFIR-50, Coherent) produced continuous-wave (CW) THz radiation. The THz wave was combined with the exciting optical wave by a parabolic mirror (PM) with a hole that allowed the UV beam to transmit through it. The two beams propagated collinearly to the sample. The focused THz beam diameter and collimated optical beam diameter were kept fixed at approximately 1.2 mm to keep the illumination area constant for 8 different THz frequencies between 1.27 and 2.74 THz. The sample was prepared by compressing pure L-Trp powder into a 13 mm diameter and 1.2 mm thick pellet using a pressure of 5 metric tons for 3 min. A computer-controlled metallic shutter controlled the *on* and *off* states of the THz radiation source. The FL spectrum from the pellet sample was obtained using a miniature fiber-optic spectrometer.

12.2.3 Experimental Results

The FL spectrum and FL intensity of Trp were monitored as a function of sample temperature, frequency, and intensity of the THz radiation source. Normalized FL spectra of the Trp sample without THz radiation and with external THz radiation of intensity 3.0 and 11.7 W/cm² are shown in Figure 12.4. The frequency of THz radiation for both intensity levels was 2.55 THz. The spectra were normalized by the FL signal corresponding to a peak emission wavelength of 329 nm.

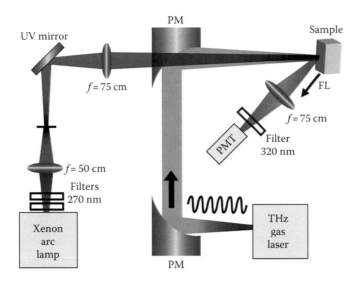

FIGURE 12.3 Schematic of the experimental setup. The UV beam at 270 nm was isolated from the broadband light generated by a xenon-UV-enhanced arc lamp to excite the sample. A tunable far-IR gas laser produced CW THz radiation. CW UV and CW THz radiation were combined using a PM with a hole that allowed UV radiation to pass through it. The FL emitted from the sample was isolated using an interference filter (central wavelength of 320 nm and bandwidth of 10 nm) and collected using a lens and detected by a PMT.

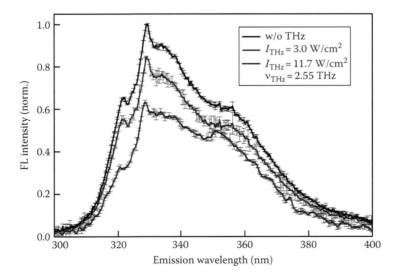

FIGURE 12.4 FL spectra of Trp pellet sample with and without the presence of THz radiation. The THz frequency was kept constant at 2.55 THz for two different THz intensities of 3.0 and 11.7 W/cm^2.

The ratio of the spectra obtained with and without THz radiation, for those two THz intensity levels, is shown in Figure 12.5. This figure demonstrates that while quenching is uniform at low THz intensity (3.0 W/cm^2), it became nonuniform over the emitted spectrum range at high THz intensities (11.7 W/cm^2). Maximum quenching occurred at 320 nm even though the peak emission wavelength of the sample was 329 nm.

For all subsequent experiments, interference filter with central wavelengths at 320 nm was used to isolate FL signal. The signal was detected using a photomultiplier tube (PMT) in a front-facing geometry

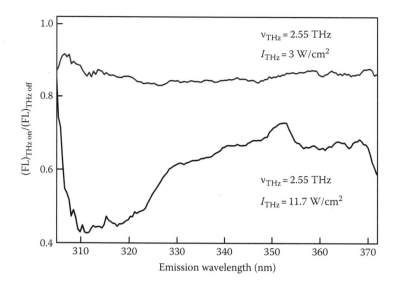

FIGURE 12.5 Ratio of the FL spectra of Trp obtained with and without the presence of THz radiation, corresponding to the spectra in Figure 12.5. THz-induced FL quenching was nonuniform for THz radiation intensity of 11.7 W/cm². The quenching remained nearly uniform for THz radiation intensity of 3 W/cm².

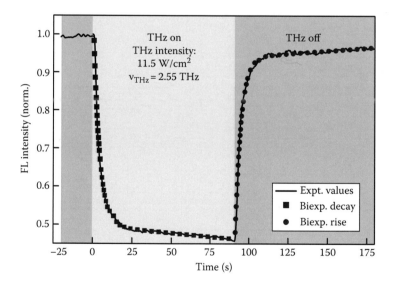

FIGURE 12.6 Normalized FL intensity of Trp in the presence of THz radiation as a function of illumination time. The THz source was turned on at $t = 0$ s and turned off at $t = 90$ s. The frequency and intensity of THz radiation were kept constant at 2.55 THz and 11.5 W/cm², respectively. The black line corresponds to the experimental data, which follow a biexponential decaying and a biexponential rising function.

and fed to the lock-in amplifier. Figure 12.6 illustrates the effect of illuminating Trp with an average intensity of 11.5 W/cm² at 2.55 THz. Initially, only UV radiation with intensity of 1 mW/cm² was incident upon the sample ($t = -20$ s to $t = 0$ s) and FL fluctuation was ~1%. At $t = 0$ s, the THz source was turned on and the FL decayed by 54% in 20 s. Once the FL reached a new equilibrium value ($t = 90$ s), the THz source was turned off and the FL recovered back to its original value over 20 s.

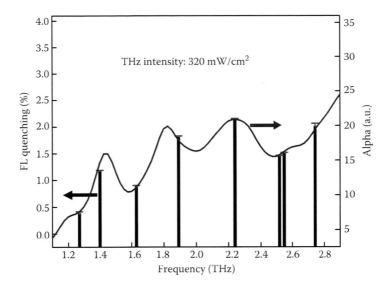

FIGURE 12.7 The FL quenching induced by THz radiation of varying frequencies. THz intensity was fixed at 320 mW/cm² for all frequencies. The THz absorption spectrum of the same sample obtained using broadband THz TDS is also shown. The FL modulation varies with changes in THz absorption coefficient highlighting the vibrational resonances.

The decaying/rising curve follows a biexponential decay/rise function:

$$FL(t) = FL_o + a_1 * \exp(-t/\tau_1) + a_2 * \exp(-t/\tau_2) \tag{12.1}$$

with $FL_o = 0.44$ (0.97), $a_1 = 0.56$ (−0.53), τ_1 3.96 (3.61), $a_2 = 0.06$ (−0.04), and τ_2 88.32 (50.29) for decaying (rising) curve.

The FL quenching was also studied as a function of the frequency of THz radiation by tuning the parameters of the THz gas laser. THz intensity for all different frequencies was kept constant at 320 mW/cm². As illustrated in Figure 12.7, the FL quenching induced by THz radiation was frequency dependent and correlated with the THz absorption coefficient of Trp.

The vertical black lines correspond to THz-induced FL quenching and the blue curve is the THz absorption coefficient. The absorption coefficient of Trp was obtained by performing broadband THz time domain spectroscopy (TDS) in a separate experiment. The THz-induced quenching closely follows the changes in the absorption coefficient, which strongly suggests the resonant interaction of THz radiation with the sample.

The FL quenching exhibited a linear dependence on THz intensity, as shown in Figure 12.8, where the THz intensity was controlled with a pair of wire grid polarizers. Difference in the slopes of the linear curves confirms the frequency-dependent response of the sample.

12.2.4 Coherent versus Noncoherent Response

12.2.4.1 Temperature-Dependent Fluorescence Modulation

One might expect the thermal effects of high-power CW radiation on the sample to lead to FL quenching. This concern was addressed by actively controlling the temperature of a sample from −180°C to 100°C and studying THz-induced FL quenching as a function of temperature. A cryogenic system (MMR Co.), constructed with a window made of poly-4-methylpentene-1 (TPX) material to facilitate transmission of the UV, IR, and THz radiation, was used to control the sample's temperature. The maximum intensity of THz radiation at the focus after transmitting through the window of the cryogenic system was 5.5 W/cm². Figure 12.9 illustrates that the sample's FL quenching induced by THz radiation increased as the sample

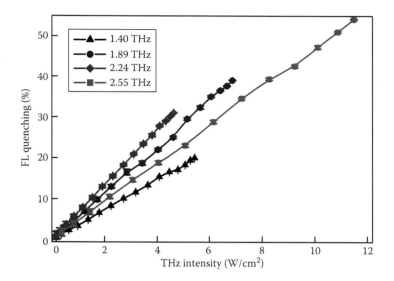

FIGURE 12.8 FL quenching as a function of THz intensity for different THz frequencies: 1.4, 1.89, 2.24, and 2.55 THz. The FL quenching depends linearly on THz intensity with different slopes for different THz frequencies.

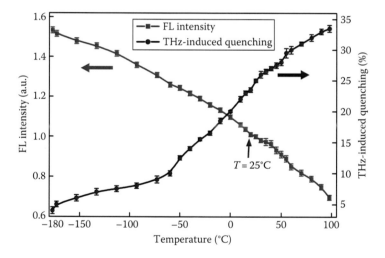

FIGURE 12.9 The FL intensity (in the absence of THz) and THz-induced FL quenching as a function of sample temperature. The FL intensity was normalized by the FL at 24°C. The frequency and intensity of THz radiation source were 2.55 THz and 5.5 W/cm², respectively.

temperature was increased. The temperature-dependent FL, in the absence of THz radiation, is also shown on the same graph, as indicated by a curve with square symbols. The FL intensity (in the absence of THz radiation) was normalized by the FL at room temperature ($T = 25°C$). The corresponding THz-induced quenching, given by a *circle symbol* curve, indicated a quenching of 25% at $T = 25°C$. The data for temperature dependent FL intensity indicates that a temperature increase ΔT of 65°C is required to produce the same level of FL quenching (25%) as was induced by THz radiation at 25°C.

The frequency of the THz radiation source for Figure 12.9 was 2.55 THz and the intensity was 5.5 W/cm². Using the same cryogenic system, IR/thermal images of the sample were recorded with a microbolometer camera to measure the temperature change caused by the THz radiation at various sample temperatures.

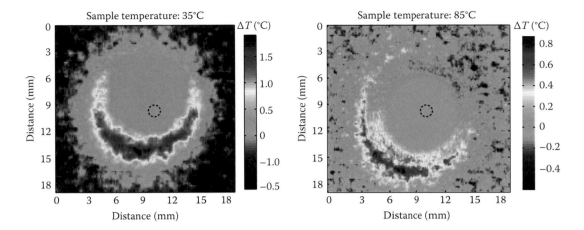

FIGURE 12.10 Difference of thermal images of the sample, with and without THz, obtained using a microbolometer camera. The temperature of the sample was actively controlled using the cryogenic system. The black-dotted region in the thermal images corresponds to the region where THz beam was focused. These images highlight that when the sample's temperature was actively controlled, heat generated by the THz beam quickly diffused to the edges and did not lead to deposition of heat where the THz beam was focused.

For each value of the sample temperature, the thermal images were obtained at $t = 0$ s (i.e., when the THz source was off) and at $t = 90$ s (i.e., after 90 s of THz radiation illumination). These two images were compared to obtain the difference in the temperature distribution and net increase in the temperature of the sample induced by THz radiation.

Figure 12.10a and b illustrates the difference images obtained for sample when it was kept at constant temperatures of 35°C and 85°C, respectively. The black-dotted region corresponds to the region where THz and optical beams were focused, meaning that black-dotted region corresponds to the active area of the sample that contributed to FL.

These IR images indicate that the sample's temperature did not change significantly under the influence of THz radiation when the sample was being actively cooled/heated by the cryogenic system. These images also indicate that heat generated by the THz beam diffused to the edges and was not deposited at the region of focus. Thereby, THz radiation was not leading to significant heating of the sample and the observed FL quenching was not due to absorptive heating.

12.2.4.2 Evidence of Nonthermal Response

The temperature-dependent FL quenching study clearly indicates the nonthermal response of the sample to incident THz radiation, as the thermal images indicate no change in sample temperature induced by THz, but there is an increase in the amount of THz-induced quenching with increasing sample temperature. The temperature change induced by THz radiation when the sample was not mounted on a cryogenic stage, that is, when the sample's temperature was not controlled externally, was also determined. The temperature change caused by THz radiation was estimated from the IR images of the sample by comparing images obtained with and without THz radiation.

Figure 12.11a provides the increase in temperature as a function of THz illumination time, and Figure 12.11b provides the FL as a function of ΔT. The temperature dependence of FL, in the absence of THz radiation, was obtained by controlling the temperature of the sample using a cryogenic system. The x-axis of Figure 12.11b is the temperature rise, meaning that $\Delta T = 0$°C corresponds to temperature, $T = 25$°C. Comparing these two figures, the change in FL as a function of time, assuming purely thermal response of the sample to THz radiation, is as shown in Figure 12.12.

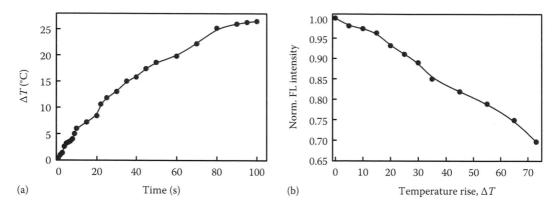

FIGURE 12.11 (a) The rise in temperature of the sample caused by localized heating from incident THz radiation. The increase in sample temperature was estimated from the thermal images. (b) The normalized FL intensity of the sample in the absence of THz radiation as a function of sample temperature. The x-axis corresponds to sample temperature rise compared to room temperature. $\Delta T = 0°C$ corresponds to sample temperature of $T = 25°C$.

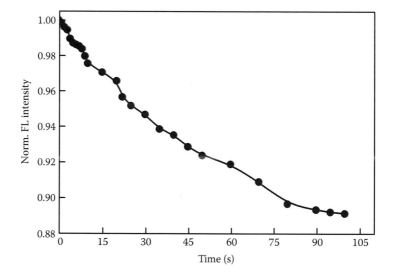

FIGURE 12.12 The expected change in FL intensity as a function of time based on an increase in the temperature of the sample due to local heating induced by THz radiation. This change is much smaller and slower compared to the observed experimental quenching as shown in Figure 12.6.

This is a clear evidence of nonthermal or coherent response of the sample to incident THz radiation. The fluctuations in the FL due to localized heating of the sample by THz radiation, as shown in Figure 12.12, are a slower process compared to the biexponential FL decay observed experimentally and as shown in Figure 12.6. As illustrated in Figure 12.11a, the sample's temperature, over a time period of 90 s, increased by 25°C ± 0.2°C when it was exposed to 2.55 THz radiation with intensity of 11.5 W/cm², which is still not significant enough to lead to the FL quenching observed in our earlier tests (54%). From the temperature-dependent FL curve (Figure 12.11b), the estimated temperature required to produce 54% quenching is approximately 140°C. The spatial profile of THz-induced heat distribution is shown in Figure 12.13, where the maximum rise in temperature is at the focus region (~1.2 mm beam diameter) of THz beam and drops to 5°C at the edges of the sample (13 mm diameter).

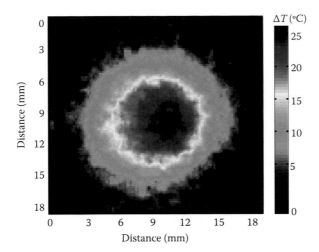

FIGURE 12.13 Difference of thermal images of the sample, with and without THz, obtained using a microbolometer camera. The intensity and frequency of THz radiation were 11.5 W/cm² and 2.55 THz, respectively. The total illumination time was 90 s, and the images at $t = 0$ s and $t = 90$ s were compared to get the temperature change. The maximum temperature rise of 25°C can be seen at focus, and it falls away from the center to the edges with change of 8°C at the edges of the sample (13 mm diameter).

Furthermore, it was observed that when a sample was heated to temperatures >90°C, it led to irreversible FL depletion of the sample in the absence of THz radiation. If THz radiation were indeed heating the sample to 140°C, the observed FL quenching would not have been reversible. In a separate *control* test, the FL emission spectrum of Trp as a function of sample temperature was also investigated. The normalized emission spectra at different temperatures, from 25°C to 95°C, exhibited uniform quenching of the FL spectrum unlike the nonuniform quenching induced by THz radiation. As demonstrated in literature, the multimodal FL emission spectrum of any molecule is due to different vibrational states corresponding to an electronic excited/ground state. The nonuniform quenching caused by THz can thereby be thought of as changes in the population density of vibrational levels due to resonant coupling of THz radiation with the vibrational modes of Trp. All these measurements clearly signify the nonthermal or coherent response of the sample to incident THz radiation.

12.2.4.3 Factors Contributing to Temperature-Dependent Quenching

The factors that can contribute to changes in temperature-dependent quenching are the changes in the THz absorption at different temperatures and the changes in thermal conductivity. The thermal conductivity of a material generally increases with increase in temperature, which means that at high temperatures, any heat deposited by THz radiation should diffuse or spread out faster leading to lower FL quenching expected from pure thermal effects. The reverse would happen at low temperatures where due to decreased thermal conductivity, the localized heating by THz radiation increases leading to increase in FL quenching. But reverse behavior of FL quenching as a function of temperature is observed.

In order to determine the changes in THz absorbance at different temperatures, the THz signal transmitted through the sample at each temperature was measured by using a pyroelectric detector. The absorption coefficient was determined by using Beer Lambert's law: $I = I_o \exp(-\alpha t)$ where t is the thickness of the sample.

The temperature dependence of the THz absorption coefficient of Trp is as shown in Figure 12.14a. Thus, even though the THz intensity illuminating the sample, I_o, was constant in the temperature-dependent FL quenching study, the absorbed THz intensity (or number of absorbed THz photons)

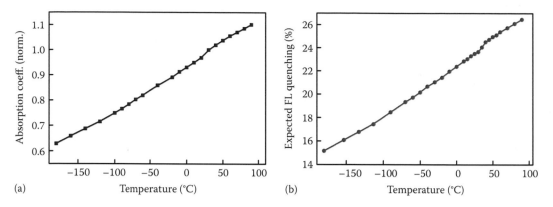

FIGURE 12.14 (a) The absorption coefficient of Trp at 2.55 THz as a function of temperature. (b) The expected FL quenching as a function of sample temperature due to temperature dependence of the THz absorption coefficient calculated using Equation 12.3.

changes at different temperatures. The change in THz intensity after propagating a distance d at temperature T can be described as

$$\Delta I = kI_o\left(1 - \exp\left(-\alpha(T)d\right)\right),\tag{12.2}$$

where

    I_o is the incident THz intensity

    $\alpha(T)$ is the temperature-dependent THz absorption coefficient

As determined in Figure 12.8, THz-induced quenching is linearly dependent on THz intensity; therefore,

$$Q(T) = kI_o\left(1 - \exp\left(-\alpha(T)d\right)\right),\tag{12.3}$$

where

    k is a constant

    d depends on the volume of molecules contributing to FL, which is limited by the penetration depth of the UV pump beam

The UV penetration depth depends on the UV molar absorbance of 5455 M/cm at 270 nm (Pace et al. 1995). The molar concentration of the sample was calculated from

$$C = \frac{W_{pellet}}{V_{pellet} * M_{Trp}}\tag{12.4}$$

$$V_{pellet} = \pi r^2 t,\tag{12.5}$$

where W_{pellet}, M_{Trp}, t, and r are the weight of the pellet, the molecular weight of Trp, and the thickness and radius of the pellet, respectively. Using C from Equation 12.4 and the value of UV molar absorbance of Trp, the UV penetration depth was approximately 2.86 μm. Thereby, the expected change in FL quenching compared to the quenching at room temperature due to the changes in absorption coefficient can be calculated using Equation 12.5 and is shown in Figure 12.14b.

This indicates that temperature dependence of quenching does not arise purely from the changes in THz absorption coefficient. The observed FL quenching can be corrected for the temperature dependence of absorption coefficient and the corrected quenching is shown in Figure 12.15. The temperature-dependent FL quenching exhibits a transition point at around −50°C. The slope of the curves and the transition temperature depend on the frequency of incident THz radiation.

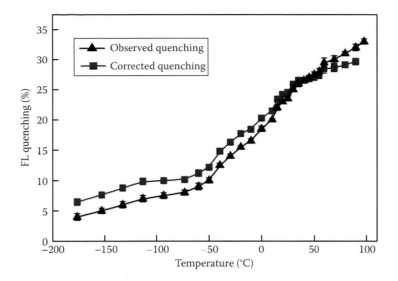

FIGURE 12.15 THz-induced FL quenching as a function of sample temperature after incorporating the temperature dependence of the THz absorption coefficient of Trp. The intensity of THz radiation was 5.5 W/cm² and the frequency was 2.55 THz.

The radiation with frequencies of 1.4 and 1.89 THz is resonant with the torsional resonance modes of Trp, and the THz absorbance increases with the decreased sample temperature. The observed change in FL quenching and expected change in FL quenching of Trp when exposed to 1.89 THz radiation are shown in Figure 12.16.

Based on Equation 12.3, the FL quenching should be higher at low temperature due to higher THz absorbance at low temperatures. In contrast to the change in FL quenching at 2.55 THz, the FL quenching under 1.89 THz radiation is nearly constant for temperatures ranging from −180°C to −120°C. The normalized FL quenching as a function of temperature after being corrected for absorbance changes is

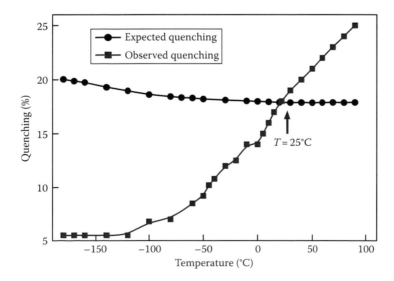

FIGURE 12.16 The observed FL quenching and expected FL quenching for the frequency of incident THz of 1.89 THz and intensity of 4.0 W/cm² as a function of sample temperature. The black (circles) curve corresponds to expected change in FL quenching due to temperature dependence of the THz absorption coefficient and calculated using Equation 12.3.

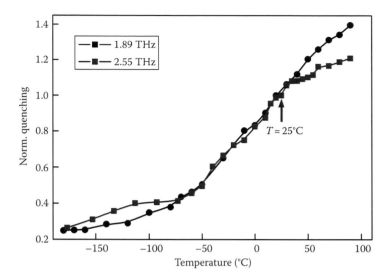

FIGURE 12.17 The normalized FL quenching as a function of temperature after being corrected for absorbance changes for two different THz radiation frequencies. The FL quenching was normalized by the FL quenching at $T = 25°C$.

shown in Figure 12.17 for two different THz radiation frequencies. The FL quenching was normalized by FL quenching at $T = 25°C$.

These results are strong indication of nonthermal quenching mechanism. When the sample is cooled to below room temperature, the molecules possess less thermal energy for lattice vibrations and the molecular bonds strengthen. Furthermore, the molecular population distribution in energy levels changes with temperature that may change the coupling of THz radiation with Trp molecules. These changes may be responsible for the observed temperature dependence of FL quenching.

12.2.5 Pathways for Modulation

12.2.5.1 THz-Radiation-Induced Nonradiative Transfer

The different pathways for FL modulation can be visualized as shown in Figure 12.18. The dark black lines correspond to the electronic states and the light black lines correspond to the vibrational levels for each electronic state. S_0 is the ground electronic state and S_1 is the first excited electronic state.

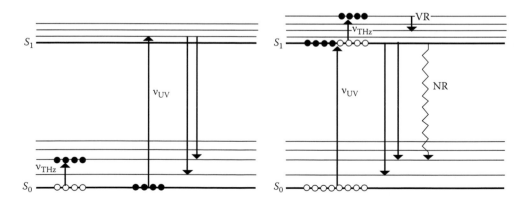

FIGURE 12.18 Model of the coupling by THz radiation and UV radiation between different energy levels of Trp molecules.

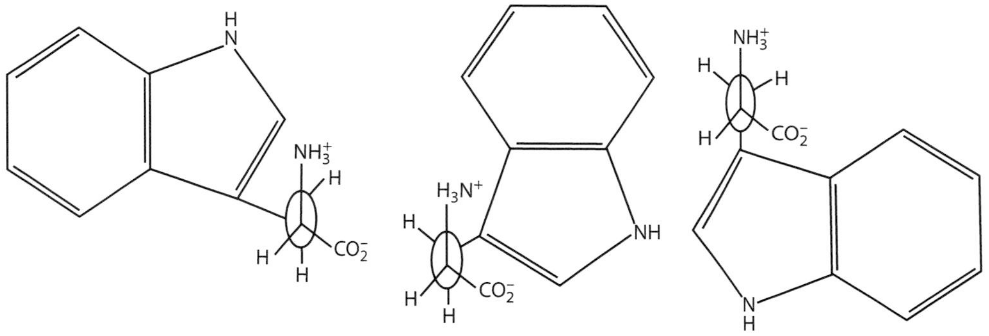

FIGURE 12.19 Different rotational isomers or rotamers of Trp molecules. (Adapted with permission from Petrich, J. W., Chang, M. C., McDonald, D. B., and Fleming, G. R., *J. Am. Chem. Soc.*, 105, 3824, 1983. © 1983, American Chemical Society.)

The schematic on the left corresponds to the scenario where THz radiation affects the molecules in the ground state. Due to the vibrational coupling of the molecules with incident THz radiation, the molecules from the lowest vibrational level of the S_0 state get transferred to higher-lying levels in the vibrational levels manifold of the ground state. This thereby decreases the number of molecules, which can be transferred to the excited electronic S_1 state by the UV beam. Thus, the emitted FL decreases due to a decreased population of Trp molecules in the ground state. The schematic on the right corresponds to the scenario where THz radiation affects the molecules in the excited state. Trp molecules are excited to a higher-lying electronic state, S_1, by the UV beam. The THz radiation then couples with the molecules and moves them to higher-lying vibrational energy levels of the S_1 electronic state. From these high-lying vibrational states, molecules may relax using nonradiative pathways.

12.2.5.2 THz-Radiation-Induced Conformational Transitions

It has been demonstrated in the past studies that Trp molecules can exist in different rotational isomers as depicted in Figure 12.19.

It may be possible that under resonant coupling with THz radiation, Trp molecules undergo cascaded transitions. It can transition from ground state to higher vibrational states through a series of intermediate levels. This is usually referred to as ladder climbing (Xie et al. 2001). Upon acquiring enough energy, it may be possible that the molecule reaches a conformational transition state and transitions into a different conformational state. Past studies involving solid-state nuclear magnetic resonance (NMR) techniques have indicated that large-amplitude motions due to torsional vibrations may take place in solid-state amino acids and proteins (Torchia 1984), which may require time scales of milliseconds to seconds. We believe that such slow motions induced by THz radiation in Trp are responsible for the observed slow FL quenching.

12.3 THz-Radiation-Induced Fluorescence Modulation in Proteins

In this section, THz-induced FL modulation study is extended to whey protein and green fluorescent protein (GFP). Whey protein is a commercially important dairy-based protein that contains a mixture of the globular proteins, alpha-lactalbumin, and beta-globulin, which are made up of 162 and 123 residues, respectively (Kinsella and Whitehead 1989). The food industry has been exploring new and alternative techniques, including using pulsed electric fields, which can help retain whey's quality attributes that may be lost during thermal processing (Kinsella and Whitehead 1989). GFP is a widely used protein in the biosciences as a marker for gene expression and protein targeting in live cells and organisms

(Kain et al. 1995; White and Stelzer 1999). Its use has revolutionized the field of FL microscopy and is also opening new avenues in the field of biosensors and photochemical memories (Misteli and Spector 1997; Shimomura et al. 1962; Tsien 1998).

12.3.1 Fluorescence Quenching in Whey Protein

Whey protein isolate powder was purchased from Davisco Foods International Inc. Whey protein isolate contains higher protein concentration compared to whey protein concentrate. The protein powder, as indicated by the manufacturer, was milk-based 100% natural protein with no added lactose, sugars, or fats. A 1 mm thick solid sample with a diameter of 13 mm was prepared by pressing whey protein powder into a pellet using a pressure of 5 metric tons for 3 min.

12.3.1.1 Experimental Setup

The schematic of the experimental setup using a Gunn diode as THz source and femtosecond UV pulses as optical excitation source is illustrated in Figure 12.20. Femtosecond (fs) optical pulses at a central wavelength of 800 nm from a Ti:sapphire regenerative amplifier laser (Hurricane, Spectra-Physics) were frequency tripled to 267 nm using a set of nonlinear β-barium borate (BBO) crystals. The 267 nm pulses were then incident upon whey protein samples causing optical excitation. FL spectra of protein sample were obtained with a miniature fiber-optic spectrometer (Ocean Optics). Interference filters were used to isolate the peak emission wavelength as determined from the spectrometer data. This filtered signal was detected using a PMT in front-facing geometry (Eisinger and Flores 1979) and fed to lock-in amplifier. The signal was collected in front-facing geometry because the solid samples attenuate FL signal in forward direction. CW THz radiation at 0.2 THz was produced using a Gunn diode (Virginia diodes).

The THz wave was combined with the exciting optical pulse by a specialty PM, and the two beams propagated collinearly to the sample. A computer-controlled metallic shutter controlled the *on* and *off* states of the THz radiation source. The beam size of collimated optical beam and focused THz beam was approximately 3 mm. The THz beam size as determined by a pyroelectric camera (Spiricon-III) is shown in Figure 12.21. The maximum intensity of the 0.2 THz source at the focus was 140 mW/cm².

12.3.1.2 Results

12.3.1.2.1 *Effects of 0.2 THz Radiation Source*

FL spectrum from whey protein obtained using miniature fiber-optic spectrometer (Ocean Optics) is shown in Figure 12.22. When the whey protein sample was exposed to radiation from a 0.2 THz Gunn

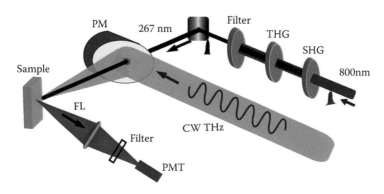

FIGURE 12.20 Schematic of experimental setup. Femtosecond 800 nm pulse frequency tripled (267 nm) by nonlinear crystals was used for the optical excitation of samples. Sample FL was collected using a PMT. UV pulses and CW THz radiation are combined using a PM with a hole that allows UV pulses to pass through it. SHG, second harmonic generation crystal; THG, third harmonic generation crystal.

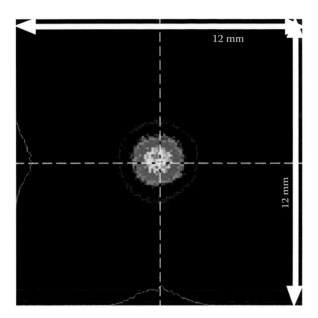

FIGURE 12.21 The image of the THz beam shape and size at the focus obtained with Spiricon pyroelectric camera. The estimated diameter of THz beam is 3 mm.

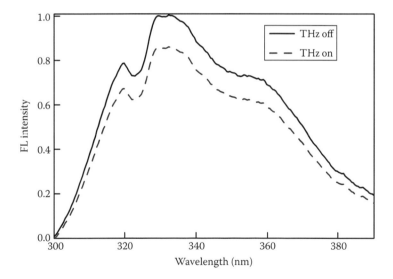

FIGURE 12.22 The FL spectrum of whey protein with and without the presence of THz radiation. The excitation wavelength was 267 nm with excitation intensity of 6 MW/cm². The frequency and intensity of THz radiation were 0.2 THz and 140 mW/cm², respectively.

diode, intrinsic FL from whey protein was quenched. The FL quenching induced by THz radiation was uniform over the entire emission spectral range.

The normalized FL emission spectrum from whey remained unchanged under different UV excitation levels. The dependence of FL intensity on optical excitation power is shown in Figure 12.23. The emitted FL intensity was linearly proportional to the excitation UV intensity up to 12 MW/cm². For UV excitation intensities >12 MW/cm², the emitted FL intensity saturated. This indicates that FL was

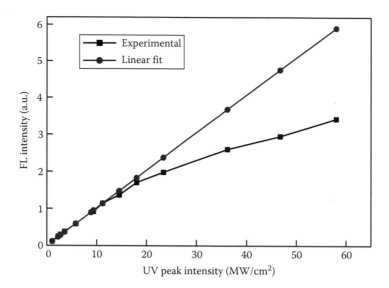

FIGURE 12.23 The power dependence of the emitted FL of whey protein as a function of UV excitation beam. The wavelength of excitation beam was 267 nm.

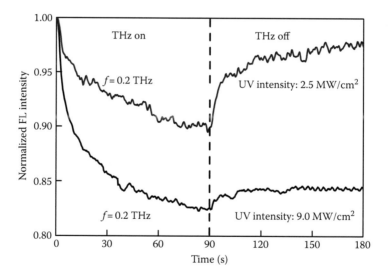

FIGURE 12.24 The normalized FL intensity as a function of time for whey protein excited using two different UV intensities. THz frequency and intensity were kept constant at 0.2 THz and 140 mW/cm², respectively. The THz source was turned on at $t = 0$ s and turned off at $t = 90$ s.

a linear process for excitation intensities below 12 MW/cm². Thereby, the maximum optical intensity used to study the influence of optical intensity on THz-induced FL quenching was limited to 9 MW/cm².

When whey protein was illuminated with 0.2 THz radiation, the intrinsic FL decreased substantially and reached equilibrium after approximately 60 s. When THz radiation was turned off, FL recovered within 10 s as shown in Figure 12.24. When the sample was excited with average UV intensity of 9 MW/cm² (average intensity of 720 µW/cm²), the FL quenching caused by THz was 18% and was irreversible. The quenching was almost reversible and reduced to 10% when peak UV intensity was reduced to 2.5 MW/cm² (average intensity of 200 µW/cm²).

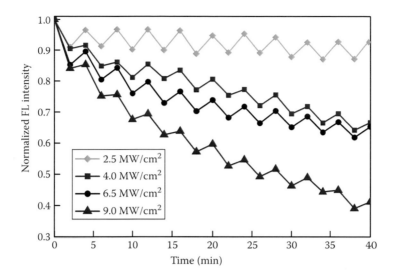

FIGURE 12.25 Irreversible and reversible FL quenching in whey protein over multiple cycles under different UV intensities. THz frequency and intensity were kept constant at 0.2 THz and 140 mW/cm², respectively. FL intensity was recorded after every 2 min wait period of *on/off* cycle.

To determine the reversibility of the quenching at different optical intensities, FL from the whey sample was recorded in 2 min intervals with the THz radiation alternating between *on* and *off* states. As shown in Figure 12.25, this process caused a loss of 60% FL over 20 cycles when this sample was excited with peak optical intensity of 9 MW/cm². In addition, this FL quenching was irreversible, as the FL did not recover even if the sample was allowed to relax for 24 h. This loss of FL is significantly larger than the ~6% quenching of FL observed over the same time period in the absence of THz radiation, resulting from photobleaching caused by high UV power. If the peak UV intensity was decreased to 2.5 MW/cm², the quenching reduced to 10% ± 0.5% and became reversible, with the initial FL recoverable over all 20 cycles. At intermediate power levels the FL recovered only partially, as shown in Figure 12.25.

12.3.1.2.2 Aqueous Protein Sample

The aqueous samples of whey were made in distilled water with a weight/volume concentration of 10%. A small volume of this solution was pipetted into a 100 micron demountable cuvette. Unlike the solid sample, the aqueous sample was transparent to UV radiation and FL emission was isotropic. Using the same technique as described earlier, FL intensity was investigated as a function of THz illumination time. Figure 12.26 illustrates the decay of FL as function of time of illumination. After correcting for the Fresnel losses at the front surface of the cuvette, THz intensity incident on the sample solution was 120 mW/cm². Both the FL emission and THz-induced FL quenching were isotropic as illustrated by Figure 12.26 (Inset).

12.3.1.2.3 Comparison of Different THz Frequencies

For comparison of different THz frequencies produced by gas laser, a setup described in Section 12.2.3 was used. Whey protein sample was excited using CW UV beam at 270 nm, and the THz gas laser produced different THz frequencies ranging from 1.08 to 2.55 THz. The gas laser source was later replaced by a Gunn-diode source to produce 0.2 THz radiation. The intensity of UV beam was kept constant at 1 mW/cm² for all THz frequencies. The dependence of FL quenching on THz intensity, for each THz frequency line, was studied by using a pair of wire grid polarizers to control the THz intensity incident on the samples. It was found that FL quenching depends linearly on the applied THz intensity for each

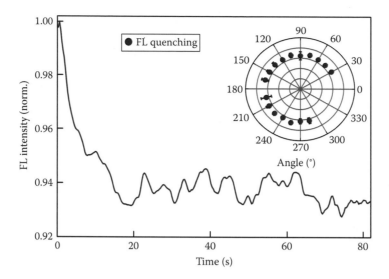

FIGURE 12.26 The normalized FL intensity as a function of time for aqueous whey protein (w/v conc. of 10%) excited using peak intensity of UV beam of 9 MW/cm². THz frequency and intensity were kept constant at 0.2 THz and 120 mW/cm², respectively. The THz source was turned on at $t = 0$ s and turned off at $t = 90$ s. Inset: isotropic pattern of THz-induced FL quenching.

frequency. Thereby, for studying the THz frequency dependence for THz-induced FL quenching, the intensity of all THz frequency lines was maintained constant at 140 mW/cm² (the maximum intensity for 0.2 THz source). It was observed that under the same THz intensity, the THz-induced FL quenching decreases as the THz frequency increases as shown in Table 12.2.

Major differences between THz-induced FL quenching in Trp and whey protein are as follows:

1. Trp amino acid does not exhibit FL quenching when illuminated with 0.2 THz radiation even though the emission from whey protein is due to Trp residues.
2. THz-induced FL quenching in whey protein decreases with increasing THz frequency, while the quenching in Trp amino acids correlates with THz absorption coefficient for Trp molecules.
3. Using the same optical excitation intensities, the THz-induced FL quenching in whey protein is irreversible for 0.2 THz radiation and becomes reversible for high THz frequencies. For 0.2 THz radiation, the FL quenching is reversible up to THz intensities of 80 mW/cm², whereafter it becomes irreversible.

Thereby, the nonthermal effects of THz radiation on whey proteins dominate at low frequencies.

TABLE 12.2 Frequency Dependence of THz-Induced FL Quenching in Whey Protein at a Fixed THz Intensity of 140 mW/cm²

Frequency (THz)	FL Quenching (%)
0.2	10 ± 0.4
1.08	2.5 ± 0.25
1.2	1.42 ± 0.25
1.4	0.8 ± 0.2
1.63	0.52 ± 0.15
1.89	0.38 ± 0.15
2.55	0.26 ± 0.15

12.3.2 Fluorescence Modulation in Green Fluorescent Protein

Using the same technique as described earlier for whey protein, FL modulation was also studied for GFP under the presence of THz radiation. GFP working stock samples with concentration of 5% (w/v) were made in phosphate-buffered saline. The sample was stored at 4°C before the experimental measurements were performed. A small volume (50 μL) of the aqueous samples was pipetted into a 0.1 mm path length demountable quartz cuvette (NSG Precision Cells).

An optical beam with a wavelength of 460 nm was isolated from the broadband light generated by a xenon-enhanced arc lamp source. A 0.2 THz radiation source was obtained from Gunn diode and higher frequencies were obtained from gas laser. The same experimental setup as described for whey protein was used. The intensity of all THz frequency lines incident on the sample, after incorporating the Fresnel losses at the interface of the front surface of the cuvette, was fixed at 120 mW/cm². The transmission coefficient of the quartz cuvette and the THz absorption coefficient of GFP in THz region were determined using THz TDS.

The spectrum of green fluorescent obtained using a miniature fiber-optic spectrometer (Ocean Optics) is shown in Figure 12.27.

The GFP sample exhibited an FL depletion of 3% ± 0.2% when exposed to 0.2 THz. However, GFP exhibited an FL enhancement of 5% ± 0.3% when exposed to an equal intensity of 2.55 THz radiation, as shown in Figure 12.28.

The rise and fall time of FL, defined as the time it takes for the system to equilibrate after the THz field perturbation is applied, was approximately 24 s for GFP. This corresponds to the folding and unfolding times reported in past studies (Makino et al. 1997).

The effect of high-frequency THz radiation (1.4–2.55 THz) on GFP is different from the low-frequency THz radiation (0.2–1.2 THz) (Table 12.3). Interestingly, either FL depletion or FL enhancement led to a net increase of 1% ± 0.2% in the FL of GFP after one *on/off* cycle. The increase saturated to 3% ± 0.7% after ten *on/off* cycles due to the competing process of photobleaching of GFP by a UV excitation beam (Table 12.3).

12.3.3 Investigation of Thermal Effects

Since FL from biological molecules is highly temperature dependent, we investigated the thermal effects of the THz wave on the whey protein samples. First, the protein FL was monitored at different sample

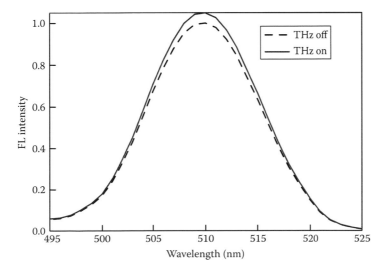

FIGURE 12.27 The FL spectrum of GFP with and without the presence of THz radiation obtained using an Ocean Optics spectrometer. The frequency of THz radiation was 2.55 THz and the FL of GFP was enhanced in the presence of THz radiation.

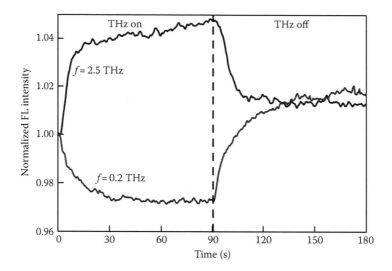

FIGURE 12.28 The normalized FL intensity as a function of time for GFP for 0.2 THz and 2.55 THz radiation with THz intensity of 120 mW/cm². The THz source was turned on at *t* = 0 s and turned off at *t* = 90 s.

TABLE 12.3 Frequency Dependence of THz-Induced FL Modulation in GFP at a Fixed THz Intensity of 120 mW/cm²

Frequency (THz)	FL Modulation (%)
0.2	−(3 ± 0.2)
1.08	−(2.5 ± 0.3)
1.2	−(1.0 ± 0.6)
1.4	0.5 ± 0.3
1.63	1.5 ± 0.5
1.89	3.5 ± 0.25
2.55	5 ± 0.3

Negative value of modulation refers to FL quenching and positive to FL enhancement.

temperatures, using a cryogenic system (MMR). These results were utilized to determine the amount of temperature change required to produce quenching levels similar to that observed in our earlier measurements. The temperature-dependent FL plot in Figure 12.29 indicates that a temperature change of 35°C is required to produce the 18% quenching observed in our experiments.

We then applied a THz wave to the sample to induce temperature changes, which were measured with a microbolometer camera and a thermocouple. The maximum increase in average temperature of the sample at the THz focus, as verified by the thermocouple and camera, was only ~1.6°C as shown in Figure 12.30. The heat diffuses away from the center to the edges of the sample, where the temperature change is only 0.4°C.

THz radiation also increased the temperature of the GFP sample by ~1°C. The FL of GFP decreases with increased temperature, but only by ~0.5% for every 1°C. This heating was insufficient to cause the observed FL quenching in the 0.2 THz data and fails to explain the FL enhancement observed with the 2.55 THz source.

The previously mentioned results clearly indicate that the quenching is mediated by nonthermal interactions of THz radiation with proteins. Another verification of the nonthermal nature of the interaction is that no quenching was observed for pure Trp amino acid at 0.2 THz even though FL at 320 nm for whey protein is due to its constituent Trp residues. This indicates that the presence

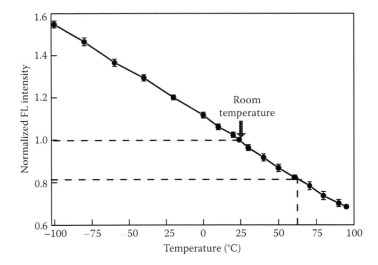

FIGURE 12.29 The FL intensity for whey protein as a function of temperature without any THz source. The FL intensity is normalized by the FL at room temperature (24°C).

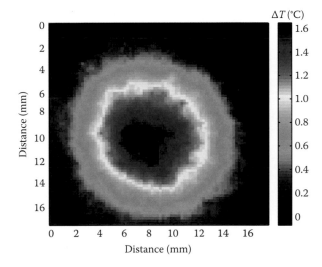

FIGURE 12.30 Difference of thermal images of the sample, with and without THz, obtained using a microbolometer camera. The maximum temperature rise of 1.6°C can be seen at focus, and it falls away from the center to the edges with change of 0.4°C at the edges of sample (13 mm diameter).

of structure or interactions with other residues in a protein might be contributing to strong FL quenching at 0.2 THz. Many proteins exhibit collective vibrations (Markelz et al. 2000; Plusquellic et al. 2007) within the THz frequency range, and we speculate that the frequency-dependent response in proteins could be due to resonance coupling of THz radiation with the intrinsic vibrational modes of the protein. Different THz frequencies may influence different modes of the protein. Such coupling may be leading to changes in the protein conformation mode and thereby influencing FL. THz absorbance exhibits a nonlinear response for solvated biomolecules where bulk water, hydration layer, and protein behave as three different components. The absorption by the protein molecules is a complex function of THz frequency and thereby it is difficult to decipher the actual mechanism of quenching. It is highly possible that molecules absorbing THz radiation undergo localized changes in their molecular conformations, which affect their intrinsic FL.

Previous studies have indicated that electrostatic interactions between the Trp ring electron density and the neighboring protein structures can affect the transition energy of Trp, which leads to differences in the FL quantum yield for different proteins (Alcala et al. 1987; Burstein et al. 1973; Vivian and Callis 2001). We believe that the externally applied THz field may lead to an uneven distribution of charges near the Trp ring and neighboring intermolecular bonds. This leads to increased probability of electron transfer and a decrease in FL. It is highly plausible that the excitation of neighboring non-Trp residues under high optical intensities may be responsible for irreversible quenching under high UV excitations.

12.4 Alternate Techniques for Detecting the Effects of Intense THz Radiation

In this section, pump–probe techniques other than FL that can be used to probe the effects of external THz radiation are discussed.

12.4.1 Time Domain Spectroscopy

The effect of CW THz radiation on whey protein sample in the absence of optical excitation was also investigated by using a compact THz time domain spectrometer (Mini-Z, Zomega Co.). Figure 12.31 illustrates the experimental setup employing this pump–probe technique.

The broadband THz pulses, from 0 to 2 THz, generated by the Mini-Z system were focused onto the sample, and the THz waveform reflected back from the sample surface was collected by the detection module of the spectrometer. CW THz radiation at a frequency of 2.55 THz and intensity of 5.5 W/cm² was focused on to the sample using a high-density polyethylene lens with focal length of 75 mm at an angle of 30° onto the same sample position.

The THz waveforms recorded after $t = 20$ and $t = 40$ s of externally applied radiation were compared to THz waveforms obtained in the absence of external THz radiation ($t = 0$ s). Figure 12.32 illustrates that the amplitude of the reflected THz signal decreased after the sample was irradiated with CW THz radiation. The signal returned back to its original value, as at $t = 0$, when the external CW THz radiation source was turned off.

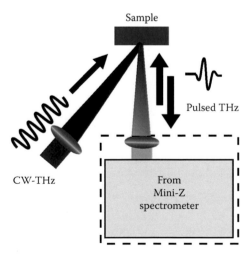

FIGURE 12.31   Schematic of the setup illustrating the THz pump–THz probe setup where the CW THz radiation from gas laser acts as a pump beam. The compact THz time domain spectrometer (Mini-Z, Zomega Co.) operating in reflection geometry detects the THz waveforms in the absence and presence of CW THz radiation.

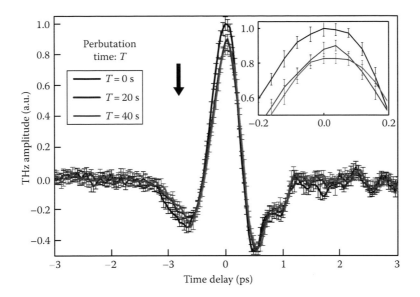

FIGURE 12.32 The THz waveforms obtained in reflection geometry using the commercial Mini-Z spectrometer. The external CW THz radiation (f = 2.5 THz and Intensity = 5.5 W/cm²) was turned on t = 0 s, and the waveforms were recorded at t = 0 s, t = 20 s, and t = 40 s. The amplitude of the reflected signal decreased as the interaction time of the external perturbation CW field increased. The signal recovered to its t = 0 s value when CW THz was turned off. Inset: zoomed in view of the waveform at t = 0 s, t = 20 s, and t = 40 s as a function of time delay from −0.2 to 0.2 s.

The spectra of the sample at different times were obtained by performing fast Fourier transform (FFT) on obtained reflected waveforms and are shown in Figure 12.33. This implies that the sample undergoes some change in the presence of external THz radiation that changes its THz absorption properties, manifesting itself in the reflected THz signal.

This demonstrates that THz TDS can be used as an alternate method for detecting the changes produced by CW radiation. With even faster acquisition speeds and higher signal-to-noise ratio, it may be possible to detect the changes happening on time scales 1 s or less with THz TDS.

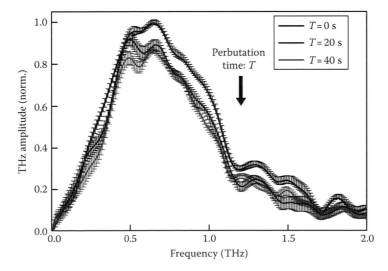

FIGURE 12.33 The THz spectra obtained by performing FFT on the obtained reflected waveforms. The changes in the spectra of reflected THz signal signify the change in the structure of protein molecules.

12.4.2 Absorption Spectroscopy

Absorption spectroscopy refers to a technique that investigates the absorption properties of the sample as a function of wavelength or frequency. As discussed in the first hypothesis for modeling THz induced FL quenching in Section 12.2.5.1, THz radiation can possibly move the Trp molecules to higher-lying vibrational states of the electronic ground state. This implies that the number of molecules absorbing UV radiation decreases under the presence of external THz radiation. Further, the change in conformation can also affect the absorption of molecules.

In order to monitor the changes in absorption, an aqueous solution of Trp was made in distilled water and the spectrum of the transmitted signal was obtained. The obtained spectrum indicated a 5% increase in the transmitted signal at 270 nm, which implies a decreased absorption of the 270 nm beam. A similar level of quenching of emitted FL signal at 320 nm by an aqueous Trp sample was observed.

The sample was studied in the aqueous form because the penetration depth of the UV beam was approximately 2 μm for the solid pellet sample. The properties of the molecules and their vibration spectra differ in the solid and liquid phase. Thereby, for fairer comparison, the FL (320 nm) signal and the reflected UV signal (270 nm) from the solid pellet sample should be compared.

12.4.3 Circular Dichroism

Circular dichroism (CD) spectroscopy refers to a technique that investigates the differences in the absorption of left-handed circularly polarized versus right-handed circularly polarized light. The non-zero CD signal arises due to asymmetry of the structure. CD spectroscopy is very sensitive to changes in the secondary or tertiary structure of proteins. Secondary structure of the protein can be determined by CD spectroscopy in the far-UV region (190–250 nm) of the electromagnetic spectrum. In a recent study by Kim et al. on the protein–water dynamics during protein folding (Kim et al. 2008), the changes in transmitted THz signal upon protein folding were monitored. This time scale was compared to time scales obtained with Trp FL, small-angle x-ray scattering, and CD. It was demonstrated that while changes in Trp FL upon protein folding happen on a time scale of seconds, the CD signal and THz signal provide information of the changes in protein dynamics on a millisecond time scale. Thereby, CD may serve as an alternate or supplement technique to FL to investigate the effects of external THz radiation.

Other optical techniques that can be potentially used to monitor the effects of external THz radiation are NMR, Raman spectroscopy, and powder x-ray diffraction.

12.5 Discussion

This chapter focused on the resonant FL quenching of Trp molecules induced by CW THz radiation. This study demonstrates that resonant interaction of CW THz radiation with Trp leads to reversible depletion of FL of Trp molecules. The reversibility suggests that THz radiation was not damaging or decomposing the sample. The effect of THz source power and frequency on the FL depletion confirmed that quenching was due to resonant coupling of THz radiation with molecules. Control experiments with pure heating and no external THz radiation confirm that changes induced by THz radiation are nonthermal in nature. We speculate that FL quenching may be due to molecular rearrangements arising from resonant coupling of THz radiation with molecules. Thereby, FL from Trp molecules can be used as a sensitive probe to understand the interactions of THz radiation with biomolecules. This method can also be used to study the interactions between molecules induced by THz radiation in a *label-free regime* using the intrinsic FL from Trp molecules. The concept of THz-induced FL modulation was also applied to whey protein and GFP. Whey protein exhibited FL quenching in the presence of THz radiation for all investigated THz frequencies (0.2–2.55 THz). While the FL quenching in whey protein was reversible for low UV excitation intensities, it became irreversible under high UV intensities. GFP exhibited FL quenching for low THz frequencies (0.2, 1.08, and 1.2 THz) and FL enhancement under THz irradiation for high THz frequencies (1.4, 1.63, 1.89, and 2.55 THz). The possible mechanism suggested for the frequency-dependent response is the vibrational resonant coupling between THz radiation and protein. A seminal paper by Fröhlich in 1968 suggests such resonant couplings between electromagnetic radiation and biomolecules. The exact mechanism behind the FL modulation is difficult to determine, but the frequency-dependent response and temperature controlled measurements of FL modulation are very strong indications that mechanism is not mediated through thermal response. This work might have potential applications towards *ground-state-depleted FL microscopy* where modulation in the FL of a specific illuminated region is required (Hell and Kroug 1995). The FL enhancement caused by THz radiation in GFP may also become useful in the field of FL microscopy, where GFP is widely used as fluorescent tag to study live cells or labeling as a reporter gene (Kain et al. 1995; White and Stelzer 1999).

References

Albinsson, B., M. Kubista, B. Norden, and E. W. Thulstrup. 1989. Near-ultraviolet electronic transitions of the tryptophan chromophore: Linear dichroism, fluorescence anisotropy, and magnetic circular dichroism spectra of some indole derivatives. *The Journal of Physical Chemistry* 93: 6646–6654.

Albinsson, B. and B. Norden. 1992. Excited-state properties of the indole chromophore: Electronic transition moment directions from linear dichroism measurements: Effect of methyl and methoxy substituents. *The Journal of Physical Chemistry* 96: 6204–6212.

Alcala, J. R., E. Gratton, and F. G. Prendergast. 1987. Fluorescence lifetime distributions in proteins. *Biophysical Journal* 51: 597–604.

Attallah, N. A. and G. F. Lata. 1968. Steroid-protein interactions studied by fluorescence quenching. *Biochimica et Biophysica Acta (BBA)-Protein Structure* 168: 321–333.

Bartel, T., P. Gaal, K. Reimann, M. Woerner, and T. Elsaesser. 2005. Generation of single-cycle THz transients with high electric-field amplitudes. *Optics Letters* 30: 2805–2807.

Burstein, E. A., N. S. Vedenkina, and M. N. Ivkova. 1973. Fluorescence and the location of tryptophan residues in protein molecules. *Photochemistry and Photobiology* 18: 263–279.

Callis, P. R. 1991. Molecular orbital theory of the 1Lb and 1La states of indole. *The Journal of Chemical Physics* 95: 4230–4240.

Callis, P. R. and B. K. Burgess. 1997. Tryptophan fluorescence shifts in proteins from hybrid simulations: An electrostatic approach. *The Journal of Physical Chemistry B* 101: 9429–9432.

Cook, D. J. and R. M. Hochstrasser. 2000. Intense terahertz pulses by four-wave rectification in air. *Optics Letters* 25: 1210–1212.

Creed, D. 1984. The photophysics and photochemistry of the near-uv absorbing amino acids–i. Tryptophan and its simple derivatives. *Photochemistry and Photobiology* 39: 537–562.

Demchenko, A. P. 1988. *Ultraviolet Spectroscopy of Proteins*. New York: Springer-Verlag.

Dlott, D. D. and M. D. Fayer. 1991. Applications of infrared free-electron lasers: Basic research on the dynamics of molecular systems. *IEEE Journal of Quantum Electronics* 27: 2697–2713.

Eftink, M. R., L. A. Selvidge, P. R. Callis, and A. A. Rehms. 1990. Photophysics of indole derivatives: Experimental resolution of La and Lb transitions and comparison with theory. *Journal of Physical Chemistry* 94: 3469–3479.

Eisinger, J. and J. Flores. 1979. Front-face fluorometry of liquid samples. *Analytical Biochemistry* 94: 15–21.

Foster, K. R., B. R. Epstein, and M. A. Gealt. 1987. "Resonances" in the dielectric absorption of DNA? *Biophysical Journal* 52: 421–425.

Fröhlich, H. 1975. The extraordinary dielectric properties of biological materials and the action of enzymes. *Proceedings of the National Academy of Sciences* 72: 4211–4215.

Gabriel, C., E. H. Grant, R. Tata et al. 1987. Microwave absorption in aqueous solutions of DNA. *Nature* 328: 145–146.

Hell, S. W. and M. Kroug. 1995. Ground-state-depletion fluorescence microscopy: A concept for breaking the diffraction resolution limit. *Applied Physics B* 60: 495–497.

Heyduk, T. 2002. Measuring protein conformational changes by FRET/LRET. *Current Opinion in Biotechnology* 13: 292–296.

Kain, S. R., M. Adams, A. Kondepudi et al. 1995. Green fluorescent protein as a reporter of gene expression and protein localization. *Biotechniques* 19: 650.

Kasha, M. 1950. Characterization of electronic transitions in complex molecules. *Discussions of the Faraday Society* 9: 14–19.

Kim, S. J., B. Born, M. Havenith, and M. Gruebele. 2008. Real-time detection of protein–water dynamics upon protein folding by terahertz absorption spectroscopy. *Angewandte Chemie International Edition* 47: 6486–6489.

Kinsella, J. E. and D. M. Whitehead. 1989. Proteins in whey: Chemical, physical, and functional properties. *Advances in Food and Nutrition Research* 33: 437–438.

Kuster, N., J. Schuderer, A. Christ, P. Futter, and S. Ebert. 2004. Guidance for exposure design of human studies addressing health risk evaluations of mobile phones. *Bioelectromagnetics* 25: 524–529.

Lakowicz, J. R. 1999. *Principles of Fluorescence Spectroscopy*. New York: Plenum Press.

Le Floc'h, N., W. Otten, and E. Merlot. 2011. Tryptophan metabolism, from nutrition to potential therapeutic applications. *Amino Acids* 41: 1195–1205.

Levitt, J. A., D. R. Matthews, S. M. Ameer-Beg, and K. Suhling. 2009. Fluorescence lifetime and polarization-resolved imaging in cell biology. *Current Opinion in Biotechnology* 20: 28–36.

Loew, G. H. and D. L. Harris. 2000. Role of the heme active site and protein environment in structure, spectra, and function of the cytochrome P450s. *Chemical Reviews* 100: 407–420.

Maier, M., C. Blakemore, and M. Koivisto. 2000. The health hazards of mobile phones. *British Medical Journal* 320: 1288–1289.

Makino, Y., K. Amada, H. Taguchi, and M. Yoshida. 1997. Chaperonin-mediated folding of green fluorescent protein. *Journal of Biological Chemistry* 272: 12468–12474.

Markelz, A. G., A. Roitberg, and E. J. Heilweil. 2000. Pulsed terahertz spectroscopy of DNA, bovine serum albumin and collagen between 0.1 and 2.0 THz. *Chemical Physics Letters* 320: 42–48.

Martin, M. M. and L. Lindqvist. 1975. The pH dependence of fluorescein fluorescence. *Journal of Luminescence* 10: 381–390.

Misteli, T. and D. L. Spector. 1997. Applications of the green fluorescent protein in cell biology and biotechnology. *Nature Biotechnology* 15: 961–964.

Ohmae, E., T. Kurumiya, S. Makino, and K. Gekko. 1996. Acid and thermal unfolding of *Escherichia coli* dihydrofolate reductase. *Journal of Biochemistry* 120: 946–953.

Pace, C. N., F. Vajdos, L. Fee, G. Grimsley, and T. Gray. 1995. How to measure and predict the molar absorption coefficient of a protein. *Protein Science* 4: 2411–2423.

Petrich, J. W., M. C. Chang, D. B. McDonald, and G. R. Fleming. 1983. On the origin of nonexponential fluorescence decay in tryptophan and its derivatives. *Journal of the American Chemical Society* 105: 3824–3832.

Plusquellic, D. F., K. Siegrist, E. J. Heilweil, and O. Esenturk. 2007. Applications of terahertz spectroscopy in biosystems. *Chemphyschem: A European Journal of Chemical Physics and Physical Chemistry* 8: 2412–2431.

Repacholi, M. H. 2001. Health risks from the use of mobile phones. *Toxicology Letters* 120: 323–331.

Shimomura, O., F. H. Johnson, and Y. Saiga. 1962. Extraction, purification and properties of aequorin, a bioluminescent protein from the luminous hydromedusan, *Aequorea*. *Journal of Cellular and Comparative Physiology* 59: 223–239.

Song, P.-S. and W. E. Kurtin. 1969. Photochemistry of the model phototropic system involving flavines and indoles. III. A spectroscopic study of the polarized luminescence of indoles. *Journal of the American Chemical Society* 91: 4892–4906.

Torchia, D. A. 1984. Solid state NMR studies of protein internal dynamics. *Annual Review of Biophysics and Bioengineering* 13: 125–144.

Tsien, R. Y. 1998. The green fluorescent protein. *Annual Review of Biochemistry* 67: 509–544.

Udenfriend, S. 1971. *Fluorescence Assay in Biology and Medicine*. New York: Academic Press.

Vivian, J. T. and P. R. Callis. 2001. Mechanisms of tryptophan fluorescence shifts in proteins. *Biophysical Journal* 80: 2093–2109.

Waider, J., N. Araragi, L. Gutknecht, and K.-P. Lesch. 2011. Tryptophan hydroxylase-2 (TPH2) in disorders of cognitive control and emotion regulation: A perspective. *Psychoneuroendocrinology* 36: 393–405.

Wallach, D. F. and P. H. Zahler. 1966. Protein conformations in cellular membranes. *Proceedings of the National Academy of Sciences of the United States of America* 56: 1552.

White, J. and E. Stelzer. 1999. Photobleaching GFP reveals protein dynamics inside live cells. *Trends in Cell Biology* 9: 61–65.

Xie, A., A. F. G. van der Meer, and R. H. Austin. 2001. Excited-state lifetimes of far-infrared collective modes in proteins. *Physical Review Letters* 88: 018102.

Xie, X., J. Dai, and X.-C. Zhang. 2006. Coherent control of THz wave generation in ambient air. *Physical Review Letters* 96: 075005.

Yamamoto, Y. and J. Tanaka. 1972. Polarized absorption spectra of crystals of indole and its related compounds. *Bulletin of the Chemical Society of Japan* 45: 1362–1366.

Yang, L., Y. Xu, Y. Su et al. 2005. FT-IR spectroscopic study on the variations of molecular structures of some carboxyl acids induced by free electron laser. *Spectrochimica Acta Part A: Molecular and Biomolecular Spectroscopy* 62: 1209–1215.

Yang, L., Y. Xu, Y. Su et al. 2007. Study on the variations of molecular structures of some biomolecules induced by free electron laser using FTIR spectroscopy. *Nuclear Instruments and Methods in Physics Research Section B: Beam Interactions with Materials and Atoms* 258: 362–368.

13

Biological Effects of Broadband Terahertz Pulses

Lyubov V. Titova
University of Alberta

Frank A. Hegmann
University of Alberta

Olga Kovalchuk
University of Lethbridge

13.1 Introduction

Recent progress in terahertz (THz) technology has resulted in the development of broadband, ultrafast laser-based THz pulse sources, as well as coherent, phase-sensitive THz detectors (Baxter and Guglietta 2011; Jepsen et al. 2011; Lee 2009; Mittleman 2013; Tonouchi 2007), and has spurred numerous applications of broadband, picosecond-duration THz pulses in medicine and biology. In particular, novel biomedical imaging modalities based on broadband THz pulses are showing much promise for improved noninvasive diagnosis of cancer (Ashworth et al. 2009; Fitzgerald et al. 2002; Woodward et al. 2003; Yu et al. 2012), assessment of burns (Arbab et al. 2011), and intraoperative tumor margin identification (Ashworth et al. 2008). THz radiation is nonionizing, is sensitive to cellular water content, suffers from significantly less scattering in tissue than visible light, and can provide submillimeter imaging resolution (Siegel 2004; Zhang 2002). Imaging with broadband THz pulses, termed THz pulsed imaging, or TPI, offers important benefits over continuous-wave, monochromatic THz imaging approaches (Pickwell-MacPherson and Wallace 2009). In addition to yielding structural images, it permits the collection of spectroscopic information within a broad spectral range, usually 0.1–3 THz, which in turn enables detection of variation in water content, ion concentrations, and other subtle differences in the composition of biological samples (Masson et al. 2006; Zhang 2002). Many important biomolecules have unique conformation-state-dependent spectral fingerprints in the THz range (Cherkasova et al. 2009; Falconer and Markelz 2012; Fischer et al. 2002; Kim et al. 2008; Markelz et al. 2002), and much research is currently dedicated to identifying intrinsic THz spectroscopic biomarkers for label-free, noninvasive detection of various types of cancer and other diseases (Joseph et al. 2009; Yu et al. 2012). Exploiting the short duration of THz pulses and applying time-domain analysis allows information from surfaces as well as subsurface layers of the imaged tissue to be obtained with depth resolution reaching 40 µm in skin (Woodward et al. 2003). Finally, implementing near-field TPI combines advantages offered by THz time-domain techniques with high spatial resolution and thus shows promise for novel noninvasive, accurate diagnostic tools as well as high-throughput, label-free pathological analysis of excised tissues (Federici et al. 2002).

This remarkable progress in the development of medical TPI applications underscores the need to carefully examine the influence of pulsed THz radiation on human tissue in order to establish clear safety standards for the future implementation of these novel techniques in a clinical setting. However, our current understanding of the biological effects of THz radiation, and particularly of broadband THz pulses, such as their genotoxicity and effects on cell activity and cell integrity, is still limited.

With typical photon energies in the range from 0.5 to 15 meV, THz radiation is nonionizing, and therefore the mechanisms by which it interacts with cells and tissues are fundamentally different from those involved in interactions of living matter with high-energy ionizing radiation (i.e., UVB, UVC, x-rays, gamma rays). While the energy of THz photons is too low to break chemical bonds, several theoretical works and experiments have shown that many of the intrinsic resonances in conformation-state-dependent vibrational spectra of cellular biomolecules, such as amino acids, proteins, and DNA, occur in the THz frequency range (Cherkasova et al. 2009; Chitanvis 2006; Fischer et al. 2002; Kim et al. 2008; Markelz et al. 2002; Prohofsky et al. 1979). Simulation studies have suggested that resonance-type linear and nonlinear interactions of THz electromagnetic fields with DNA may significantly alter DNA dynamics and, under certain conditions, even induce localized openings (bubbles) in the DNA strands by coupling to breathing vibrational modes of the hydrogen bonds between two complementary DNA strands (Alexandrov et al. 2010; Bergues-Pupo et al. 2013; Chitanvis 2006; Maniadis et al. 2011). THz radiation can also couple to the vibrations of hydrogen bonds that form between the side chains of cellular proteins and the water molecules of the hydration shell and thus can affect the dynamic relaxation properties of proteins, which in turn may impact cellular functions (Born et al. 2009; Kim et al. 2008; Pal and Zewail 2004).

The growing theoretical and experimental evidence suggests that THz radiation may indeed influence biological systems and functions. In particular, THz pulses of picosecond duration are most likely to elicit molecular and cellular responses through resonant-type, nonlinear interactions as the effectiveness of THz radiation coupling to vibrational modes of hydrogen bond networks increases with the peak THz power. Whereas the average power of picosecond THz pulse beams is usually quite low (in the µW or mW range), the peak powers can be as high as 1 MW, sufficient to penetrate into the interior of cells and cell nuclei even in the highly attenuating liquid environment (Xu et al. 2006).

Unlike continuous-wave THz radiation, intense THz pulses are also characterized by extremely high peak electric fields that can reach MV/cm (Blanchard et al. 2011; Hirori et al. 2011; Hoffmann and Fülöp 2011; Junginger et al. 2010). With rise times on femtosecond time scales, many orders of magnitude faster than typical charging times of cellular and nuclear membranes of most mammalian cells (~100 ns), THz pulse electric fields pass through the membrane into the cytoplasm and inside the cell nucleus and can affect biological function (Schoenbach et al. 2008). Moreover, it has been suggested that coherent THz modes play an important role in biological self-organization, and therefore exploring the interactions between THz radiation (especially intense THz pulses) and biological systems may provide a new avenue for studying biological systems (Weightman 2012).

The majority of experimental investigations of the cellular and molecular effects of THz radiation focused almost exclusively on the effects of continuous-wave, monochromatic THz radiation (Hintzsche et al. 2011; Wilmink and Grundt 2011; Wilmink et al. 2010a,b). Whereas some studies using continuous-wave THz sources did not uncover any biological effects, other studies have revealed, under specific exposure conditions, changes in cell membrane permeability and induction of genotoxicity in human lymphocytes (Korenstein-Ilan et al. 2009) as well as spindle disturbances in human–hamster hybrid cells (Hintzsche et al. 2011). Exposing Jurkat cells and human dermal fibroblasts to continuous-wave radiation at 2.52 THz was also found to trigger dramatic upregulation of cellular stress response and inflammation genes and, after prolonged exposure (30–40 min), cell death (Wilmink et al. 2010a). Importantly, in most of these continuous-wave THz exposure experiments, cellular effects were attributed to heat shock. Heating due to strong absorption by water in biological materials may be significant in the case of continuous-wave THz excitation (Kristensen et al. 2010). For this reason, until recently, exposure risks associated with THz radiation were evaluated based on simulations of thermal effects due to absorption of THz radiation by water (Berry et al. 2003; Fitzgerald et al. 2002; Kristensen et al. 2010; Wilmink and Grundt 2011).

However, average thermal effects are minimal for picosecond-duration THz pulses with low repetition rates and average powers (Kristensen et al. 2010). Recent experiments have conclusively demonstrated that intense THz pulses can elicit cellular and molecular changes in exposed cells and tissues in the absence of thermal changes, as confirmed by the unchanged levels of temperature-change-sensitive genes and proteins (Alexandrov et al. 2011, 2013; Kim et al. 2013; Titova et al. 2013b).

13.2 In Vitro Studies of Biological Effects of THz Pulses on Cells

The first experiments aimed at uncovering the biological effects of broadband THz pulses were carried out using in vitro cell culture models (Alexandrov et al. 2011, 2013; Bock et al. 2010; Clothier and Bourne 2003; Williams et al. 2013). Cell culture refers to cells removed from tissue and proliferated under appropriate conditions in a medium that supplies the essential nutrients. The major advantage of using in vitro cell culture is the consistency and reproducibility of experiments performed on nearly identical cells, preferably derived from the same clonal population. In order to clearly distinguish the effects of exposure to broadband THz pulses on cells in a culture from spurious effects such as thermal shock or desiccation, it is crucial to maintain the cells in the appropriate physiological conditions, such as temperature, medium pH, and gas (O_2, CO_2) composition, for the entire duration of the THz exposure (Williams et al. 2013).

13.2.1 Influence of THz Pulses on Human Epithelial and Stem Cells

Due to the strong absorption of THz radiation by the hydrogen bonding network of water in biological tissue, the most promising medical uses of pulsed THz imaging are in diagnosis of skin cancer, assessment of skin burns, and sensing of corneal hydration. Therefore, exploring the biological effects of broadband THz pulses on human skin and ocular cells and tissues is not only interesting from the standpoint of understanding the nonthermal effects of interactions between ultrashort pulses of electromagnetic radiation and living matter but is also critical for establishing acceptable exposure parameters for biomedical pulsed THz imaging applications.

In 2003, Clothier and Bourne (Clothier et al. 2003) studied the impact of prolonged exposure to picoseconds-duration THz pulses on the differentiation, activity, and viability of primary human keratinocytes. Keratinocytes account for over 90% of the cells in the epidermis, the outermost layer of the skin. Through a programmed process of differentiation, during which keratinocytes migrate to the surface, stop dividing, and undergo morphological changes, they synthesize major structural components of the epidermal barrier that protects against pathogens, heat, UV radiation, and water loss (Eckert and Rorke 1989; Eckert et al. 2013). As significant disruptions of the differentiation process may result in deleterious consequences such as inflammatory skin diseases and skin cancer (Eckert et al. 2013), evaluation of the effects of broadband THz pulses on keratinocyte differentiation is important. Triggering differentiation of the primary keratinocytes in vitro is achieved via activation of transglutaminases by increasing the culture medium calcium concentration above 0.1 mM. For their experiments, Clothier and Bourne used primary keratinocytes isolated from skin of several donors and cultured for 48 h in keratinocyte growth medium with 0.06 mM calcium to form a confluent layer and exposed them to picosecond-duration, 0.2–3.0 THz bandwidth THz pulses through the base of the culture plates. Based on the experimental details provided in the paper, the estimated energy density per pulse was ~3 pJ/cm^2 (3×10^{-6} μJ/cm^2), while the total exposure did not exceed 0.45 mJ/cm^2 over 30 min. The standard resazurin reduction assay was used to assess the impact of THz exposure on cell viability, while monitoring the changes in fluorescein cadaverine uptake was used to detect changes in differentiation capacity. The viability and differentiation capacity of the exposed cells were analyzed immediately after exposure, as well as 3, 6, and 8 days after that. No discernible effects were reported on the differentiation, activity, or viability of primary human keratinocytes in vitro at any time after the exposure, compared with the unexposed controls.

Recently, Williams et al. (2013) explored the influence of significantly more intense THz pulses produced by the ALICE (Daresbury Laboratory, United Kingdom) synchrotron source on two types of human ocular cell lines, corneal epithelial cells and retinal pigment epithelial cells, as well as on human embryonic stem cells in vitro. ALICE is an intense source of broadband THz radiation, delivering picosecond-duration pulses with a repetition rate of 41 MHz in ten 100 μs duration pulse trains per second. The energy density per pulse was varied from 3 to 10 nJ/cm^2 (0.003–0.01 μJ/cm^2) with exposure times ranging from 2 to 6 h, so that the total exposure was in the 0.15–5.4 J/cm^2 range.

The morphology of the exposed epithelial cells was analyzed using live cell phase contrast microscopy, and no morphological changes were observed for the retinal pigment epithelial cells following 3 h exposure to intense THz pulses, as the morphology of the exposed cells was identical to that of the unexposed controls. Likewise, it was demonstrated that the THz radiation did not affect the proliferation of the retinal pigment or corneal epithelial cells exposed to either a single exposure over various times or repeated 3 h long exposures over several days. In addition to epithelial cells, human embryonic stem cells were also exposed to the intense THz pulses. While embryonic stem cells would not be exposed to THz pulses in the proposed medical imaging applications, they are highly responsive to physical and chemical environmental stimuli and therefore provide a uniquely sensitive test bed for assessing the possible cellular effects of intense THz pulses in vitro. Using the ALICE source, Williams et al. found that intense THz pulse exposure did not affect the attachment, morphology, proliferation, and differentiation of the human embryonic stem cells. The authors concluded that cells maintained under ideal conditions (at 37°C in a cell culture CO_2 incubator) are capable of compensating for any effects caused by exposure to THz radiation with the peak power levels employed in these studies. However, this study did not explore changes in protein and gene expression in the exposed cells.

13.2.2 Effects of Intense Broadband THz Pulses on Gene Expression in Mammalian Stem Cells

As mentioned earlier, stem cells are highly susceptible to even very small changes in their environment and thus are uniquely suited for studying the cellular effects of various exogenous stimuli. As discussed in the previous section, Williams et al. found that the morphology of human embryonic stem cells appeared unaffected by prolonged exposure to THz pulses with pulse energy densities up 0.01 μJ/cm^2 (Williams et al. 2013). In a different series of experiments, Bock et al. and Alexandrov et al. exposed mouse mesenchymal stem cells (MSCs) to broadband THz pulses with pulse energy densities around 1 μJ/cm^2 and demonstrated for the first time broadband THz-pulse-induced gene expression changes in mammalian cells (Alexandrov et al. 2011, 2013; Bock et al. 2010). In these experiments, THz pulses were generated by frequency mixing the fundamental and second-harmonic laser fields at 800 nm and 400 nm, respectively, in pressurized argon, as illustrated in Figure 13.1. MSCs in a petri dish were exposed to broadband (~1–15 THz, centered at 10 THz, as also shown in Figure 13.1a), subpicosecond THz pulses. Control mMSC culture was placed next to the irradiated sample (screened from the THz radiation), and the temperature of the cultures was kept at 26°C–27°C for both the control and irradiated sample (Figure 13.1b). Differential changes in gene expression between the THz-pulse-exposed samples and their respective controls were analyzed using GeneChip microarrays immediately after the

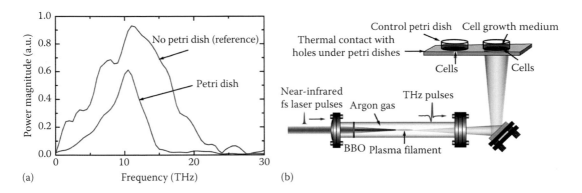

FIGURE 13.1 Exposure of MSCs to intense broadband THz pulses. (a) Spectrum of the THz pulse measured with and without a petri dish. (b) Schematic diagram of the experimental setup displaying the irradiated and control MSC cultures. Both the irradiated sample and its control were held at the same temperature. (Reprinted from Bock, J. et al., *PLoS ONE*, 5, e15806, 2010. With permission.)

exposure. Importantly, since each differentiation stage of the MSCs is characterized by the distinct gene expression pattern, the mouse MSC cultures were pretreated with culture medium inducing adipose phenotype and synchronized to be at the same differentiation time point immediately before exposure to THz pulses. These studies showed that prolonged exposure to broadband THz pulses accelerates differentiation of MSCs toward the adipose phenotype, as the cells exposed for 9 or 12 h exhibited gene expression patterns associated with adipocytes (fat cells) (Alexandrov et al. 2013; Bock et al. 2010). Most notably, reverse transcription polymerase chain reaction (RT-PCR) analysis demonstrated that the genes PPARG, adiponectin, GLUT4, and FABP4, which are transcriptionally active in differentiated adipocytes and not in pluripotent stem cells, were upregulated in MSCs exposed for 12 h (Alexandrov et al. 2013; Figure 13.2) or 9 h (Bock et al. 2010), compared to the unexposed control cells. At the same

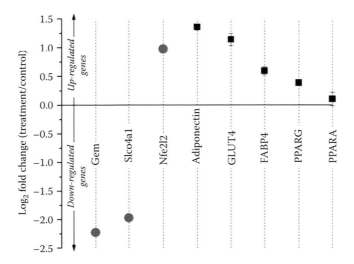

FIGURE 13.2 Selected genes differentially expressed following 12 h exposure of mouse MSCs to broadband THz pulses. Log_2 fold changes for Gem, Slco4a1, and Nfe2l2 (circles) were determined using microarray analysis, and the corresponding p-values are 6.91×10^{-19}, 4.47×10^{-15}, and 1.03×10^{-04}, respectively (Alexandrov et al. 2013). Log_2 fold changes for the other genes (squares) were calculated using relative RNA levels determined by RT-PCR measurements (Alexandrov et al. 2013). Error bars indicate corresponding standard deviations and in some cases are smaller than the symbol size. (Courtesy of A. Usheva.)

time, genes that encode for heat shock, as well as stress response proteins such as HSP90, HSP105, and others, were unaffected, underscoring the nonthermal nature of cellular changes induced by exposure to broadband, subpicosecond THz pulses (Alexandrov et al. 2013).

In addition to the genes PPARG, adiponectin, GLUT4, and FABP4, identified in RT-PCR measurements as upregulated by broadband THz pulses, 20 differentially expressed genes were discovered by microarray analysis of the cells exposed for 12 h. In particular, the downregulation of Gem and Slco4a1 and the upregulation of Nfe2l2 are most interesting (Figure 13.2). Both Gem and Slco4a1 are typically transcriptionally active in pluripotent stem cells and become silenced as cell differentiation toward the adipose phenotype progresses. On the other hand, Nfe2l2 is involved in adipogenesis and is overexpressed in adipose tissue. These three differentially expressed genes, in addition to PPARG, adiponectin, GLUT4, and FABP4, suggest that prolonged exposure to intense, broadband THz pulses accelerates differentiation of mouse MSCs. Interestingly, the impact of broadband THz pulse exposure was also found to depend on the level of stem cell differentiation (Alexandrov et al. 2013; Bock et al. 2010). While the changes in the gene expression showing accelerated differentiation toward adipose phenotype were observed in cells exposed either 48 h or 120 h following the treatment by the differentiation medium, only MSCs that were further along in their differentiation program exhibited THz-pulse-induced morphological changes typical of mature adipocytes, such as clear lipid droplet-like inclusions in the cellular cytoplasm.

The effects of broadband THz pulse exposure also depend on the exposure parameters, as can be seen by comparing gene expression levels in mouse MSCs exposed for 2 h to those exposed for 12 h, both 48 h after initiation of differentiation. Exposure for 2 h did not affect the expression of PPARG, adiponectin, GLUT4, and FABP4, and upregulated Slco4a1. However, Slco4a1 was downregulated in the 12 h exposure. During the 2 h exposure, cells are still pluripotent, while 12 h exposures are sufficiently long for substantial progress in differentiation to occur. In both cases, THz pulse exposure affected the expression of transcriptionally active genes at a given differentiation stage of stem cells (Alexandrov et al. 2013).

Alexandrov et al. suggested that the effects of THz pulses on gene expression are only catalytic and might be determined by the availability of transcription factors (suppressors or enhancers), the position of the dynamically active promoters in the chromatin structure, as well as the nature of the active transcription pathways (Alexandrov et al. 2013). It is possible that broadband THz pulses can induce conformational changes of transcription factors, which are proteins that control gene expression activity (Cherkasova et al. 2009). Conformation changes would affect the binding of transcription factors to DNA and would either favor transcription or interfere with it. Another possible mechanism for the catalytic action of THz radiation may be driven by nonlinear resonant coupling of the THz field to breathing modes of DNA in the regions where the core promoters of the affected genes are located (Alexandrov et al. 2010, 2013; Bock et al. 2010). The core promoter is the region on the DNA strand that is located upstream from the gene and directs the accurate initiation of gene transcription by RNA polymerase II (Smale and Kadonaga 2003). After binding to the core promoter region of the DNA, the RNA polymerase initiates gene transcription through a series of reactions that result, among other changes, in the separation of the DNA strands to form a transcription bubble, which is an unwound section of DNA approximately 13 base pairs in length. This localized DNA opening is necessary to facilitate access to the transcribed gene, and transient, thermally induced strand separation motions of the double-stranded DNA, known as DNA breathing, play an important role in its formation and, therefore, in transcription initiation (Alexandrov et al. 2009, 2012). Using extended Peyrard–Bishop–Dauxois (EPBD) Langevin molecular dynamics simulations, Alexandrov et al. and Bock et al. studied intrinsic double-stranded DNA breathing dynamics in the core promoter regions of several genes. They found substantial variability in their propensity for local breathing, characterized by the following three dynamic criteria: length, amplitude, and lifetime of the bubble (Alexandrov et al. 2013; Bock et al. 2010). While the core promoter regions of some genes demonstrate the dynamic pattern characterized by long-lived and large-amplitude bubbles (i.e., >5 ps lifetime and >1.5 Å amplitude over ~10 base pairs) at the transcription start sites, the dynamic activity in the promoter regions of other genes is almost evenly distributed and noticeably weaker (Alexandrov et al. 2009). The effect of broadband THz pulse exposure on the expression of particular genes appears to be correlated with the local breathing dynamics of their core promoter regions. The upregulated genes, including PPARG, Gem, and Slco4a1, are found to have large transient bubbles at the transcription start sites in their core promoter regions, while the promoter region for PPARA, which belongs to the same group of nuclear receptor proteins as PPARG but is unaffected by the THz pulses (Figure 13.2), is characterized by a low propensity for local breathing. These relationships between the susceptibility of certain genes to the THz-pulse-induced transcription changes and the DNA breathing dynamics of their core promoters suggest that THz radiation can couple to vibrations of hydrogen bonds between the DNA strands, enhance existing transcription bubbles, and even create new bubbles, thus affecting the gene transcription.

13.3 Biological Effects of Intense Picosecond THz Pulses on Human Skin Tissue

In vitro cell culture models provide an important testing ground for exploring the cellular and molecular effects of exogenous stimuli such as broadband THz pulses. The most important advantage of in vitro experiments is that by limiting the investigation to the individual cells, they greatly simplify the complex task of identifying the biological effects of THz pulses on living organisms. Furthermore, the use of clonal cell lines provides for consistency and reproducibility in experimental results. As discussed in the previous section, in vitro studies yielded the first experimental demonstration of broadband THz-pulse-induced changes in gene expression in mammalian stem cells (Alexandrov et al. 2013; Bock et al. 2010). However, extrapolating the conclusions of in vitro experiments using cell cultures to living organisms in their intact state can be challenging. On the other hand, in vivo studies with animal models, which are presumed to be the gold standard for assessing the biological effects of various environmental stimuli, suffer from their own set of drawbacks such as ethical considerations, difficulty in interpretation of results due to the complexity of the model, and significant variability between the test subjects. Three-dimensional tissue models that recreate the structural architecture as well as the cell–cell and cell–extracellular matrix interactions provide a practical alternative to both in vitro and in vivo experiments.

In a series of recent studies, Titova et al. analyzed the effects of intense, broadband, picosecond-duration THz pulses on DNA integrity and global gene expression in a 3D human skin tissue model (Titova et al. 2013a–c). For these experiments, THz pulses with 1 kHz repetition rate and pulse energy variable up to 1 µJ were generated by optical rectification of tilted-pulse-front 800 nm pulses from an amplified Ti/sapphire laser source in LiNbO$_3$ (Blanchard et al. 2011; Hoffmann and Fülöp 2011), as shown in Figure 13.3a and as discussed in Chapter 3. The THz pulse waveform (Figure 13.3b), which was recorded using free-space electro-optic sampling, has an amplitude spectrum that is peaked at 0.5 THz with a bandwidth of 0.1–2 THz (Figure 13.3c).

The EpiDermFT (MatTek) full thickness human skin tissue model reconstructs normal skin tissue structure and consists of normal, human-derived epidermal keratinocytes and dermal fibroblasts that form a multilayered, highly differentiated model of the human dermis and epidermis (Figure 13.3d). It is mitotically and metabolically active, preserves the arrangement and communication of cells in skin tissue in vivo (Boelsma et al. 2000; Sedelnikova et al. 2007), and thus provides an excellent platform for assessing the effects of intense THz pulses on human skin. The THz spot size at the focus was 1.5 mm diameter (Figure 13.4e), yielding a peak incident THz electric field of approximately 220 kV/cm at the surface of the tissue for 1 µJ THz pulses. For comparison, exposure for 2 min to pulsed UVA light was used (400 nm, 0.1 ps duration, with pulse energy of either 0.080 µJ or 0.024 µJ). While UVA is not directly absorbed by DNA, it stimulates the production of reactive oxygen species that in turn cause indirect damage to DNA and other cellular components (Agar et al. 2004; Cadet et al. 2005; Wang et al. 2001). After exposure, the tissues were incubated at 37°C for 30 min, followed by a snap freeze on dry ice. THz exposures were carried out at biologically low ambient temperature (21°C), and the estimated

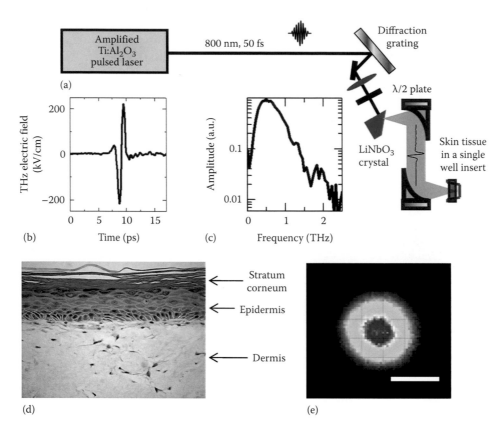

FIGURE 13.3 Exposure of artificial human skin tissues to intense THz pulses. (a) Schematic of the tilted-pulse-front THz pulse source and EpiDermFT tissue in a single-well insert placed at the focus of THz beam. (b) Waveform of a 1.0 µJ THz pulse and (c) corresponding amplitude spectrum. (d) Histology of the EpiDermFT tissue sample (400× magnification, courtesy of MatTek Corporation) that reveals epidermis containing basal, spinous, granular keratinocytes, and stratum corneum. The dermis contains numerous viable fibroblasts. (e) THz beam profile at the sample location as imaged by a pyroelectric infrared camera. The $1/e^2$ diameter of the THz beam is 1.5 mm (scale bar—1 mm). (Reprinted from Titova, L. V. et al., *Biomed. Opt. Express*, 4, 559, 2013a. With permission.)

temperature increase due to THz exposure was less than 0.7°C (Kristensen et al. 2010). Furthermore, none of the heat shock protein encoding genes were differentially expressed in the THz- or UVA-exposed tissues, and thus all the observed cellular and molecular effects are presumed to be nonthermal.

13.3.1 Evidence for DNA Damage and Activation of Damage Response

As discussed in Section 13.2.2, molecular dynamics simulations have shown that high-intensity THz radiation with frequencies resonant with the local breathing dynamics of double-stranded DNA may amplify spontaneously forming localized openings in the double strands (Alexandrov et al. 2010). Picosecond-duration, intense THz pulses have instantaneous THz fields of hundreds of kV/cm, which might be sufficient to not only temporarily disrupt DNA dynamics but also cause DNA damage by resonantly enhancing rapid, large-amplitude motions of the DNA strands. Of the various types of DNA damage, double-strand breaks (DSBs), in which two complementary strands of the DNA double helix are damaged simultaneously, are potentially the most dangerous, and, if unrepaired, can lead to cell death or cancer (Sedelnikova et al. 2003). DSBs can occur in cells due to many different endogenous factors, such as oxidative damage by reactive oxygen species, as well as exogenous factors such as exposure to x-ray (Hoeijmakers 2001) or UV radiation (Agar et al. 2004; Cadet et al. 2005; Wang et al. 2001).

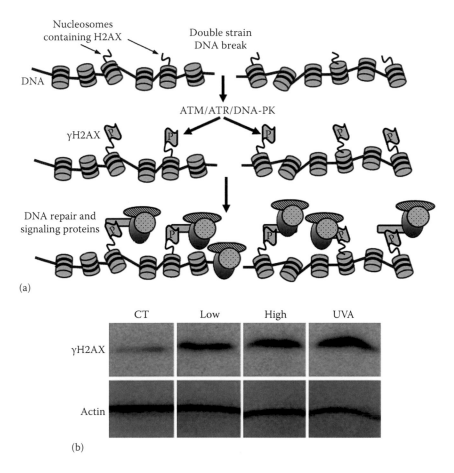

FIGURE 13.4 (a) Schematic illustration of induction of phosphorylated H2AX foci following DSB. Once the lesion is detected, kinase enzymes such as ATM, ATR, and DNA-PK are activated and phosphorylate histone H2AX proteins in the vicinity of the DSB. γH2AX helps recruit and retain DNA repair and signaling proteins to the DSB site. (b)Induction of γH2AX in artificial human skin tissue equivalents following 10 min exposure to either high (1.0 μJ) or low (0.1 μJ) energy THz pulses or 2 min exposure to UVA pulses (400 nm, 0.080 μJ), as compared to control (CT) samples. Actin was used as a loading control. Each experiment included pooled lysates from three tissues for each exposure condition with equal representation for each tissue. (Reprinted from Titova, L. V. et al., *Biomed. Opt. Express*, 4, 559, 2013a. With permission.)

Regardless of the exact mechanism of DSB formation, cells respond to them by promptly initiating the phosphorylation of the histone protein H2AX, as illustrated schematically in Figure 13.4a. DNA inside the cells is organized into a compact form known as chromatin. The basic unit of chromatin, called nucleosome, is composed of DNA wrapped around the cores made of histone proteins to form a structure that resembles beads on a string (Bonner et al. 2008; Downs and Jackson 2003). One of the histone proteins, H2AX, a variant of histone H2A, plays a critical role in recognizing DSBs and initiating repair processes (Bassing et al. 2002; Fragkos et al. 2009; Sedelnikova et al. 2003). Hundreds or even thousands of H2AX molecules in the chromatin surrounding the break site become phosphorylated at the Serine 139 position within minutes of DSB formation, and phosphorylation reaches a maximum approximately 30 min later (Dickey et al. 2011). The resulting phosphorylated H2AX (γH2AX) promotes chromatin remodelling to bring the broken DNA ends closer together and plays an important role in the formation of DNA repair foci at the sites of DSBs (Bassing et al. 2002; Downs and Jackson 2003; Fragkos et al. 2009; Sedelnikova et al. 2003). Because H2AX phosphorylation is abundant, rapid, correlates well with the number of DBSs, and can be readily detected by standard western immunoblotting, it is widely used as a surrogate marker for DNA damage and a reliable DSB detection tool (Bonner et al. 2008; Downs and Jackson 2003).

To assess the occurrence of DSBs in THz-exposed and UV-exposed tissues compared to unexposed controls, Titova et al. used western blot analysis to compare the levels of γH2AX protein (Titova et al. 2013a). (Western blotting is a molecular biology technique that detects the presence of a specific protein in a complex mixture extracted from cells. The target protein is separated from others by size using gel electrophoresis and is detected by binding to a specifically directed antibody.) Exposure to intense THz pulses with either high (1.0 μJ) or low (0.1 μJ) pulse energy for 10 min led to a significant induction of H2AX phosphorylation compared to control samples (Figure 13.4b). This is an indication that intense THz pulses cause DNA DSBs to occur in human skin cells. UVA (400 nm), which creates reactive oxygen species and causes oxidative DNA damage (Agar et al. 2004; Cadet et al. 2005; Wang et al. 2001, 2010), also leads to significant increases in γH2AX levels in exposed tissues (Figure 13.4b), which may be indicative of DSB induction.

In response to DSBs, proteins that participate in DNA repair are activated (Fillingham et al. 2006; Mirzayans et al. 2012). Of them, protein p53 is particularly important. A potent tumor suppressor and cell-cycle regulator, p53 has been termed "the guardian of the genome" (Bolderson et al. 2009; Menendez et al. 2009). It activates cell-cycle checkpoints to give the cell time to repair or, if the damage is too extensive, induces apoptotic cell death (Attardi and DePinho 2004; Bolderson et al. 2009; Fragkos et al. 2009; Menendez et al. 2009; Mirzayans et al. 2012; Figure 13.5a). In their study, Titova et al. observed increased levels of protein p53 in THz-exposed tissues (Figure 13.5b). Once the p53 protein is activated

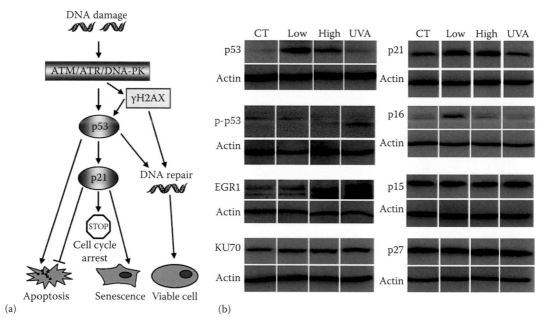

FIGURE 13.5 (a) Simplified diagram of the p53/p21 DNA damage response pathway. In response to DNA damage, kinase proteins ATM, ATR, and DNA-PK activate the tumor suppressor protein p53, which in turn induces protein p21 and similar cell-cycle regulatory proteins, as well as activates a pathway leading to apoptosis in the case that the damage is too extensive. Protein p21, in conjunction with other cell-cycle regulators, implements cell-cycle arrest to allow time for DNA repair. It also suppresses apoptosis and triggers senescence (Attardi and DePinho 2004; Downs and Jackson 2003; Mirzayans et al. 2012). (b) Expression of tumor suppressor and cell-cycle regulatory proteins in artificial human skin tissue equivalents following 10 min exposure to either high (1.0 μJ) or low (0.1 μJ) energy THz pulses or two-minute exposure to 0.080 μJ UVA (400 nm) pulses, as compared to unexposed control (CT) samples. Actin was used as a loading control. (Reprinted from Titova, L. V. et al., *Biomed. Opt. Express*, 4, 559, 2013a. With permission.)

in response to DNA damage, its function and stabilization are regulated by phosphorylation (Meek and Anderson 2009). The level of the phosphorylated variant of the p53 protein, p-p53, phosphorylated at Serine 15, was barely affected in THz-exposed tissues compared to controls. On the other hand, the pattern was reversed in UVA-exposed tissue, as the detected level of p53 was low while the induction of its phosphorylated variant, p-p53, was significant (Figure 13.5b). These variations in phosphorylation levels of protein p53 following THz-pulse-induced and UVA-induced DNA damage might provide clues to the differences in cellular damage response pathways activated by these exogenous stimuli.

Another tumor suppressor and a key player in DNA damage response is protein p21, a universal inhibitor of cell-cycle progression and a p53 target (Mirzayans et al. 2012; Rowland and Peeper 2005). In addition to its role in cell-cycle control, p21 controls gene expression, suppresses apoptosis, and induces cell senescence, which is a sustained proliferation block (Mirzayans et al. 2012; Figure 13.5a). The induction of p21 was found to be significantly more pronounced in THz-exposed tissues compared to UVA-exposed tissues (Figure 13.5b).

The levels of three other proteins, p16, p15, and p27, belonging to the same family of cyclin-dependent kinase inhibitors as protein to p21, were also measured in THz- and UVA-exposed tissues. All of them are cell-cycle regulators and tumor suppressors, acting to provide cells crucial time for DNA repair (Sherr and Roberts 1995). The levels of p16 and p27 were elevated in THz-exposed tissues but not in UVA-exposed tissues. At the same time, however, the levels of p15 proteins were unaltered by either THz pulses or UVA. Nevertheless, the observed concerted upregulation of p53, p16, and p21 is a hallmark of DNA damage response in cells (Sperka et al. 2012).

Another in vivo tumor suppressor that is induced by p53 in response to genotoxic stress and that assists p53 in controlling cellular growth and cell cycle is the early growth response 1 (EGR1) protein (Krones-Herzig et al. 2005; Zwang et al. 2011). It was observed that EGR1 was also slightly upregulated by THz exposure, albeit weaker than by UVA. Finally, a significant upregulation of KU70 in THz- and UVA-exposed tissues was observed (Figure 13.5b). This protein is essential for nonhomologous end joining, a primary mechanism of DSB repair in mammalian cells (Hoeijmakers 2001; Khanna and Jackson 2001; Mahaney et al. 2009), and has been shown to be induced in cells exposed to x-ray (Reynolds et al. 2012), gamma ray (Chen et al. 2012), and UV radiation (Rastogi et al. 2010). Its upregulation in THz-pulse-exposed tissues indicates activation nonhomologous end joining process to repair DSBs.

In summary, the study by Titova et al. has shown that exposure to intense THz pulses induces phosphorylation of H2AX, indicative of the formation of DNA DSBs, and at the same time profoundly activates DNA damage response in human skin tissue (Titova et al. 2013a). Future studies are needed to understand the precise molecular mechanisms for the induction of DNA damage and changes in cellular signaling by intense picosecond THz pulses.

13.3.2 Effects on Gene Expression

Experiments using mammalian stem cell cultures, discussed earlier in Section 13.2.2, have demonstrated that exposure to broadband, picosecond-duration THz pulses influenced differentiation of mammalian stem cells and elicited parameter-specific changes in the expression levels of genes that were transcriptionally active at a given differentiation time point (Alexandrov et al. 2011, 2013; Bock et al. 2010). In human skin tissue, THz-pulse-induced activation of damage repair proteins, as discussed in Section 13.3.1, suggests that intense, picosecond THz pulses affect gene expression. An extensive study of the THz-pulse-induced changes in global gene expression in human tissue using microarray analysis was carried out by Titova et al. (2013c). In this study, full thickness normal skin tissues (EpiDermFT, Figure 13.3b) were exposed to intense THz pulses or 400 nm pulses (UVA) following the same protocol that was used to study the effects of intense THz pulses on DNA integrity.

Global gene expression analysis demonstrated that intense THz pulses have a profound effect on gene expression in human skin tissue. The expression levels of hundreds of genes were altered in tissues exposed to either 1.0 or 0.1 μJ THz pulses for 10 min, compared to unexposed controls. As illustrated

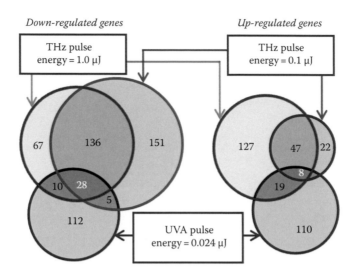

FIGURE 13.6 Venn diagrams summarizing differentially expressed genes in EpiDermFT tissues exposed to either 1.0 μJ or 0.1 μJ THz or UVA (400 nm, 0.024 μJ) pulses. Genes with a false discovery rate (FDR) adjusted *p*-value < 0.05 and log$_2$ fold change > 0.6 (1.5× change) were considered differentially expressed. Left diagram: downregulated genes. Right diagram: upregulated genes. (Reprinted from Titova, L. V. et al., *Sci. Rep.*, 3, 2363, 2013c. With permission.)

in the Venn diagrams in Figure 13.6 for the upregulated and downregulated genes, there is a significant overlap between genes affected by both THz pulse energies, while little overlap exists between THz- and UVA-induced gene expression profiles. These differences in THz-pulse-induced and UVA-induced changes in gene expression, as well as in postexposure levels of cell-cycle regulatory and DNA repair proteins that were discussed earlier, underscore a fundamental difference in how low photon energy (~4 meV) THz radiation and high photon energy (3.1 eV) UVA radiation interact with living cells.

The expression of a significant number of genes (219 in total) was affected by both THz pulse energy exposure regimes. These genes, many of which play important roles in the molecular etiology of skin cancer and inflammatory skin diseases, as well as in apoptotic signaling pathways, are of particular interest as potential THz-pulse-exposure biomarkers. Of them, 164 are downregulated and 55 are upregulated as a result of THz pulse exposure (Figure 13.6).

One of the most important findings of this study is the highly nonuniform distribution of the THz-pulse-sensitive genes throughout the genome. The most pronounced changes were observed in the expression of genes belonging to the epidermal differentiation complex (EDC), the 1.6 Mb locus on human chromosome 1q21 that contains 57 genes responsible for terminal differentiation of keratinocytes and regulation of epidermal barrier function, as well as for skin immune and inflammation responses (de Guzman Strong et al. 2010; Kypriotou et al. 2012; Mischke et al. 1996). The genes of the EDC are organized into clustered families of genes encoding calcium-binding proteins (S100), filaggrin (FLG)-like proteins, *late cornified envelope* (LCE) proteins, and small proline-rich region (SPRR) proteins, as illustrated schematically in Figure 13.7a. Nearly half of them (27) were affected by intense THz pulses, with expression of 20 of them being markedly suppressed, compared to only 12 affected by UVA pulses (Figure 13.7a and b). Enhanced expression of EDC genes results in increased keratinocyte proliferation and differentiation, commonly observed in inflammatory skin disorders (de Guzman Strong et al. 2010; Hoffjan and Stemmler 2007; Roberson and Bowcock 2010) and in skin cancers (Haider et al. 2006). For example, S100A genes S100A15 (Eckert et al. 2004) and S100A12 (Eckert et al. 2004; Semprini et al. 2002), SPRR genes SPRR2A, SPRR3, and SPRR2B (Haider et al. 2006), as well as multiple LCE3 genes (Bergboer et al. 2011), which are known to be upregulated in psoriasis(Haider et al. 2006; Roberson and Bowcock 2010), were suppressed by intense THz pulses (Figure 13.7b). Also, the enhanced expression of eight of the EDC genes that were downregulated by intense THz pulse exposure—S100A11, S100A12, SPRR1B, SPRR2B, SPRR2C, SPRR3, involucrin (IVL), and LCE3D—has been associated with cutaneous squamous cell carcinomas (SCCs) (Haider et al. 2006, Hudson et al. 2010). Downregulation of these genes by intense THz pulses may open new avenues for targeted treatments for psoriasis and SCC (Haider et al. 2006; Hudson et al. 2010).

In addition to EDC genes, exposure to intense THz pulses also altered expression levels of many other genes implicated in inflammatory dermatological diseases, as well as in cancer, including aggressive oral SCC (OSCC) and nonmelanoma skin cancer (SCC and basal cell carcinoma [BCC]; Figure 13.7c). For example, THz pulses significantly lowered the mRNA expression of four human β-defensins (DEFB103A, DEFB4, LOC728454, and DEFB1). Enhanced expression of β-defensin genes is associated with OSCC (Sawaki et al. 2001), BCC, and SCC (Gambichler et al. 2006; Haider et al. 2006; Hudson et al. 2010; Mburu et al. 2011; Muehleisen et al. 2012), as well as psoriasis (Roberson and Bowcock 2010). Interestingly, the effect of intense THz pulses on the expression levels of β-defensins was similar to that of UVA pulses. Exposure to intense THz pulses (but not UVA) also downregulated corneodesmosin (CDSN), the expression of which is often enhanced in psoriasis (Allen et al. 2001), and two genes, S100P (Arumugam et al. 2005; Kim et al. 2009; Schor et al. 2006) and CD24 (Baumann et al. 2005; Haider et al. 2006; Lee et al. 2009) whose overexpression is associated with aggressive tumor progression, metastasis, and poor prognosis in various cancers. In addition, THz pulse irradiation also downregulated four out of ten serine proteinase inhibitor (serpin) B genes located in the serpin gene cluster on chromosome 18q21.3 (Schneider et al. 1995), once again demonstrating a nonuniform distribution of intense-THz-pulse-sensitive genes throughout the genome. Two of them, serpinB3 and serpinB4, tandemly arranged on chromosome 18q21.3, are also known as SCC antigens 1 and 2, respectively. While their function in carcinogenesis has

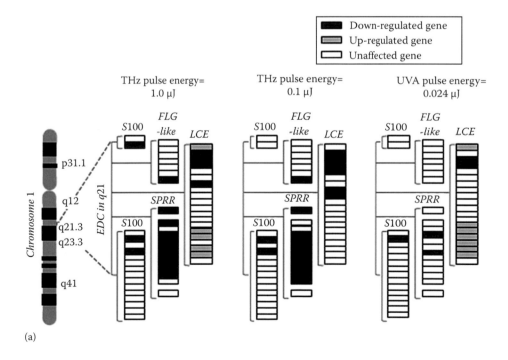

FIGURE 13.7 Differentially expressed EDC genes and selected other genes associated with nonmelanoma skin cancer or inflammatory skin diseases. (a) EDC genes that were either upregulated (gray rectangles) or downregulated (black rectangle) after exposure of skin tissues to either 1.0 or 0.1 J THz pulse energies or UVA pulses. Unaffected genes are depicted as white rectangles.

(continued)

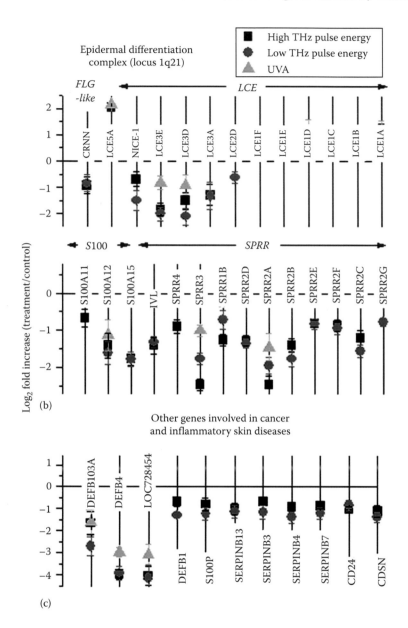

FIGURE 13.7 (continued) Differentially expressed EDC genes and selected other genes associated with nonmelanoma skin cancer or inflammatory skin diseases. (b) Log₂ fold changes for differentially expressed genes belonging to four families of EDC genes (FLG-like, LCE, S100, and SPRR). (c) Other genes (selected) involved in dermatological diseases and cancer, whose expression levels were altered by exposure to both THz pulse energy regimes. In (b) and (c), error bars indicate corresponding standard deviations and, in some cases, are smaller than the symbol size. (Reprinted from Titova, L. V. et al., *Sci. Rep.*, 3, 2363, 2013c. With permission.)

not yet been clearly established, studies find a correlation between their enhanced expression and aggressive progression of SCCs and breast carcinoma (Catanzaro et al. 2011; Murakami et al. 2010). SerpinB13, or hurpin, is also overexpressed in SCC, as well as in psoriatic lesions (Moussali et al. 2005).

    Overall, the observed changes in expression of EDC genes and several other genes implicated in dermatological diseases and epithelial cancers are opposite to disease-related changes. While these results may raise questions about possible therapeutic applications of intense THz pulses aimed at changing the expression of these disease-related genes, extensive studies tracking the THz-induced effects over a period of time following the exposure of healthy as well as diseased tissues will have to be carried out before the full potential of intense THz pulses for clinical use can be realized.

Additionally, comparison of the gene expression in directly exposed tissues (Titova et al. 2013c) and in tissue that encompassed directly exposed as well as neighboring unexposed cells (Titova et al. 2013b) suggests that intense THz pulses may also result in biological effects in the neighboring unexposed cells by what is known as the bystander effect. In the bystander effect, the directly exposed cells communicate distress to their neighbors by several mechanisms such as gap-junction intercellular communication, oxidative metabolism, and secretion of various soluble factors (Azzam et al. 2003; Mothersill and Seymour 1998; Nagasawa and Little 1999; Watson et al. 2000). The bystander effect has been documented upon exposure to ionizing radiation (Mothersill and Seymour 1998; Nagasawa and Little 1999; Ojima et al. 2011; Sedelnikova et al. 2007; Watson et al. 2000), as well as UVA and UVB light (Banerjee et al. 2005; Dahle et al. 2005; Nishiura et al. 2012). The existence of the THz-pulse-induced bystander effect, as well as the possible signaling events involved in it, will have to be addressed in future studies.

13.4 In Vivo Studies of Nonthermal Effects of Subpicosecond THz Pulses on Mouse Skin

In the first in vivo investigation of the nonthermal cellular effects of broadband THz pulses, Kim et al. applied subpicosecond THz pulses to mouse skin (Kim et al. 2013). The skin of C57BL/6J and BALB/c nude mice was exposed for one hour to subpicosecond-duration THz pulses with 0.1–2.5 THz bandwidth, repetition rate of 1 kHz, and energy density per pulse of 0.32 nJ/cm^2. Whole-genome expression profiling using mRNA microarray technology was carried out to delineate the effects of the THz exposure on gene expression 24 h postexposure. Similar to earlier studies using mouse MSCs (Alexandrov et al. 2013; Bock et al. 2010) and artificial human skin tissues (Titova et al. 2013b,c) that were discussed earlier, exposure to broadband THz pulses resulted in considerable changes in gene expression in exposed mouse skin compared to unexposed controls, upregulating 82, and downregulating 67 genes. Using gene set enrichment analysis, Kim et al. have established the set of biological functions most affected by broadband THz pulses. Of those, healing function was found to be most significantly affected, with many of the differentially expressed genes such as Bmp2, Cd44, Krt6a, Lep, Serpinc1, Sprr1b, and Thbs1 playing important roles in wound healing.

Wound healing is a complex process that progresses through four overlapping phases: hemostasis, inflammation, proliferation, and remodelling (Diegelmann and Evans 2004; Penn et al. 2012). Of the various signaling cascades that are activated in response to the disruption of an epithelial layer and drive the healing process, the transforming growth factor-beta (TGF-β) signaling pathway is one of the most important. Protein TGB-β is a cytokine that controls cellular proliferation and stimulates angiogenesis (Gharaee-Kermani and Phan 2001), thus playing a critical role in the proliferative phase of wound healing (Deonarine et al. 2007). This protein acts mainly by activating NFκB1 and Smad proteins, transcription factors that regulate expression of multiple genes involved in various aspects of cellular growth and proliferation (Ear et al. 2010; Ishinaga et al. 2007), including Serpine 1, THBS1, BMP2, CD44, and other genes that were differentially expressed in mouse skin following THz pulse exposure (Kim et al. 2013). Kim et al. confirmed activation of TGF-β signaling in mouse skin exposed to broadband THz pulses by analyzing the expression of TGF-β1 mRNA by real-time RT-PCR. They found that the relative level of TGF-β1 mRNA increased significantly to the levels observed in the in vivo wound model within 1 h after exposure in both C57BL/6J and BALB/c nude mice skin and decreased back to control levels within 24 h. While the gene expression profile characteristic of wound healing was observed in THz-exposed mouse skin, the exposed area did not show any histological evidence of wound damage. Hypothesizing that THz-pulse-induced activation of TGF-β signaling might impact healing, Kim et al. have also studied the effects of broadband THz pulses on the closing of 4 mm punch wounds in the skin of C57BL/6J mice. They found that after 10 days, the punch wounds were completely closed in sham-exposed skin, while the closure of the wounds exposed to THz pulses daily for 1 h was significantly delayed. At the same time, the expression of TGF-β1 protein, which typically increases as the healing process progresses to the proliferation and remodelling stages (Deonarine et al. 2007; Penn et al. 2012), was significantly elevated at 5 and 9 days after wounding in THz-pulse-exposed wound tissue as compared to unexposed wounds. The authors concluded that increased expression of TGF-β and activation of its downstream target genes perturbs the wound healing process in vivo and thus delays wound closing, consistent with earlier reports that overexpressed TGF-β inhibits reepithelialization with scarring (Shah et al. 1999; Yang et al. 2001).

While the significant differences in exposure parameters (0.32 nJ/cm^2 used by Kim et al. versus $6-60$ μJ/cm^2 used by Titova et al. [2013a, c]) and in sampling time points (24 h following 1 h exposure versus 30 min after 10 min exposure) make direct comparison of the results of these studies difficult, several interesting questions can be raised based on these studies. TGF-β, which was activated in response to THz pulse exposure in mouse skin, controls cellular proliferation and can play various roles depending on the conditions in the tissue (Massagué 2012). In wounds, it enhances proliferation to facilitate remodelling of damaged epithelium (Penn et al. 2012). However, in normal epithelial tissues and in early stage neoplasms, it acts as a potent proliferation inhibitor, thus playing a role of a tumor suppressor (Cui et al. 1994; Massagué 2012; Piccolo 2008). Does THz-pulse-induced activation of TGF-β play a role in transcriptional changes in genes implicated in carcinogenesis and inflammatory skin diseases, which was observed in human skin tissues upon exposure to intense THz pulses (Titova et al. 2013c)? Furthermore, it is known that protein p53 cooperates with Smad proteins to mediate TGF-β signals and acts as a regulator of TGF-β (Cordenonsi et al. 2003; Piccolo 2008; Prime et al. 2004). Also, cell-cycle inhibitor p21 plays an important role in the TGF-β pathway (Dai et al. 2013; Voss et al. 1999). The enhanced expression of both p53 and p21 was observed in human skin tissues exposed to intense THz pulses (Titova et al. 2013a). Is THz-pulse-induced activation of these tumor suppressor proteins solely a response to DNA damage by high-intensity THz pulses, which might not be occurring at significantly lower THz energy densities as used by Kim et al. (2013)? Or did p53 and p21 also play a critical role in mediating THz-induced activation of the TGF-β signaling cascade observed in exposed mouse skin? The interplay between the various THz-pulse-sensitive genes and signaling pathways in skin tissue certainly needs to be explored in future studies, which not only could shed light on the mechanisms of THz pulse interactions with cells and tissues but also might have important implications for clinical applications of broadband THz pulses. Furthermore, the very regulation of gene expression by THz pulses needs to be investigated in detail. It is important to establish which transcription factors control gene expression upon THz radiation. This may be achieved by an in-depth bioinformatic analysis of the promoter regions of genes susceptible to THz pulse exposure, followed by chromatin immune-precipitation assays that detect the presence of transcription factors at different genomic sites.

13.5 Summary

In the past decade, only a handful of experimental investigations aimed at elucidating the effects of broadband THz pulses on living cells and tissues have been carried out, as discussed in this chapter. The vast differences in the biological systems studied, including in vitro cell cultures, artificial human skin tissues, and in vivo animal models, as well as differences in THz exposure conditions, make drawing parallels between those studies challenging. Importantly, in all of the studies that reported observation of THz-pulse-induced effects, the observed effects were nonthermal with negligible temperature increases upon THz pulse exposure (Alexandrov et al. 2013; Bock et al. 2010; Kim et al. 2013) and unchanged expression levels of heat shock protein-coding genes (Alexandrov et al. 2013; Kim et al. 2013; Titova et al. 2013c). Profound changes in gene expression profiles in response to broadband THz pulse exposure were reported in mouse MSCs (Alexandrov et al. 2013; Bock et al. 2010), human skin tissues (Titova et al. 2013c), and in the skin of mice exposed in vivo (Kim et al. 2013). Intense THz pulses were also shown to significantly induce phosphorylation of H2AX in human skin tissue, indicative of DNA damage, and, at the same time, caused increases in the levels of multiple cell-cycle regulatory and tumor suppressor proteins (Titova et al. 2013a). These cellular effects occurred in response to broadband THz pulses with energy densities per pulse of 0.32 nJ/cm^2 used in experiments on mouse skin in vivo (Kim et al. 2013), 1 µJ/cm^2 in studies using mouse MSCs (Alexandrov et al. 2013; Bock et al. 2010), and 6–60 µJ/cm^2 used to expose artificial human skin tissues (Titova et al. 2013a–c). No visible morphological changes in the cells or histological evidence of tissue damage were reported in these experiments, with a notable exception of mouse MSCs that were exposed to broadband THz pulses at an advanced stage in their differentiation program (Bock et al. 2010). It is therefore conceivable that cellular and epigenetic changes were also taking place in human epithelial and embryonic stem cells exposed by Williams et al. to intense THz pulses with energy densities per pulse reaching 10 nJ/cm^2 produced by the ALICE synchrotron source (Williams et al. 2013). In that study, the authors reported no change in the attachment, morphology, proliferation, and differentiation of the exposed cells but did not analyze the gene expression profiles or the levels of cellular proteins. Therefore, the results of the studies carried out to date do not yet allow establishing THz pulse energy density thresholds and other exposure parameters for the onset of THz-pulse-induced changes to gene expression and cellular function.

Furthermore, the mechanisms by which intense THz pulses affect gene expression and induce DNA damage are not known. As it has been suggested by Alexandrov et al., intense THz pulse radiation may enhance existing transcription bubbles or create new open states in the DNA double helix, thus influencing transcription initiation or binding of transcription factors (Alexandrov et al. 2010, 2013; Bock et al. 2010). Alternatively, the changes in gene expression may constitute a cellular response to THz-pulse-induced DNA damage (Titova et al. 2013a) or to damage or changes in conformation states of intracellular proteins (Cherkasova et al. 2009). Future studies will require detailed and comprehensive analysis of the exact cellular and molecular effects of broadband THz pulses.

References

Agar, N. S., G. M. Halliday, R. S. Barnetson et al. 2004. The basal layer in human squamous tumors harbors more UVA than UVB fingerprint mutations: A role for UVA in human skin carcinogenesis. *Proceedings of the National Academy of Sciences of the United States of America* 101: 4954–4959.

Alexandrov, B. S., Y. Fukuyo, M. Lange et al. 2012. DNA breathing dynamics distinguish binding from nonbinding consensus sites for transcription factor YY1 in cells. *Nucleic Acids Research* 40: 10116–10123.

Alexandrov, B. S., V. Gelev, A. R. Bishop, A. Usheva, and K. Ø. Rasmussen. 2010. DNA breathing dynamics in the presence of a terahertz field. *Physics Letters A* 374: 1214–1217.

Alexandrov, B. S., V. Gelev, Y. Monisova et al. 2009. A nonlinear dynamic model of DNA with a sequence-dependent stacking term. *Nucleic Acids Research* 37: 2405–2410.

Alexandrov, B. S., M. L. Phipps, L. B. Alexandrov et al. 2013. Specificity and heterogeneity of terahertz radiation effect on gene expression in mouse mesenchymal stem cells. *Scientific Reports* 3: 1184.

Alexandrov, B. S., K. Ø. Rasmussen, A. R. Bishop et al. 2011. Non-thermal effects of terahertz radiation on gene expression in mouse stem cells. *Biomedical Optics Express* 2: 2679.

Allen, M., A. Ishida-Yamamoto, J. McGrath et al. 2001. Corneodesmosin expression in psoriasis vulgaris differs from normal skin and other inflammatory skin disorders. *Laboratory Investigation* 81: 969–976.

Arbab, M. H., T. C. Dickey, D. P. Winebrenner et al. 2011. Terahertz reflectometry of burn wounds in a rat model. *Biomedical Optics Express* 2: 2339.

Arumugam, T., D. M. Simeone, K. Van Golen, and C. D. Logsdon. 2005. S100P promotes pancreatic cancer growth, survival, and invasion. *Clinical Cancer Research* 11: 5356–5364.

Ashworth, P. C., P. O'Kelly, A. D. Purushotham et al. 2008. An intra-operative THz probe for use during the surgical removal of breast tumors. Presented at *33rd International Conference on Infrared, Millimeter and Terahertz Waves*, Pasadena, CA, pp. 1–3.

Ashworth, P. C., E. Pickwell-MacPherson, E. Provenzano et al. 2009. Terahertz pulsed spectroscopy of freshly excised human breast cancer. *Optics Express* 17: 12444–12454.

Attardi, L. D. and R. A. DePinho. 2004. Conquering the complexity of p53. *Nature Genetics* 36: 7–8.

Azzam, E. I., S. M. de Toledo, and J. B. Little. 2003. Oxidative metabolism, gap junctions and the ionizing radiation-induced bystander effect. *Oncogene* 22: 7050–7057.

Banerjee, G., N. Gupta, A. Kapoor, and G. Raman. 2005. UV induced bystander signaling leading to apoptosis. *Cancer Letters* 223: 275–284.

Bassing, C. H., K. F. Chua, J. Sekiguchi et al. 2002. Increased ionizing radiation sensitivity and genomic instability in the absence of histone H2AX. *Proceedings of the National Academy of Sciences* 99: 8173–8178.

Baumann, P., N. Cremers, F. Kroese et al. 2005. CD24 expression causes the acquisition of multiple cellular properties associated with tumor growth and metastasis. *Cancer Research* 65: 10783–10793.

Baxter, J. B. and G. W. Guglietta. 2011. Terahertz spectroscopy. *Analytical Chemistry* 83: 4342–4368.

Bergboer, J. G. M., G. S. Tjabringa, M. Kamsteeg et al. 2011. Psoriasis risk genes of the late cornified envelope-3 group are distinctly expressed compared with genes of other LCE groups. *The American Journal of Pathology* 178: 1470–1477.

Bergues-Pupo, A. E., J. M. Bergues, and F. Falo. 2013. Modeling the interaction of DNA with alternating fields. *Physical Review E* 87: 022703.

Berry, E., G. C. Walker, A. J. Fitzgerald et al. 2003. Do in vivo terahertz imaging systems comply with safety guidelines? *Journal of Laser Applications* 15: 192.

Blanchard, F., G. Sharma, L. Razzari et al. 2011. Generation of intense terahertz radiation via optical methods. *IEEE Journal of Selected Topics in Quantum Electronics* 17: 5–16.

Bock, J., Y. Fukuyo, S. Kang et al. 2010. Mammalian stem cells reprogramming in response to terahertz radiation. *PLoS ONE* 5: e15806.

Boelsma, E., S. Gibbs, C. Faller, and M. Ponec. 2000. Characterization and comparison of reconstructed skin models: Morphological and immunohistochemical evaluation. *Acta Dermatovenereologica-Stockholm-* 80: 82–88.

Bolderson, E., D. J. Richard, B.-B. S. Zhou, and K. K. Khanna. 2009. Recent advances in cancer therapy targeting proteins involved in DNA double-strand break repair. *Clinical Cancer Research* 15: 6314–6320.

Bonner, W. M., C. E. Redon, J. S. Dickey et al. 2008. γH2AX and cancer. *Nature Reviews Cancer* 8: 957–967.

Born, B., S. J. Kim, S. Ebbinghaus, M. Gruebele, and M. Havenith. 2009. The terahertz dance of water with the proteins: The effect of protein flexibility on the dynamical hydration shell of ubiquitin. *Faraday Discussions* 141: 161–173.

Cadet, J., E. Sage, and T. Douki. 2005. Ultraviolet radiation-mediated damage to cellular DNA. *Mutation Research/Fundamental and Molecular Mechanisms of Mutagenesis* 571: 3–17.

Catanzaro, J. M., J. L. Guerriero, J. Liu et al. 2011. Elevated expression of squamous cell carcinoma antigen (SCCA) is associated with human breast carcinoma. *PLoS ONE* 6: e19096.

Chen, H., Y. Bao, L. Yu et al. 2012. Comparison of cellular damage response to low-dose-rate ^{125}I seed irradiation and high-dose-rate gamma irradiation in human lung cancer cells. *Brachytherapy* 11: 149–156.

Cherkasova, O. P., V. I. Fedorov, E. F. Nemova, and A. S. Pogodin. 2009. Influence of terahertz laser radiation on the spectral characteristics and functional properties of albumin. *Optics and Spectroscopy* 107: 534–537.

Chitanvis, S. M. 2006. Can low-power electromagnetic radiation disrupt hydrogen bonds in dsDNA? *Journal of Polymer Science Part B: Polymer Physics* 44: 2740–2747.

Clothier, R. H. and N. Bourne. 2003. Effects of THz exposure on human primary keratinocyte differentiation and viability. *Journal of Biological Physics* 29: 179–185.

Cordenonsi, M., S. Dupont, S. Maretto et al. 2003. Links between tumor suppressors: p53 is required for TGF-β gene responses by cooperating with Smads. *Cell* 113: 301–314.

Cui, W., C. J. Kemp, E. Duffie, A. Balmain, and R. J. Akhurst. 1994. Lack of transforming growth factor-β1 expression in benign skin tumors of p53null mice is prognostic for a high risk of malignant conversion. *Cancer Research* 54: 5831–5836.

Dahle, J., O. Kaalhus, T. Stokke, and E. Kvam. 2005. Bystander effects may modulate ultraviolet A and B radiation-induced delayed mutagenesis. *Radiation Research* 163: 289–295.

Dai, M., A. Al-Odaini, N. Fils-Aimé et al. 2013. Cyclin D1 cooperates with p21 to regulate TGFβ-mediated breast cancer cell migration and tumor local invasion. *Breast Cancer Research* 15: 1–14.

de Guzman Strong, C., S. Conlan, C. B. Deming et al. 2010. A milieu of regulatory elements in the epidermal differentiation complex syntenic block: Implications for atopic dermatitis and psoriasis. *Human Molecular Genetics* 19: 1453–1460.

Deonarine, K., M. Panelli, M. Stashower et al. 2007. Gene expression profiling of cutaneous wound healing. *Journal of Translational Medicine* 5: 11.

Dickey, J. S., F. J. Zemp, A. Altamirano et al. 2011. H2AX phosphorylation in response to DNA double-strand break formation during bystander signalling: Effect of microRNA knockdown. *Radiation Protection Dosimetry* 143: 264–269.

Diegelmann, R. F. and M. C. Evans. 2004. Wound healing: An overview of acute, fibrotic and delayed healing. *Frontier in Bioscience* 9: 283–289.

Downs, J. A. and S. P. Jackson. 2003. Cancer: Protective packaging for DNA. *Nature* 424: 732–734.

Ear, T., C. F. Fortin, F. A. Simard, and P. P. McDonald. 2010. Constitutive Association of TGF-β–Activated Kinase 1 with the IκB Kinase Complex in the Nucleus and Cytoplasm of Human Neutrophils and Its Impact on Downstream Processes. *The Journal of Immunology* 184: 3897–3906.

Eckert, R. L., G. Adhikary, C. A. Young et al. 2013. AP1 transcription factors in epidermal differentiation and skin cancer. *Journal of Skin Cancer* 2013: 537028.

Eckert, R. L., A. M. Broome, M. Ruse et al. 2004. S100 proteins in the epidermis. *Journal of Investigative Dermatology* 123: 23–33.

Eckert, R. L. and E. A. Rorke. 1989. Molecular biology of keratinocyte differentiation. *Environmental Health Perspectives* 80: 109–116.

Falconer, R. J. and A. G. Markelz. 2012. Terahertz spectroscopic analysis of peptides and proteins. *Journal of Infrared, Millimeter, and Terahertz Waves* 33: 973–988.

Federici, J. F., O. Mitrofanov, M. Lee et al. 2002. Terahertz near-field imaging. *Physics in Medicine and Biology* 47: 3727.

Fillingham, J., M.-C. Keogh, and N. J. Krogan. 2006. γ H2AX and its role in DNA double-strand break repair This paper is one of a selection of papers published in this Special Issue, entitled 27th International West Coast Chromatin and Chromosome Conference, and has undergone the Journal's usual peer review process. *Biochemistry and Cell Biology* 84: 568–577.

Fischer, B. M., M. Walther, and P. U. Jepsen. 2002. Far-infrared vibrational modes of DNA components studied by terahertz time-domain spectroscopy. *Physics in Medicine and Biology* 47: 3807.

Fitzgerald, A. J., E. Berry, N. N. Zinovev et al. 2002. An introduction to medical imaging with coherent terahertz frequency radiation. *Physics in Medicine and Biology* 47: R67.

Fragkos, M., J. Jurvansuu, and P. Beard. 2009. H2AX is required for cell cycle arrest via the p53/p21 pathway. *Molecular and Cellular Biology* 29: 2828–2840.

Gambichler, T., M. Skrygan, J. Huyn et al. 2006. Pattern of mRNA expression of β-defensins in basal cell carcinoma. *BMC Cancer* 6: 163.

Gharaee-Kermani, M. and S. H. Phan. 2001. Role of cytokines and cytokine therapy in wound healing and fibrotic diseases. *Current Pharmaceutical Design* 7: 1083–1103.

Haider, A. S., S. B. Peters, H. Kaporis et al. 2006. Genomic analysis defines a cancer-specific gene expression signature for human squamous cell carcinoma and distinguishes malignant hyperproliferation from benign hyperplasia. *Journal of Investigative Dermatology* 126: 869–881.

Hintzsche, H., C. Jastrow, T. Kleine-Ostmann et al. 2011. Terahertz radiation induces spindle disturbances in human-hamster hybrid cells. *Radiation Research* 175: 569–574.

Hirori, H., F. Blanchard, and K. Tanaka. 2011. Single-cycle THz pulses with amplitudes exceeding 1 MV/cm generated by optical rectification in LiNbO3. *Applied Physics Letters* 98: 091106.

Hoeijmakers, J. H. J. 2001. Genome maintenance mechanisms for preventing cancer. *Nature* 411: 366–374.

Hoffjan, S. and S. Stemmler. 2007. On the role of the epidermal differentiation complex in ichthyosis vulgaris, atopic dermatitis and psoriasis. *British Journal of Dermatology* 157: 441–449.

Hoffmann, M. C. and J. A. Fülöp. 2011. Intense ultrashort terahertz pulses: Generation and applications. *Journal of Physics D: Applied Physics* 44: 083001.

Hudson, L. G., J. M. Gale, R. S. Padilla et al. 2010. Microarray analysis of cutaneous squamous cell carcinomas reveals enhanced expression of epidermal differentiation complex genes. *Molecular Carcinogenesis* 49: 619–629.

Ishinaga, H., H. Jono, J. H. Lim et al. 2007. TGF-β induces p65 acetylation to enhance bacteria-induced NF-κB activation. *The EMBO Journal* 26: 1150–1162.

Jepsen, P. U., D. G. Cooke, and M. Koch. 2011. Terahertz spectroscopy and imaging–modern techniques and applications. *Laser & Photonics Reviews* 5: 124–166.

Joseph, C. S., A. N. Yaroslavsky, M. Al-Arashi et al. 2009. Terahertz spectroscopy of intrinsic biomarkers for non-melanoma skin cancer. Presented at *Proceedings of SPIE*: 72150I-I-10.

Junginger, F., A. Sell, O. Schubert et al. 2010. Single-cycle multiterahertz transients with peak fields above 10 MV/cm. *Optics Letters* 35: 2645–2647.

Khanna, K. K. and S. P. Jackson. 2001. DNA double-strand breaks: Signaling, repair and the cancer connection. *Nature Genetics* 27: 247–254.

Kim, J. K., K. H. Jung, J. H. Noh et al. 2009. Targeted disruption of S100P suppresses tumor cell growth by down-regulation of cyclin D1 and CDK2 in human hepatocellular carcinoma. *International Journal of Oncology* 35: 1257.

Kim, K.-T., J. H. Park, S. J. Jo et al. 2013. High-power femtosecond-terahertz pulse induces a wound response in mouse skin. *Scientific Reports* 3: 2296.

Kim, S. J., B. Born, M. Havenith, and M. Gruebele. 2008. Real-time detection of protein-water dynamics upon protein folding by terahertz absorption spectroscopy. *Angewandte Chemie International Edition* 47: 6486–6489.

Korenstein-Ilan, A., A. Barbul, P. Hasin et al. 2009. Terahertz radiation increases genomic instability in human lymphocytes. *Radiation Research* 170: 224–234.

Kristensen, T. T. L., W. Withayachumnankul, P. U. Jepsen, and D. Abbott. 2010. Modeling terahertz heating effects on water. *Optics Express* 18: 4727–4739.

Krones-Herzig, A., S. Mittal, K. Yule et al. 2005. Early growth response 1 acts as a tumor suppressor in vivo and in vitro via regulation of p53. *Cancer Research* 65: 5133–5143.

Kypriotou, M., M. Huber, and D. Hohl. 2012. The human epidermal differentiation complex: Cornified envelope precursors, S100 proteins and the 'fused genes' family. *Experimental Dermatology* 21: 643–649.

Lee, J.-H., S.-H. Kim, E.-S. Lee, and Y.-S. Kim. 2009. CD24 overexpression in cancer development and progression: A meta-analysis. *Oncology Reports* 22: 1149–1156.

Lee, Y. S. 2009. *Principles of Terahertz Science and Technology*. New York: Springer.

Mahaney, B., K. Meek, and S. Lees-Miller. 2009. Repair of ionizing radiation-induced DNA double-strand breaks by non-homologous end-joining. *Biochemical Journal* 417: 639–650.

Maniadis, P., B. S. Alexandrov, A. R. Bishop, and K. Ø. Rasmussen. 2011. Feigenbaum cascade of discrete breathers in a model of DNA. *Physical Review E* 83: 011904.

Markelz, A., S. Whitmire, J. Hillebrecht, and R. Birge. 2002. THz time domain spectroscopy of biomolecular conformational modes. *Physics in Medicine and Biology* 47: 3797–3805.

Massagué, J. 2012. TGFβ signalling in context. *Nature Reviews Molecular Cell Biology* 13: 616–630.

Masson, J.-B., M.-P. Sauviat, J.-L. Martin, and G. Gallot. 2006. Ionic contrast terahertz near-field imaging of axonal water fluxes. *Proceedings of the National Academy of Sciences of the United States of America* 103: 4808–4812.

Mburu, Y. K., K. Abe, L. K. Ferris, S. N. Sarkar, and R. L. Ferris. 2011. Human β-defensin 3 promotes NF-κB-mediated CCR7 expression and anti-apoptotic signals in squamous cell carcinoma of the head and neck. *Carcinogenesis* 32: 168–174.

Meek, D. W. and C. W. Anderson. 2009. Posttranslational modification of p53: Cooperative integrators of function. *Cold Spring Harbor Perspectives in Biology* 1: a000950.

Menendez, D., A. Inga, and M. A. Resnick. 2009. The expanding universe of p53 targets. *Nature Reviews Cancer* 9: 724–737.

Mirzayans, R., B. Andrais, A. Scott, and D. Murray. 2012. New insights into p53 signaling and cancer cell response to DNA damage: Implications for cancer therapy. *Journal of Biomedicine and Biotechnology* 2012: 170325.

Mischke, D., B. P. Korge, I. Marenholz, A. Volz, and A. Ziegler. 1996. Genes encoding structural proteins of epidermal cornification and S100 calcium-binding proteins form a gene complex ("epidermal differentiation complex") on human chromosome 1q21. *Journal of Investigative Dermatology* 106: 989–992.

Mittleman, D. M. 2013. Frontiers in terahertz sources and plasmonics. *Nature Photonics* 7: 666–669.

Mothersill, C. and C. Seymour. 1998. Cell-cell contact during gamma irradiation is not required to induce a bystander effect in normal human keratinocytes: Evidence for release during irradiation of a signal controlling survival into the medium. *Radiation Research* 149: 256–262.

Moussali, H., M. Bylaite, T. Welss et al. 2005. Expression of hurpin, a serine proteinase inhibitor, in normal and pathological skin: Overexpression and redistribution in psoriasis and cutaneous carcinomas. *Experimental Dermatology* 14: 420–428.

Muehleisen, B., S. B. Jiang, J. A. Gladsjo et al. 2012. Distinct innate immune gene expression profiles in non-melanoma skin cancer of immunocompetent and immunosuppressed patients. *PLoS ONE* 7: e40754.

Murakami, A., C. Fukushima, K. Yositomi et al. 2010. Tumor-related protein, the squamous cell carcinoma antigen binds to the intracellular protein carbonyl reductase. *International Journal of Oncology* 36: 1395–1400.

Nagasawa, H. and J. B. Little. 1999. Unexpected sensitivity to the induction of mutations by very low doses of alpha-particle radiation: Evidence for a bystander effect. *Radiation Research* 152: 552–557.

Nishiura, H., J. Kumagai, G. Kashino et al. 2012. The bystander effect is a novel mechanism of UVA-induced melanogenesis. *Photochemistry and Photobiology* 88: 389–397.

Ojima, M., A. Furutani, N. Ban, and M. Kai. 2011. Persistence of DNA double-strand breaks in normal human cells induced by radiation-induced bystander effect. *Radiation Research* 175: 90–96.

Pal, S. K. and A. H. Zewail. 2004. Dynamics of water in biological recognition. *Chemical Reviews* 104: 2099–2124.

Penn, J. W., A. O. Grobbelaar, and K. J. Rolfe. 2012. The role of the TGF-β family in wound healing, burns and scarring: A review. *International Journal of Burns and Trauma* 2: 18–28.

Piccolo, S. 2008. p53 regulation orchestrates the TGF-β response. *Cell* 133: 767–769.

Pickwell-MacPherson, E. and V. P. Wallace. 2009. Terahertz pulsed imaging—A potential medical imaging modality? *Photodiagnosis and Photodynamic Therapy* 6: 128–134.

Prime, S. S., M. Davies, M. Pring, and I. C. Paterson. 2004. The role of TGF-β in epithelial malignancy and its relevance to the pathogenesis of oral cancer (part II). *Critical Reviews in Oral Biology & Medicine* 15: 337–347.

Prohofsky, E. W., K. C. Lu, L. L. Van Zandt, and B. F. Putnam. 1979. Breathing modes and induced resonant melting of the double helix. *Physics Letters A* 70: 492–494.

Rastogi, R. P., A. Kumar, M. B. Tyagi, and R. P. Sinha. 2010. Molecular mechanisms of ultraviolet radiation-induced DNA damage and repair. *Journal of Nucleic Acids* 16: 592980.

Reynolds, P., J. A. Anderson, J. V. Harper et al. 2012. The dynamics of Ku70/80 and DNA-PKcs at DSBs induced by ionizing radiation is dependent on the complexity of damage. *Nucleic Acids Research* 40: 10821–10831.

Roberson, E. and A. M. Bowcock. 2010. Psoriasis genetics: Breaking the barrier. *Trends in Genetics* 26: 415–423.

Rowland, B. D. and D. S. Peeper. 2005. KLF4, p21 and context-dependent opposing forces in cancer. *Nature Reviews Cancer* 6: 11–23.

Sawaki, K., N. Mizukawa, T. Yamaai et al. 2001. High concentration of beta-defensin-2 in oral squamous cell carcinoma. *Anticancer Research* 22: 2103–2107.

Schneider, S. S., C. Schick, K. E. Fish et al. 1995. A serine proteinase inhibitor locus at 18q21. 3 contains a tandem duplication of the human squamous cell carcinoma antigen gene. *Proceedings of the National Academy of Sciences* 92: 3147–3151.

Schoenbach, K. H., S. Xiao, R. P. Joshi et al. 2008. The effect of intense subnanosecond electrical pulses on biological cells. *IEEE Transactions on Plasma Science* 36: 414–422.

Schor, T., A. Paula, F. M. Carvalho et al. 2006. S100P calcium-binding protein expression is associated with high-risk proliferative lesions of the breast. *Oncology Reports* 15: 3–6.

Sedelnikova, O. A., A. Nakamura, O. Kovalchuk et al. 2007. DNA double-strand breaks form in bystander cells after microbeam irradiation of three-dimensional human tissue models. *Cancer Research* 67: 4295–4302.

Sedelnikova, O. A., D. R. Pilch, C. Redon, and W. M. Bonner. 2003. Histone H2AX in DNA damage and repair. *Cancer Biology and Therapy* 2: 233–235.

Semprini, S., F. Capon, A. Tacconelli et al. 2002. Evidence for differential S100 gene over-expression in psoriatic patients from genetically heterogeneous pedigrees. *Human Genetics* 111: 310–313.

Shah, M., D. Revis, S. Herrick et al. 1999. Role of elevated plasma transforming growth factor-beta1 levels in wound healing. *The American Journal of Pathology* 154: 1115–1124.

Sherr, C. J. and J. M. Roberts. 1995. Inhibitors of mammalian G1 cyclin-dependent kinases. *Genes & Development* 9: 1149–1163.

Siegel, P. H. 2004. Terahertz technology in biology and medicine. *IEEE Transactions on Microwave Theory and Techniques* 52: 2438–2447.

Smale, S. T. and J. T. Kadonaga. 2003. The RNA polymerase II core promoter. *Annual Review of Biochemistry* 72: 449–479.

Sperka, T., J. Wang, and K. L. Rudolph. 2012. DNA damage checkpoints in stem cells, ageing and cancer. *Nature Reviews Molecular Cell Biology* 13: 579–590.

Titova, L. V., A. K. Ayesheshim, A. Golubov et al. 2013a. Intense THz pulses cause H2AX phosphorylation and activate DNA damage response in human skin tissue. *Biomedical Optics Express* 4: 559.

Titova, L. V., A. K. Ayesheshim, A. Golubov et al. 2013b. Intense picosecond THz pulses alter gene expression in human skin tissue in vivo. Presented at *Proceedings of SPIE* 8585: 85850Q-Q-10, San Francisco, CA.

Titova, L. V., A. K. Ayesheshim, A. Golubov et al. 2013c. Intense THz pulses down-regulate genes associated with skin cancer and psoriasis: A new therapeutic avenue? *Scientific Reports* 3: 2363.

Tonouchi, M. 2007. Cutting-edge terahertz technology. *Nature Photonics* 1: 97–105.

Voss, M., B. Wolff, N. Savitskaia et al. 1999. TGFbeta-induced growth inhibition involves cell cycle inhibitor p21 and pRb independent from p15 expression. *International Journal of Oncology* 14: 93–194.

Wang, H.-T., B. Choi, and M.-s. Tang. 2010. Melanocytes are deficient in repair of oxidative DNA damage and UV-induced photoproducts. *Proceedings of the National Academy of Sciences* 107: 12180–12185.

Wang, S. Q., R. Setlow, M. Berwick et al. 2001. Ultraviolet A and melanoma: A review. *Journal of the American Academy of Dermatology* 44: 837–846.

Watson, G. E., S. A. Lorimore, D. A. Macdonald, and E. G. Wright. 2000. Chromosomal instability in unirradiated cells induced in vivo by a bystander effect of ionizing radiation. *Cancer Research* 60: 5608–5611.

Weightman, P. 2012. Prospects for the study of biological systems with high power sources of terahertz radiation. *Physical Biology* 9: 053001.

Williams, R., A. Schofield, G. Holder et al. 2013. The influence of high intensity terahertz radiation on mammalian cell adhesion, proliferation and differentiation. *Physics in Medicine and Biology* 58: 373–391.

Wilmink, G. and J. Grundt. 2011. Invited review article: Current state of research on biological effects of terahertz radiation. *Journal of Infrared, Millimeter, and Terahertz Waves* 32: 1074–1122.

Wilmink, G. J., B. L. Ibey, C. L. Roth et al. 2010a. Determination of death thresholds and identification of terahertz (THz)-specific gene expression signatures. *Presented at Proceedings of SPIE* 7562, San Francisco, CA: 75620K.

Wilmink, G. J., B. D. Rivest, B. L. Ibey et al. 2010b. Quantitative investigation of the bioeffects associated with terahertz radiation. Presented at *Proceedings of SPIE* 7562: 75620L-L-10, San Francisco, CA.

Woodward, R. M., V. P. Wallace, D. D. Arnone, E. H. Linfield, and M. Pepper. 2003. Terahertz pulsed imaging of skin cancer in the time and frequency domain. *Journal of Biological Physics* 29: 257–259.

Xu, J., K. W. Plaxco, and S. J. Allen. 2006. Absorption spectra of liquid water and aqueous buffers between 0.3 and 3.72 THz. *The Journal of Chemical Physics* 124: 036101.

Yang, L., T. Chan, J. Demare et al. 2001. Healing of burn wounds in transgenic mice overexpressing transforming growth factor-$\hat{\text{I}}^2$1 in the epidermis. *The American Journal of Pathology* 159: 2147–2157.

Yu, C., S. Fan, Y. Sun, and E. Pickwell-MacPherson. 2012. The potential of terahertz imaging for cancer diagnosis: A review of investigations to date. *Quantitative Imaging in Medicine And Surgery* 2: 33–45.

Zhang, X.-C. 2002. Terahertz wave imaging: Horizons and hurdles. *Physics in Medicine and Biology* 47: 3667–3677.

Zwang, Y., A. Sas-Chen, Y. Drier et al. 2011. Two phases of mitogenic signaling unveil roles for p53 and EGR1 in elimination of inconsistent growth signals. *Molecular Cell* 42: 524–535.

III

Terahertz Biomedical Applications

14

Terahertz Dynamic Imaging of Skin Drug Absorption

Kyung Won Kim
Dana-Farber Cancer Institute

Joon Koo Han
College of Medicine Seoul National University

Joo-Hiuk Son
University of Seoul

14.1 Introduction

Terahertz (THz) spectroscopy using the THz time-domain spectroscopy (TDS) technique enables identification of chemical composition owing to the spectral absorption properties of materials (Kawase et al. 2007; Son 2009; Watanabe et al. 2004). THz techniques are continuously applied in pharmaceutical research (Reid et al. 2010). For example, a THz spectroscopic technique can be used to detect illegal drugs such as amphetamine or to identify the spectral features of antibiotics such as doxycycline and sulfapyridine even when they are mixed with other materials (Redo-Sanchez et al. 2011).

Since the development of THz spectroscopic imaging technique in the mid-1990s, there has been an increasing interest in its biomedical applications for the detection of the skin, oral, and breast cancer (Brun et al. 2010; Fitzgerald et al. 2006; Ji et al. 2009; Son 2009). These biomedical applications have included surface imaging such as that of the skin, oral mucosa, or excised tissue, because the penetration depth of THz beam into biological tissues is limited. In that sense, the skin is the most promising target of THz imaging, and several commercial THz imaging devices have been developed for evaluation of skin diseases (Wallace et al. 2004).

Theoretically, a combination of spectroscopy and the imaging capability of THz-TDS technique may enable combination of drug identification and imaging of drug components. Considering that the skin is practically the best target of THz imaging, imaging of skin drug absorption might be a good biomedical application of THz imaging. The use of transdermal drug delivery systems has grown over the past three decades since the first commercially available scopolamine transdermal patch for motion sickness was approved by the US Food and Drug Administration (FDA) in 1979 (Nachum et al. 2006; Tfayli et al. 2007). The global market for transdermal drug delivery was estimated at $12.7 billion in 2005 and is expected to grow to $31.5 billion in 2015 (Tanner and Marks 2008). The ultimate goal of transdermal drug delivery research is to develop a noninvasive device that delivers the right dose of the active agent at the right rate according to a patient's need. Novel, noninvasive approaches to enhance and control drug transport across the skin are under intensive investigation (Guy 2010). In this chapter, we review the basic concepts of transdermal drug delivery and the feasibility of THz imaging as an imaging tool for research on transdermal drug delivery.

14.2 Transdermal Drug Delivery

14.2.1 Overview of Transdermal Drug Delivery

Transdermal drug delivery has several advantages over drug administration via oral, intravenous, and intramuscular routes (Table 14.1). It can provide a sustained and controlled release of the medication so that blood concentration of the drug can be kept within the therapeutic window for extended periods of time. It is a noninvasive form or drug administration and allows easy control of drug administration. The administration of the drug can be stopped simply by removing the transdermal drug delivery system, such as a patch. Furthermore, transdermal drug delivery can help avoid hepatic metabolism, which may affect the efficacy of a drug when it is administered through intravenous and oral routes. Thus, compared to systemic administration, transdermal drug delivery can help reduce the dose of a drug and systemic side effects (Guy 2010).

Transdermal drug delivery has several drawbacks and limitations. Owing to the physical and chemical barriers of the skin, only small molecules (e.g., those with a molecular weight of less than 500 Da), which are small enough to penetrate the skin via the intercellular pathway, can be delivered by this method. In addition, a drug should be lipophilic to penetrate the skin and should have some aqueous solubility to be dissolved in the blood. Locally irritating or sensitizing drugs cannot be used in clinical practice (Thomas and Finnin 2004).

The most common transdermal drug delivery system is an adhesive skin patch that contains a specific dose of medication and/or a penetration enhancer (Berner and John 1994). There are several types of patches, such as single-layer drug-in-adhesive, multilayer drug-in-adhesive, reservoir system, and matrix system, but their basic mechanism is similar in that the drug is released into the skin, through the adhesive, from the drug-containing part such as a membrane, matrix, or reservoir. The flux of the drug across the skin is influenced by the diffusivity of the drug and the concentration gradient of the drug within the skin, in accordance with Fick's first law, which postulates that the flux goes from regions of high concentration to regions of low concentration with a magnitude that is proportional to the concentration gradient (Figure 14.1) (Surber et al. 1990). The diffusivity of a drug varies according to the chemical structure of the drug, which might show variable interactions with components of the skin. Furthermore, conventional transdermal drug delivery systems, such as patches, creams, and gels, have some limitations. To improve drug penetration, various techniques have been developed, including chemical penetration enhancers and physical enhancement methods such as iontophoresis, sonophoresis, electroporation, and microneedles (Thomas and Finnin 2004).

Development of penetration-enhancing techniques is one of the important issues in the manufacturing of transdermal drug delivery systems. Numerous compounds have been evaluated for use as

TABLE 14.1   Advantages of Transdermal Drug Delivery

A controlled delivery of drug for a long time
Weaker fluctuation in the blood concentration of drugs (allows avoiding the drug spike concentrations after intravenous or oral administration)
Noninvasive and patient-friendly method
Improved patient compliance
Easily terminated by removal of the system
Avoidance of hepatic metabolism, thus enabling dose reduction
Avoidance of gastric side effects and systemic side effects

Fick's first law

$$J = D \cdot A \cdot K \frac{\Delta C}{h} = P \cdot \Delta C$$

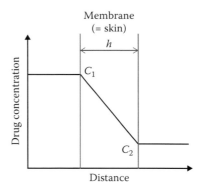

J: Flow rate, the amount of drug that will flow through a unit area during a unit time interval (Unit: $\mu g/cm^2/h$)

D: Difffusivity or diffusiono coefficient (Unit: cm^2/s), reflecting the velocity of the diffusion drug

A: Surface area

K: Partition coefficient, reflecting hydrophilicity

ΔC: Concentration difference of drug ($= C_2 - C_1$)

h: Membrane thickness

P: Permeability (unit: cm/s)

FIGURE 14.1   Fick's first law.

chemical penetration enhancers, such as sulfoxides (e.g., dimethyl sulfoxide [DMSO]), azones (e.g., laurocapram), pyrrolidones (e.g., 2-pyrrolidone), alcohols (e.g., ethanol), or glycols (e.g., propylene glycol) (Pathan et al. 2009). Penetration enhancers may act by reducing the barrier resistance of the stratum corneum or by increasing the drug's diffusivity through skin proteins. Various mechanical enhancement methods are very promising, but a detailed explanation of such methods is beyond the scope of this chapter.

During the last three decades, transdermal drug delivery systems have gained considerable attention from researchers and pharmaceutical companies, and many drugs have been approved for transdermal administration (Table 14.2). This progress is based on a better understanding of the skin barrier function and the physicochemical and pharmacokinetic characteristics of the drug. However, the pool of potential candidates for transdermal drug delivery has not dramatically expanded, with some distinct challenges remaining (Guy 2010). One of these challenges is that in the currently used in vitro and in vivo methods, a long time is required to test a chemical and perform extensive chemical analytic testing such as HPLC-based testing. Real-time in vivo evaluation methods for skin drug absorption, which can reduce the testing time and in which dedicated chemical analytic testing can be skipped, might enable mass screening of the candidate chemicals for transdermal drug delivery.

TABLE 14.2   Commonly Used Drugs Approved for Transdermal Administration

Drug Compound	Target Diseases or Conditions
Scopolamine	Motion sickness
Nitroglycerin	Angina pectoris
Clonidine	Hypertension
Estradiol	Hormone replacement therapy
Fentanyl	Analgesia
Nicotine	Smoking cessation
Testosterone	Hypogonadism
Lidocaine	Local anesthetic
Oxybutynin	Incontinence
Selegiline	Depression
Methylphenidate	Attention deficit hyperactivity disorder
Buprenorphine	Analgesia
Rivastigmine	Dementia
Rotigotine	Parkinson's disease
Granisetron	Antiemetic

14.2.2 Anatomy and Role of the Skin

The skin is the largest organ covering the whole body, and it plays a vital role in maintaining life. The skin has several functions. The most important function of the skin is to form a barrier to protect the body from physical injury as well as from microorganisms, ultraviolet radiation, and toxic agents. The skin also has very dynamic functions in maintaining the homeostasis of the body, including regulation of body's temperature, water content, and skin respiration, as well as controlling the inward and outward passage of water, electrolytes, and various substances.

The skin is a multilayered organ that is composed of three layers: the epidermis, the dermis, and the hypodermis (Figure 14.2). Understanding the structure and function of each of the skin layers is vital for both transdermal drug delivery research and THz imaging research (Table 14.3). The epidermis is a stratified squamous cornifying epithelium, consisting of several layers of cells. The keratinocytes proliferate in the stratum basale, migrate through the epidermis toward the surface, and finally transform into corneocytes forming the stratum corneum. The stratum corneum, the outermost layer of the epidermis, is the most important layer in terms of the transdermal drug delivery, because it forms a barrier that protects the underlying skin tissue from chemicals (Haftek et al. 2011).

The stratum corneum is composed of 10–20 layers of corneocytes, stacked up in a brick-like fashion. Each corneocyte is a nonviable dead cell with a flat, platelike structure (Bouwstra et al. 2003). The thickness of stratum corneum is approximately between 20 and 170 μm, depending on an individual's age, anatomical location, and hydration (Egawa et al. 2007). The main component of the cell membrane of corneocyte is ceramides, a family of waxy lipid molecules composed of sphingosine and fatty acids, which make the cellular membrane the most insoluble structure of the corneocyte. Each corneocyte contains a dense network of keratins, which can hold large amounts of water and keep the skin hydrated by preventing water evaporation. Corneocytes are held together by specialized protein structures called corneodesmosomes (Hatta et al. 2006). Surrounding the corneocytes in the extracellular space are stacked layers of lipid, the so-called intercellular lipid matrix. This intercellular lipid matrix also acts as a barrier for water, drugs, and other substances (Hatta et al. 2006). In terms of components, the stratum corneum contains substantial amount of lipid (5%–15% of its dry weight) and protein (mainly keratin, 75%–85% of its dry weight) and variable water content (generally 15%–25% at the surface of a skin but can absorb three times its weight). This structural complexity of stratum corneum influences the transdermal drug delivery by altering drug diffusivity or drug–skin interaction.

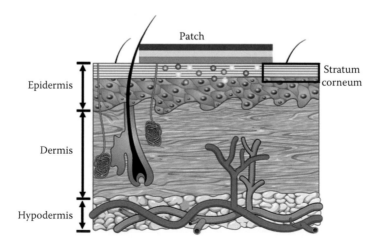

FIGURE 14.2 Structure of the skin. The skin is composed of three layers: the epidermis, the dermis, and the hypodermis. The most superficial layer of the epidermis is stratum corneum.

TABLE 14.3   Skin Structure

Layer	Characteristics
Nonviable epidermis (stratum corneum)	• The outermost layer of the skin, the actual physical barrier • The strongest barrier in transdermal drug delivery • Thickness ranging from 20 to 170 μm • Composed of corneocytes and intercellular lipid matrix • Lipid (5%–15% of dry weight), proteins (mainly keratin, 75%–85% of dry weight), and variable water content (15%–25% at the skin surface)
Viable epidermis	• This layer is located between the stratum corneum and the dermis • Composed of living cells named keratinocytes • Thickness ranging from 0.5 to 1.5 mm • The water content is about 70%
Dermis	• Middle layer beneath the epidermis • Mainly composed of a matrix of loose connective tissue embedded in an amorphous ground substance • Thickness ranging from 0.3 to 3 mm • Contains blood vessels and lymphatics, into which the drug is absorbed systemically
Hypodermis (subcutaneous fat layer)	• Mostly fatty tissue containing a few blood vessels and lymphatics

The viable epidermis is composed of living cells, mainly keratinocytes. The thickness of epidermis ranges from 0.5 to 1.5 mm (Brannon 2007). The water content of stratum corneum increases from about 15% to 25% (g water/g tissue) at the skin surface to a constant level of about 70% in the viable epidermis (Warner et al. 1988). Between the epidermis and dermis is the basement membrane, a very thin connective tissue mainly consisting of type IV collagen.

The dermis, the midlayer of the skin, is usually thicker than the epidermis, ranging from 0.3 mm on the eyelids to 3 mm on the back, palms, and soles (Brannon 2007). The major component of dermis is connective tissue such as collagen and elastic fiber. It also contains nerves, dermal vasculature, lymphatics, sweat glands, and hair roots. In the dermis layer, the penetrating drug is absorbed into systemic circulation (Pathan et al. 2009).

The hypodermis, also called the subcutaneous fat, is the innermost layer and is mostly composed of fat. The subcutaneous tissue varies in its thickness according to the age, sex, location, and the nutritional status. Most of the drug that permeates through the skin enters the systemic circulation before reaching the hypodermis, although the fatty tissue could serve as a depot for the drug (Pathan et al. 2009).

The rate-determining layer for most topical drugs is the stratum corneum. Regarding the pathway of transdermal drug delivery, drug permeation can occur by diffusion via intercellular permeation through the intercellular lipid matrix of stratum corneum, by transcellular permeation through intracellular keratin domains of stratum corneum, or via the hair follicle, sebaceous, and sweat glands. Most molecules are believed to penetrate through the skin via intercellular route (Pathan et al. 2009; Tanner and Marks 2008). Permeability through the skin depends on the lipophilicity of the chemical and the thickness of the outer layer of epidermis, as well as on factors such as molecular weight and concentration of the substance. Even though the physical nature of the skin's barrier is now well-understood, the understanding of the detailed penetration mechanisms of various candidate drugs and biological interaction between drugs and skin components remains a real challenge.

14.2.3 Methods for the Evaluation of Skin Drug Absorption

Accurate and reliable data regarding the diffusion kinetics of the transdermal drug are necessary in order to understand and optimize the skin absorption of transdermal drugs. To this end, guidance notes on skin absorption studies were developed by the Organisation for Economic Co-operation and Development (OECD) Expert Group on Dermal Absorption (OECD 2011). The methods for measuring percutaneous absorption can be divided into two categories: in vitro methods and in vivo methods.

Currently, a well-established method that has been adopted by the OECD guidelines is an in vitro skin absorption test, which uses diffusion cells that measure the diffusion of chemicals across the skin to a receptor fluid reservoir. These in vitro methods can provide a great deal of valuable information such as the penetration of chemicals, expressed as a percentage of the dose or as a rate. The most common method is Franz cell diffusion test combined with high-performance liquid chromatography (HPLC) analysis. The basic Franz cell diffusion system is illustrated in Figure 14.3 (Kim et al. 2012b). The diffusion cell is composed of a donor chamber and a receptor chamber between which the skin is positioned. The receptor chamber contains buffer solution, which is kept at a constant temperature by a circulating water bath. After a test drug is applied to the donor chamber, which is in contact with the skin, aliquots of receptor fluid are serially sampled at predefined times in order to evaluate penetration kinetics. The sampled receptor fluid is assayed by the HPLC to determine the concentration of drug that has permeated through the skin (Franz 1975, 1978). Even though the Franz cell diffusion method is an accurate, reliable, and well-established method, it only provides information related to the diffusion of the transdermal drug. It cannot provide information regarding the in vivo state. In addition, this method requires many steps, such as diffusion cell experiments and HPLC analysis.

There are several different in vivo methods for determining transdermal drug absorption: (1) measurement of the radioactivity of blood and of the radioactivity of the radiolabeled drug applied on the skin, (2) measurement of the parent chemical and/or its metabolite level in blood, (3) microdialysis technique, and (4) stratum corneum tape stripping (Kezic 2008). The most important advantage of the in vivo methods is that they allow evaluation of transdermal drug absorption and skin toxicity in the physiologically and metabolically intact systems. There are several disadvantages associated with the use of in vivo methods. Live animals need to be used with these methods, which may raise ethical issues. The differences in permeability of animal skin and human skin may hamper the application of the results of in vivo methods in human research or clinical practice. A radiolabeled material is necessary to facilitate reliable results, which may cause radiation exposure. Determining the early absorption phase is difficult with in vivo methods (OECD 2004). Because of these drawbacks, several noninvasive, time-resolved,

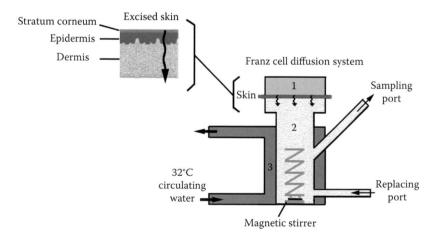

FIGURE 14.3 Schematic drawing of excised skin mounted in the Franz cell diffusion system. The donor chamber (1) above the skin contains the applied topical agent. The chamber below the skin is the receptor chamber (2) from which the samples are taken through the sampling port. The receptor chamber is surrounded by a water jacket (3), maintained at 32°C. A magnetic stirrer and stirring helix are magnetically rotated at the bottom of the receptor chamber. The topical drug, which is applied to the stratum corneum side of the skin, permeates into the dermis side and then crosses the skin. (Reproduced from Kim, K.W. et al., *Opt. Exp.*, 20, 9476, 2012b. With permission.)

3D imaging techniques are under development, including spectroscopic imaging techniques (Jiang et al. 2008; Tfayli et al. 2007) and multiphoton microscopic imaging (Tsai et al. 2009).

Spectroscopic imaging techniques are a promising tool for transdermal drug delivery research, because they allow simultaneous, real-time determination of both the spatial information in the form of imaging and the chemical properties of a drug, while preserving the integrity of the drug and the subject (Jiang et al. 2008). A variety of spectroscopic methods have been applied in in vivo studies of transdermal drug delivery, including infrared (IR) spectroscopic imaging and Raman spectroscopic imaging (Jiang et al. 2008; Tfayli et al. 2007). IR spectroscopy and Raman spectroscopy are vibrational spectroscopic techniques that provide information on the molecular composition and structure of a sample as well as the interactions that occur within it. In both techniques, the radiation light causes changes in oscillating and rotational energy of the molecules.

Among the various IR spectroscopic techniques, Fourier transform infrared (FTIR) spectroscopic imaging is the most commonly used method for studying penetration kinetics of transdermal drugs. In this technique, an IR beam is emitted through an IR transparent crystal into the target skin. Then, the radiated skin absorbs IR at frequencies corresponding to the absorption spectrum of the drug, and the penetrating drug can be quantified by measuring the IR spectrum (Kezic 2008). This technique can achieve high spatial resolution and allow fast acquisition on the order of minutes (Jiang et al. 2008). However, the penetration depth is low. Further, the thermal emission decay (TED) FTIR method, which is a recently developed noncontact technique for detection of the presence and concentration of chemicals in the stratum corneum, can also evaluate the spectrum up to a stratum corneum depth of 10 µm (Kezic 2008).

Raman spectroscopy is based on laser radiation that is focused with microscope objective on a point in the skin sample. When light irradiates a molecule, most of the photons are scattered elastically, while a small amount of light is scattered inelastically. The inelastically scattered light is termed Raman scattering. Raman scattering demonstrates frequency shifts with respect to the incident light and corresponds to the vibrational energy transitions of the molecule (Tfayli et al. 2007). The Raman spectrum can be considered a spectral fingerprint of the molecule. Confocal Raman spectroscopy is a combination of Raman spectroscopy and confocal microscopy. It enables noninvasive, depth-resolved, and time-resolved imaging of the skin, which can provide information on the chemical composition of the skin and the spatial and depth distribution of its components. Owing to these characteristics, it also allows real-time in vivo monitoring of the transdermal absorption of a drug (i.e., determination of the concentration profile of the transdermal drug or estimation of the diffusion coefficient and permeability coefficient of the transdermal drug) (Kezic 2008). This technique can achieve high spatial resolution and relatively high penetration depth of several hundred micrometers (Cal et al. 2009; Kezic 2008). However, Raman spectroscopic techniques have a few limitations: (1) heating of a sample using intense laser radiation can destroy tissue; (2) the Raman effect is very weak, requiring a highly sensitive and optimized device; and (3) the method can only detect chemicals that have a specific absorption spectrum, distinct from that of the skin structure (Cal et al. 2009; Kezic 2008).

14.3 Application of Terahertz Imaging for Skin Drug Absorption

THz imaging is a novel spectroscopic imaging technique that can provide nondestructive, label-free, rapid imaging and spectroscopic information. Among various THz imaging techniques, the most established approach is the THz-TDS technique, which is based on the photoconductive switch. Femtosecond pulsed laser excitation of photoconductive switches causes rapid transients in the photoconductor carrier density and conductance (Sun et al. 2011). The THz-TDS imaging technique is a coherent technique for acquisition of the THz wave's temporal electric fields, which enables simultaneous acquisition of THz pulse amplitude and phase. Each pixel of a THz-TDS image contains full THz pulse waveforms that can be transformed from time domain to the frequency domain by Fourier transformation to provide the spectra, which enables measurements of both refractive index and absorption coefficient (Wallace et al. 2004). The absorption relies on the chemical constituents of the medium. Acquisition of information of a large spectral range by broadband detection can provide specific spectroscopic information regarding the samples.

A THz-TDS imaging in reflection mode is a time-domain technique, thus allowing detection of THz pulse reflections from distinct layered structures within the tissue. When THz pulses reach a boundary between media that have different refractive indices, the pulses are reflected back. Reflections from different layers have different optical time delays. By using the optical time delays, information regarding the depth of the layers can be obtained (Wallace et al. 2004; Woodward et al. 2002). The THz B-mode images can be generated by plotting the time-domain profile as a function of position and can be reconstructed as the vertical axis represents optical time delay and the gray scale represents the THz amplitude, such as in ultrasound B-mode scanning. A THz-TDS imaging technique can achieve a very high signal-to-noise ratio (SNR), because the background noise can be eliminated efficiently with coherent time-gated detection (Mittleman et al. 1996). The high SNR of the imaging technique enables detection and imaging of small changes in the refractive index of materials. These coherent, time-gated, and low-noise techniques of THz imaging have the potential to provide both structural and functional information because of their chemical specificity (Wallace et al. 2008).

Recently, authors demonstrated that THz reflection 2D imaging and tomographic B-mode imaging are feasible for visualizing the distribution and penetration of a topical agent and its serial dynamic changes, using DMSO containing ketoprofen, in conventional THz-TDS system (Kim et al. 2012a,b). THz dynamic reflection imaging can reflect the drug that penetrated through the skin based on the changes in the reflection signals of the skin to which the drug was transdermally applied. The drug-applied site was not initially visible on the serial THz reflection 2D images (Figure 14.4) but appeared on the images that were taken 8–16 min after the drug application. On these images, the drug-applied site was characterized by a lower reflection signal compared to that of the adjacent normal skin. As the drug that is applied on the stratum corneum of the skin penetrates into the dermis, the reflection signal of the drug-applied site gradually decreases for certain time period and then remains constant. Each pixel on a THz reflection 2D image has full spectral data of THz waveforms. As illustrated in Figure 14.5a, a serial time-domain waveform, acquired at the drug-applied site, shows two peaks, which can be explained as follows: (1) the first peak, which decreases over time, is the main factor used to determine the THz reflection image; and (2) the second peak, which

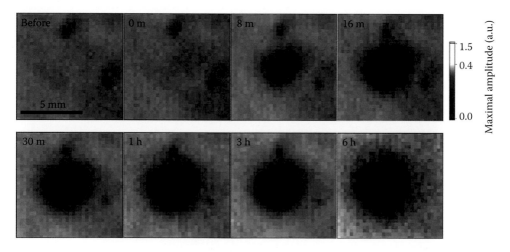

FIGURE 14.4 Serial THz reflection images of the drug-applied site. (Reproduced from Kim, K.W. et al., *Opt. Exp.*, 20, 9476, 2012b. With permission.) The drug-applied site is not visible on the initial image but appears as a dark shaded area, approximately 8 min after the drug application. The drug-applied sites appear in a darker shade (with lower reflection signal) compared to the unapplied skin. The intensity of darkness gradually increases for 1 h and then remains constant. The area of the dark shade increases with time, which may indicate diffusion and distribution changes of the topical drug in the skin.

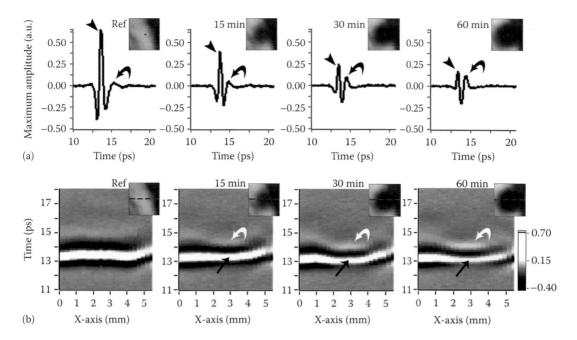

FIGURE 14.5 Time-domain waveforms and B-scan images of a drug-applied site. (Reproduced from Kim, K.W. et al., *Opt. Exp.*, 20, 9476, 2012b. With permission.) Serial time-domain waveforms (a) obtained before (Ref) and 15, 30, and 60 min after the application of the drug show the maximal amplitude of the first main peak (arrowheads) and the second small peak (curved arrow) at the drug-applied sites (at the central dot in the THz reflection image on the upper right). THz B-mode images (b) of the drug-applied site show the reconstructed perpendicular images (at the dotted line in the THz reflection image on the upper right) as the vertical axis represents the optical delay and the gray scale represents the THz amplitude. Serial THz B-mode images show that the signal and optical time width of the quartz–dermis interface (arrows) decreases with time, which corresponds to the first main peak in time-domain waveforms. The signal of the layer (curved arrows) just above the quartz plate–dermis interface, which corresponds to the small peak in time-domain waveform, increases with time.

increases with time, may indicate the presence of an interface between the drug-permeated layer and the rest of the skin dermis. THz tomographic B-mode images can be reconstructed using the optical time delay and THz reflection signals of these peaks (Figure 14.5b). This experiment is the first attempt to image the skin drug absorption on THz reflection image. However, the tomographic B-mode images reflect the optical time depth rather than the anatomic depth, warranting further investigation and development.

When compared with the standard in vitro skin absorption test—the Franz cell diffusion test—the dynamic pattern of the THz reflection signal decrease is similar to that of DMSO absorption analyzed by the Franz cell diffusion test, which indicates that THz imaging mainly reflects the penetration and distribution of DMSO component (Figure 14.6) (Kim et al. 2012b). This can be explained by the fact that the THz imaging is highly sensitive to polar materials and DMSO is a polar solvent that also strongly absorbs the THz radiation (Suen et al. 2009). These results may imply that the THz imaging is good for testing penetration of polar molecules. However, further investigation is warranted to find out the detailed characteristics of THz imaging for transdermal delivery of various drugs, because various types of topical agents might induce variable responses to THz radiation.

The potential advantages of THz imaging for skin drug absorption are as follows: (1) THz imaging is very sensitive to polar molecules, which may enable analyzing permeation kinetics of polar molecules (Suen et al. 2009); (2) THz imaging provides rapid acquisition without special preparation such as labeling, which can provide real-time information on the skin and drug in their natural states; (3) THz radiation is a safe, nonionizing radiation with very low photon energy (Bourne et al. 2008); (4) the wavelength

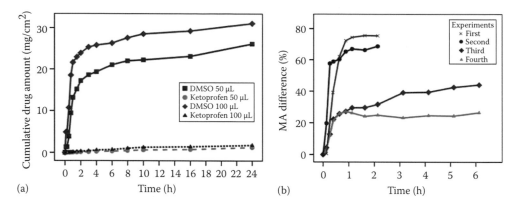

FIGURE 14.6 Profiles of the cumulative permeated drug amount obtained using the Franz cell diffusion test (a) and a diagram showing the time course of THz reflection signal (b). (Reproduced from Kim, K.W. et al. *Opt. Exp.*, 20, 9476, 2012b. With permission.) In both curves, the permeated DMSO amount and the THz signal increase rapidly during the early time period and then remain constant, suggesting that the THz imaging mainly reflects the DMSO component.

range used in the THz imaging (3–100 μm) is significantly larger than the size of scattering structures in skin tissue, which can considerably reduce the scattering effects compared with those of other spectroscopic techniques that use shorter wavelength radiation, for example, near-IR imaging (Wallace et al. 2004); and (5) THz imaging can provide high SNR of images and can estimate the depth information of layered structures within the tissue, as explained previously (Mittleman et al. 1996; Wallace et al. 2008).

The potential drawbacks of THz imaging are its relatively low spatial resolution and axial resolution, compared to those of IR spectroscopy or Raman spectroscopy. The conventional resolution is not enough for cellular or molecular imaging, which is becoming popular in medical imaging technology. The penetration depth is still a crucial issue in in vivo imaging. Penetration depths range from a few hundred micrometers for tissues with high water content to several centimeters for tissues with high fat content (Arnone et al. 1999; Pickwell and Wallace 2006). Because water or polar molecules have strong absorption across the THz range, the detection of THz signal from nonpolar molecules is limited, which may hamper the analysis of transdermal drug delivery of multicompound drugs (Pickwell-MacPherson 2010). These drawbacks are all future research topics and are currently under intensive investigation.

14.4 Prospects for Future Research

The demand for in vivo real-time imaging methods to trace chemical compounds within the skin has been increasing. New in vivo methods should enable noninvasive investigations of skin compartments that do not require tissue removal and real-time tracking of drugs under natural conditions in order to assess pharmacokinetics of topical drugs. THz imaging may meet these demands (Kim et al. 2012b). THz imaging is a coherent, time-resolved, and low-noise imaging technique that has the potential to provide both structural information and chemical specificity on skin and transdermal drugs (Wallace et al. 2008). In addition, THz imaging is a noninvasive, label-free tool to observe skin changes under natural conditions (Hattori and Sakamoto 2007). Furthermore, THz imaging may detect the different diffusion rates of each drug component in the mixed topical agents.

In order to use THz imaging for transdermal drug delivery research, the spatial and axial resolution, the penetration depth, and the acquisition rate of this method should be improved. THz imaging is still in the early stages of development, and great advances in THz generation, THz detection, and THz imaging have been made in the past few decades. We hope that in the near future, the THz imaging will be translated from the laboratory to the clinic for the transdermal drug delivery evaluation as well as for other biomedical applications.

References

Arnone, D. D., C. M. Ciesla, A. Corchia et al. 1999. Applications of terahertz (THz) technology to medical imaging. Presented at *Proceedings of SPIE* 3828: 209–219.

Berner, B. and V. A. John. 1994. Pharmacokinetic characterisation of transdermal delivery systems. *Clinical Pharmacokinetics* 26: 121–134.

Bourne, N., R. H. Clothier, M. D'Arienzo, and P. Harrison. 2008. The effects of terahertz radiation on human keratinocyte primary cultures and neural cell cultures. *Alternatives to Laboratory Animals* 36: 667–684.

Bouwstra, J. A., P. L. Honeywell-Nguyen, G. S. Gooris, and M. Ponec. 2003. Structure of the skin barrier and its modulation by vesicular formulations. *Progress in Lipid Research* 42: 1–36.

Brannon, H. 2007. Dermatology—Epidermis. http://dermatology.about.com/cs/skinanatomy/a/anatomy.htm, assessed on February 8, 2014.

Brun, M. A., F. Formanek, A. Yasuda et al. 2010. Terahertz imaging applied to cancer diagnosis. *Physics in Medicine and Biology* 55: 4615.

Cal, K., J. Stefanowska, and D. Zakowiecki. 2009. Current tools for skin imaging and analysis. *International Journal of Dermatology* 48: 1283–1289.

Egawa, M., T. Hirao, and M. Takahashi. 2007. In vivo estimation of stratum corneum thickness from water concentration profiles obtained with Raman spectroscopy. *Acta Dermato-Venereologica* 87: 4–8.

Fitzgerald, A. J., V. P. Wallace, M. Jimenez-Linan et al. 2006. Terahertz pulsed imaging of human breast tumors1. *Radiology* 239: 533–540.

Franz, T. J. 1975. Percutaneous absorption. On the relevance of in vitro data. *Journal of Investigative Dermatology* 64: 190–195.

Franz, T. J. 1978. The finite dose technique as a valid in vitro model for the study of percutaneous absorption in man. *Current Problems in Dermatology* 7: 58.

Guy, R. H. 2010. Transdermal drug delivery. In *Drug Delivery*, ed. Schäfer-Korting, M., pp. 399–410. New York: Springer.

Haftek, M., S. Callejon, Y. Sandjeu et al. 2011. Compartmentalization of the human stratum corneum by persistent tight junction-like structures. *Experimental Dermatology* 20: 617–621.

Hatta, I., N. Ohta, K. Inoue, and N. Yagi. 2006. Coexistence of two domains in intercellular lipid matrix of stratum corneum. *Biochimica et Biophysica Acta (BBA)-Biomembranes* 1758: 1830–1836.

Hattori, T. and M. Sakamoto. 2007. Deformation corrected real-time terahertz imaging. *Applied Physics Letters* 90: 261106.

Ji, Y. B., E. S. Lee, S.-H. Kim, J.-H. Son, and T.-I. Jeon. 2009. A miniaturized fiber-coupled terahertz endoscope system. *Optics Express* 17: 17082–17087.

Jiang, J., M. Boese, P. Turner, and R. K. Wang. 2008. Penetration kinetics of dimethyl sulphoxide and glycerol in dynamic optical clearing of porcine skin tissue in vitro studied by Fourier transform infrared spectroscopic imaging. *Journal of Biomedical Optics* 13: 021105.

Kawase, K., A. Dobroiu, A. Masatsugu Ya, Y. Sasaki, and C. Otani 2007. Terahertz rays to detect drugs of abuse. In *Terahertz Frequency Detection and Identification of Materials and Objects*, ed. Miles, R. E., pp. 241–250. New York: Springer.

Kezic, S. 2008. Methods for measuring in-vivo percutaneous absorption in humans. *Human & Experimental Toxicology* 27: 289–295.

Kim, K. W., H. Kim, J. Park, J. K. Han, and J.-H. Son. 2012a. Terahertz tomographic imaging of transdermal drug delivery. *IEEE Transactions on Terahertz Science and Technology* 2: 99–106.

Kim, K. W., K.-S. Kim, H. M. Kim et al. 2012b. Terahertz dynamic imaging of skin drug absorption. *Optics Express* 20: 9476–9484.

Mittleman, D. M., R. H. Jacobsen, and M. C. Nuss. 1996. T-ray imaging. *IEEE Journal of Selected Topics in Quantum Electronics* 2: 679–692.

Nachum, Z., A. Shupak, and C. R. Gordon. 2006. Transdermal scopolamine for prevention of motion sickness. *Clinical Pharmacokinetics* 45: 543–566.

OECD. 2004. Test No. 427: Skin Absorption: In Vivo Method, OECD Guidelines for the Testing of Chemicals, Section 4, OECD Publishing, Paris, France.

OECD. 2011. Guidance notes on dermal absorption, series on testing and assessment, No. 156. http://www.oecd.org/env/ehs/testing/48532204.pdf, accessed on February 8, 2014.

Pathan, I. B. and C. M. Setty. 2009. Chemical penetration enhancers for transdermal drug delivery systems. *Tropical Journal of Pharmaceutical Research* 8: 173–179.

Pickwell, E. and V. P. Wallace. 2006. Biomedical applications of terahertz technology. *Journal of Physics D-Applied Physics* 39: R301–R310.

Pickwell-MacPherson, E. 2010. Practical considerations for in vivo THz imaging. *Terahertz Science and Technology* 3: 163–171.

Redo-Sanchez, A., G. Salvatella, R. Galceran et al. 2011. Assessment of terahertz spectroscopy to detect antibiotic residues in food and feed matrices. *Analyst* 136: 1733–1738.

Reid, C. B., E. Pickwell-MacPherson, J. G. Laufer et al. 2010. Accuracy and resolution of THz reflection spectroscopy for medical imaging. *Physics in Medicine and Biology* 55: 4825.

Son, J. H. 2009. Terahertz electromagnetic interactions with biological matter and their applications. *Journal of Applied Physics* 105: 102033.

Suen, J. Y., P. Tewari, Z. D. Taylor et al. 2009. Towards medical terahertz sensing of skin hydration. *Studies in Health Technology and Informatics* 142: 364–368.

Sun, Y., M. Y. Sy, Y.-X. J. Wang et al. 2011. A promising diagnostic method: Terahertz pulsed imaging and spectroscopy. *World Journal of Radiology* 3: 55.

Surber, C., K.-P. Wilhelm, M. Hori, H. I. Maibach, and R. H. Guy. 1990. Optimization of topical therapy: Partitioning of drugs into stratum corneum. *Pharmaceutical Research* 7: 1320–1324.

Tanner, T. and R. Marks. 2008. Delivering drugs by the transdermal route: Review and comment. *Skin Research and Technology* 14: 249–260.

Tfayli, A., O. Piot, F. Pitre, and M. Manfait. 2007. Follow-up of drug permeation through excised human skin with confocal Raman microspectroscopy. *European Biophysics Journal* 36: 1049–1058.

Thomas, B. J. and B. C. Finnin. 2004. The transdermal revolution. *Drug Discovery Today* 9: 697–703.

Tsai, T.-H., S.-H. Jee, C.-Y. Dong, and S.-J. Lin. 2009. Multiphoton microscopy in dermatological imaging. *Journal of Dermatological Science* 56: 1–8.

Wallace, V., A. Fitzgerald, S. Shankar et al. 2004. Terahertz pulsed imaging of basal cell carcinoma ex vivo and in vivo. *British Journal of Dermatology* 151: 424–432.

Wallace, V. P., E. MacPherson, J. A. Zeitler, and C. Reid. 2008. Three-dimensional imaging of optically opaque materials using nonionizing terahertz radiation. *Journal of the Optical Society of America A* 25: 6.

Warner, R. R., M. C. Myers, and D. A. Taylor. 1988. Electron probe analysis of human skin: Determination of the water concentration profile. *Journal of Investigative Dermatology* 90: 218–224.

Watanabe, Y., K. Kawase, T. Ikari et al. 2004. Component analysis of chemical mixtures using terahertz spectroscopic imaging. *Optics Communications* 234: 125–129.

Woodward, R. M., B. E. Cole, V. P. Wallace et al. 2002. Terahertz pulse imaging in reflection geometry of human skin cancer and skin tissue. *Physics in Medicine and Biology* 47: 3853.

15

Terahertz for the Detection of Skin Cancer

Cecil Joseph
University of Massachusetts Lowell

Shuting Fan
The Hong Kong University of Science and Technology

Emma MacPherson
The Chinese University of Hong Kong

Vincent P. Wallace
University of Western Australia

15.1 Overview of Skin Cancer

Skin cancer is the most common form of cancer in the United States. According to the statistics provided by the Skin Cancer Foundation of the United States, more than 3.5 million cases are diagnosed annually. Each year, there are more new cases of skin cancer than the total combination of cancers of the breast, prostate, lung, and colon. Ultraviolet (UV) radiation from sun exposure is the primary cause of skin cancer. Other factors that play a role include smoking, HPV infections, some genetic syndromes, and chronic nonhealing wounds (Saladi and Persaud 2005).

There are mainly three types of skin cancer: basal cell carcinoma (BCC), squamous cell carcinoma (SCC), and malignant melanoma, each of which is named after the skin cell from where it arises.

BCC is formed in the lowest layer of the epidermis, the basal layer. It usually appears as a raised, smooth, pearly bump on the sun-exposed skin of the head, neck, or shoulders. Sometimes small blood vessels can be seen within the tumor. Crusting and bleeding in the center of the tumor are frequently observed and it is often mistaken for a sore that does not heal.

BCC is the most common form of skin cancer; an estimated 2.8 million cases are diagnosed annually in the United States, which represents approximately 80% of all nonmelanoma skin cancers (NMSCs) (Rubin et al. 2005). BCC has a comparatively low death rate and only accounts for less than 0.1% of all deaths caused by cancer (Miller and Weinstock 1994); however, it can be highly disfiguring if allowed to grow.

Squamous cell cancer originates within the middle layer of the epidermis and is the second most common type of skin cancer following BCC. Skin areas that have been exposed to the sun such as ears, face, and mouth are most likely to develop squamous cell cancer. Symptoms of this kind of skin cancer include a bump that turns into an open sore; ulceration; reddish, flat spot that is sometimes crusty; a bump that gets larger; and a sore that won't heal. When left untreated, it can spread to other parts of the body, such as the lymphatic system, bloodstream, and nerve routes.

Melanoma, which is the most aggressive, easy to spread, and fatal, originates in the pigment-producing cells (melanocytes). It can develop on any part of the body but is most commonly found on the arms, legs, and trunk. When detected early, the cure rate is considered to be very high. Symptoms include a mole, freckle, or new/existing spot that changes in size, shape, and color. It may have an irregular outline and possibly be more than one color.

Due to its seriousness, melanoma is categorized alone, and BCC, SCC, and other skin cancers are categorized under the generalized term of NMSC.

The likelihood of developing skin cancer varies with geography, ethnic origin, gender, and age. First of all, because UV radiation of sunlight is the main reason for skin cancer, people in areas that receive more sunlight have a higher chance of getting skin cancer. Second, ethnic background is also an important factor in the epidemiology of skin cancers: light-skinned Caucasians are more likely to be diagnosed with skin cancer than people with any other skin type. Research has also found that light-skinned Caucasians with skin cancer suffer a greater mortality. Australia has the highest incidence rate of skin cancer. According to the Department of Health and Aging of the Australian government, "Australians are four times more likely to develop skin cancer than any other form of cancer" (AACR and AIHW 2008); and approximately two in three Australians will be diagnosed with skin cancer before the age of 70. Men and women generally have different skin cancer rates with the incidence for males higher than for females. The Centers for Disease Control and Prevention in the United States has reported that the chance of getting skin cancer is becoming greater (CDC 2012). Reports from dermatologists in Hong Kong also indicate that the age of skin cancer patients is getting younger.

15.2 Mohs Micrographic Surgery

For well-defined, solid, cystic, and superficial BCCs and for tumors of less than 20 mm in diameter, surgical excision is the treatment of choice. A minimum margin of 4 mm is required to completely excise the tumor in more than 95% of cases (Wolf and Zitelli 1987). Ill-defined, micronodular, infiltrating, and sclerosing tumors may extend 15 mm or more beyond the clinical edge. Accurate histology reflecting all the tumor margins seems to be essential to achieve a complete cure. None of the reported range of histological techniques used is ideal (Rapini 1990), but Mohs micrographic surgery (MMS) is probably the best as it allows review of all the margins and same-day closure of the defects (Shriner et al. 1998).

The method and results of MMS are shown in Figure 15.1. Obvious tumor is debulked (removed), as indicated by the yellow and pink excisions in Figure 15.1. With a scalpel at 45° to the surface, a saucer-shaped excision is made, which includes the undersurface and entire epidermal margin, indicated by the dashed line. The tissue is divided into sections and color coded to assist in orientation for any further excisions. Horizontal frozen histology sections of the lower layer of the tissue, which encompasses all margins, are reviewed during surgery, and the tissue excised until all margins are clear.

There are a number of advantages offered by MMS. Infiltrating, sclerotic tumors can be defined accurately and the direction of the tumor spread identified, thus conserving tissue and avoiding the need for a *blind* wide excision. High cure rates of 99% at 5 years in primary tumors and 96% in recurrent tumors are widely reported using this technique (Chu and Edelson 1999). However, MMS is time consuming and expensive. Any system that could help define the histological subtype of the BCC and direction of subclinical spread preoperatively without performing a biopsy may simplify MMS to a single layer for all but the most extensive tumors. Also, by doing MMS, the surgeon can cut away all the cancer cells and spare as much healthy skin as possible, thus reducing disfigurement.

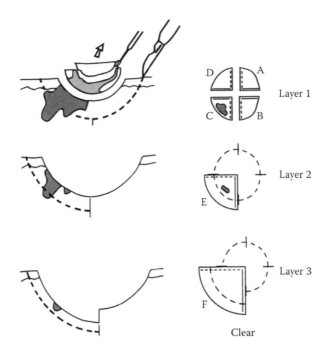

FIGURE 15.1 Method of tissue excision using MMS. (From Pye, R.J., *Horizons Med.*, 12, 339, 2000.)

15.3 Overview of Existing and Research-Based Skin Image Techniques

A large number of imaging/diagnostic techniques are currently under investigation and at different stages of development for clinical use (Mogensen and Jemec 2007; Ulrich et al. 2007). Imaging techniques for NMSC are generally compared to the *gold standard* of histology sections of the tissue sample. For noninvasive techniques, correlation of the acquired data/images to the gold standard is of critical importance in determining the sensitivity, specificity, and predictive values of the modality. In the following, we describe some of the conventional and experimental methods.

15.3.1 Clinical and Physical Examination

This is the most accessible and used test for NMSC diagnosis; however, precision and accuracy are unclear. Sensitivity and specificity values for these tests are reported based upon the individual performing the examination (dermatologists, family practitioners, and primary care physicians). Overall estimates of sensitivity range between 56%–90% and specificity of 75%–90% (Cooper and Wojnarowska 2002; Davis et al. 2005; Ek et al. 2005; Hallock and Lutz 1998; Har-Shai et al. 2001; Leffell et al. 1993; Morrison et al. 2001; Schwartzberg et al. 2005; Whited and Hall 1997; Whited et al. 1995).

15.3.2 Dermoscopy

Dermoscopy is essentially the noninvasive imaging of skin with a magnifying lens and a light source. Generally, immersion oil is applied to the lens–skin interface to improve the index match and remove scattering from the air–skin interface. Other names for this technique are dermatoscopy, epiluminescence microscopy, incident light microscopy, and skin surface microscopy. The magnification used varies between 10× and 100×. Reported sensitivities for BCC range between 86% and 96% with specificities ranging from 72% to 92% (Argenziano et al. 2004; Chin et al. 2003; Kreusch 2002; Menzies 2002; Newell et al. 2003; Otis et al. 2004; Zalaudek 2005; Zalaudek et al. 2004, 2006). Dermoscopy has also been shown to improve primary care physician's triage of suspect skin lesions significantly (Argenziano et al. 2006).

15.3.3 Optical Coherence Tomography

Optical coherence tomography (OCT) is essentially the optical analogue of ultrasound pulse–echo imaging. OCT has been applied with great success to retina imaging and is under investigation for its use in skin. A typical OCT system consists of a Michelson interferometer (see Figure 15.2). The beam is split into a reference beam and a sample beam. Controlling the optical path length travelled by the reference beam allows for controlled interference with the sample beam. OCT uses a low-coherence source so that there is interference when the two beams are matched to within the coherence length of the source. Thus, varying the distance travelled by the reference beam images different depths within the sample. Two- and three-dimensional images can then be created by moving the beam across the target (Pitris et al. 2010).

Initial OCT systems used mechanical means to vary the reference path length and this implementation is called time domain OCT. Another implementation is Fourier domain OCT (FD-OCT), which has led to a significant increase in scanning speed (10×–100×). In FD-OCT, the reference arm is held stationary, and instead, the wavelength spectrum of the reflected light is detected. Taking the Fourier transform of the recovered spectrum yields the conventional axial scan. In spectral FD-OCT, a broadband source and spectrograph are used to collect images. Another implementation is swept source OCT, which varies the source wavelength, and the resulting spectrum is captured at sequential wavelengths with a detector. This is also called optical frequency domain imaging (OFDI) (Chinn et al. 1997; Yun et al. 2003). Polarization-sensitive (PS) OCT can be used to map the birefringence of samples.

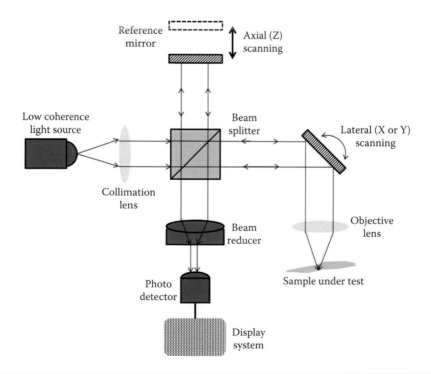

FIGURE 15.2   OCT system schematic.

Unlike many other noninvasive optical imaging techniques, OCT provides vertical cross-sectional images of tissue like conventional histology. Noninvasive imaging can be performed in real time. The axial resolution of OCT is 1–10 μm and is limited by the bandwidth of the source. The penetration depth in skin is 1–2 mm. Contrast in OCT images comes from the refractive index variation of tissue structures. Several studies have looked at NMSC using OCT. The diagnostic use is based upon identification of characteristic changes of tissue morphology in cancer as compared to normal tissue, for example, a disruption in the characteristic layering of human skin in cancer (Gambichler et al. 2007a,b; Mogensen et al. 2009a,b; Olmedo et al. 2006; Strasswimmer et al. 2004; Welzel 2001).

15.3.4 Confocal Microscopy

Confocal scanning laser microscopy is based on the confocal principle and has been applied extensively to human tissue both ex vivo and in vivo. Confocal microscopy was developed to overcome limitations of conventional far-field optical microscopes. In a conventional far-field microscope, the entire specimen volume in the beam path contributes to the collected signal; there is no distinction between signals collected from different depths within the imaged volume, and as a result, scattering causes image noise and a loss in the microscope's resolution. A confocal microscope uses an objective lens and the placement of a pinhole in the detection arm to reject out-of-focus light. The pinhole is placed in the conjugate plane of the system objective, and this removes light that is scattered from other planes within the imaged volume. Thus, the image is comprised of light coming from a highly localized imaging volume within the sample that is confined laterally by the objective (essentially diffraction limited) and axially by the pinhole. This principle forms the basis of confocal microscopy. Confocal laser scanning microscopy (CLSM) provides the ability to perform noninvasive optical sectioning on human tissue with cellular resolution (Rajadhyaksha et al. 1995, 1999).

Confocal images are generated by scanning the point source beam in two dimensions across the sample. Confocal images provide en face views of the skin surface. By translating the objective lens orthogonally to the sample, 2D en face images at different depths within the sample are generated and can be stacked vertically to represent the 3D samples. The penetration depth of light in tissue is limited by scattering. Current commercial systems use near-infrared light (830 nm) and offer a lateral resolution of 0.5–1 μm and axial resolution of 1–5 μm. The microscope has a mechanical fixture to allow for use of water immersion lenses as this minimizes refractive index mismatch at the interface and enables deeper imaging into the skin (Nehal et al. 2008). The maximum depth of imaging in skin is 300 μm.

Contrast in reflectance confocal imaging in vivo relies upon refractive index variations within normal tissue microstructures. Endogenous contrast is provided by melanin, keratin, mitochondria and cytoplasmic organelles, chromatin in the nuclei, and collagen in the dermis (Nehal et al. 2008; Rajadhyaksha et al. 1995, 1999). Correlating the morphological information achieved using CLSM with conventional histology has allowed for the identification of features of NMSC that has been used to improve the sensitivity and specificity of NMSC diagnosis using in vivo confocal imaging (Gerger et al. 2006, 2008; Nori et al. 2004).

15.3.5 Fluorescence Imaging

Fluorescence imaging offers a relatively fast noninvasive technique for imaging skin and NMSC demarcation. The concept is essentially straightforward; fluorescence signal is excited in tissue using a laser or a lamp and mapped using a camera and filters. Fluorescence imaging can be performed with or without the application of an external fluorophore. Autofluorescence generally refers to green tissue fluorescence (around 470 nm) and can be observed without the application of an external photosensitizer. Autofluorescence is generally attributed to naturally occurring fluorophores including collagen, elastin, and nicotinamide adenine dinucleotide (NADH). Contrast is observed as a decrease in the fluorescence intensity in tumors when compared to normal skin, possibly due to faster metabolism in the tumor (Brancaleon et al. 2001; Na et al. 2001; Onizawa et al. 2003; Panjehpour et al. 2002).

Fluorescence can also be exogenously induced by topical application of δ-5-aminolaevulinic acid (ALA) on the skin. This leads to an accumulation of protoporphyrin IX (Pp IX), which has been found to be higher in tumor as compared to normal. Pp IX fluorescence can be excited in the UV and blue (365 nm, 405 nm) and emits in the red (610–700 nm). The increased red fluorescence serves as a biomarker for tumor (Andersson-Engels et al. 2000; Brancaleon et al. 2001; Na et al. 2001; Onizawa et al. 2003; Panjehpour et al. 2002). Combining ALA-induced fluorescence with autofluorescence has been shown to be promising for BCC demarcation (Ericson et al. 2005; Stenquist et al. 2006).

15.3.6 High-Frequency Ultrasound

High-frequency ultrasound (HFUS) has been widely investigated for characterizing skin tumors. HFUS images are generated by the emission of pulsed ultrasound waves by a transducer and then detecting and registering the echo from the sample. An A-scan represents the intensity and time delay of the echo, and a B-scan is generated by moving the transducer laterally across the sample to produce a 2D image.

The penetration depth and resolution of the imaging system are related to the frequency. Increasing the frequency will improve the resolution but decrease the penetration depth. HFUS systems typically use 20 MHz probes. The axial resolution achieved at 20 MHz is 50 μm and the lateral resolution is 350 μm. The penetration depth in skin at this frequency is approximately 6–7 mm (Desai et al. 2007). Contrast in HFUS maps the acoustic impedance of normal and cancerous tissue. Normal dermis exhibits higher ultrasound (US) remittance (echogenic) when compared to NMSC, specifically BCC, which has higher acoustic impedance (hypoechogenic). Several studies have investigated the use of HFUS for NMSC diagnosis and demarcation (Gupta et al. 1996; Harland et al. 1993; Schmid-Wendtner and Burgdorf 2005).

15.3.7 Raman Spectroscopy

Raman spectroscopy is one of the most promising techniques for noninvasive NMSC diagnosis. Raman scattering is observed when a photon of incident light is inelastically scattered by molecules resulting in remitted light at a shifted frequency. This process is different from fluorescence as it is not a resonance process and thus occurs for all incident wavelengths. The emitted Raman spectra provide information of the vibrational states of different molecules and can be used to classify the molecule.

Several groups have shown that the characteristic Raman spectra of NMSC can be differentiated from normal tissue (Choi et al. 2005; Gniadecka et al. 1997; Lieber et al. 2008; Nijssen et al. 2002; Sigurdsson et al. 2004). A study showed that BCC sections can be distinguished using their Raman spectra with a sensitivity of 100% and a specificity of 93% (Nijssen et al. 2002). An in vivo study of NMSC using Raman microspectroscopy indicated 100% sensitivity and 91% specificity for abnormality (Lieber et al. 2008). Raman microscopy has been combined with confocal microscopy (confocal Raman microscopy) as this can assist significantly with reducing the autofluorescence contribution from tissue that interferes with the Raman signature (Choi et al. 2005).

Due to the prevalence of NMSC and the ease of surgical treatment, several other noninvasive imaging modalities are also being investigated for their potential use. These include polarized light imaging (PLI) (Yaroslavsky et al. 2003), fluorescence confocal imaging (Astner et al. 2008), multiphoton imaging (Paoli et al. 2007), electrical impedance mapping (Åberg et al. 2005), and even conventional CT, PET, and MRI (Fosko et al. 2003; Querleux 1995; Williams et al. 2001). Despite all this work, the reference standard remains to be skin biopsy with histopathological assessment.

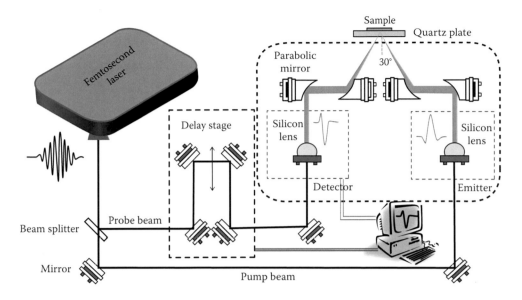

FIGURE 15.3 THz system in reflection geometry.

15.3.8 THz Imaging

Terahertz (THz) imaging, a comparatively young imaging technique, has become a hot topic for investigation in the recent 20 years. A THz time domain spectroscopy (THz TDS) system can be set up in two different geometries: transmission and reflection. In reflection geometry (Figure 15.3), the detector works by detecting the THz beam reflected back from the sample, which makes it especially suitable for skin imaging. Research has been done on skin cancer detection using THz light both in vivo and ex vivo (Woodward et al. 2003a,b). For measuring in vivo samples, reflection geometry is preferred; the THz beam is focused onto the sample with a 30° angle and is raster scanned in the sample plane (Wallace et al. 2004).

15.4 THz Imaging of BCC

Although traditional imaging techniques such as MRI and HFUS have been utilized to be diagnostic tools for skin tumor margin assessment, they both have limitations. MRI can be only used to detect tumors that extend beyond 15 mm below the skin surface. HFUS, which can provide a visualizing dimension of 80 and 200 μm and a penetration depth of 7 mm, is not capable of distinguishing benign and malignant skin lesions. Therefore, its role in determining tumor margins may also be limited.

Water accounts for 70% of the contents in human skin. Many imaging techniques including MRI and near-infrared spectroscopy have found a higher hydration level in skin tumor tissues. Some studies done on porcine and chicken skin confirmed this theory. The fact that THz electromagnetic radiation is highly absorbed by water makes it sensitive to cancer diagnosis as tumors have been shown to have an increased water content.

Wallace et al. undertook a study on the absorption of ex vivo BCC samples from 19 patients and successfully reported discrimination between diseased and healthy tissues (see Figure 15.4). Transmission geometry was used to measure the transmitted THz wave from excised skin tissues. By looking at the difference of the refractive index and absorption coefficient calculated from the spectroscopy, contrast can be found between normal tissue and tumor. Different techniques were used to model the skin tissue (Pickwell et al. 2004), and subsequently, several studies have been done in this field (Tonouchi 2007; Wallace et al. 2006).

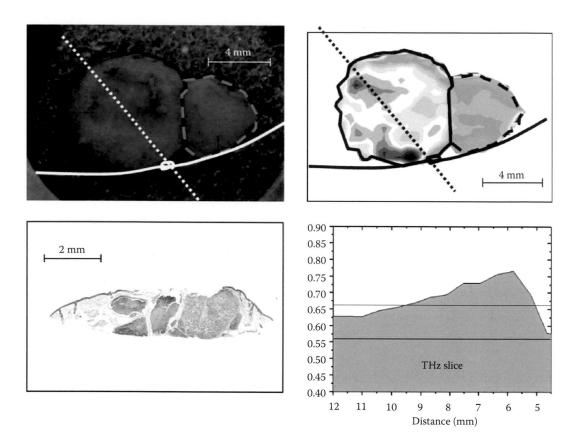

FIGURE 15.4   This THz images show the contrast between diseased and healthy tissues.

Due to the fact that not all skin abnormalities requiring diagnosis will be skin cancer, it is better that we can diagnose them before having to excise the skin tissue. Furthermore, most of BCC tumors are found on the face, so for cosmetic reasons, invasive diagnosing methods are not preferred. THz imaging can be performed in reflection mode, in which the detector receives signals reflected back from the surface of the sample. This geometry allows for in vivo sample measurement and makes THz imaging even more suitable for skin cancer diagnosis and tumor margin marking.

15.5 Image Contrast

One of the advantages of THz imaging is that it can achieve better image contrast of soft tissues than x-ray and many other imaging techniques. Many studies have proved the sensitivity of THz light in measuring the difference of different biomedical tissues. However, it is not always easy to get access to fresh tissues. Formalin fixing and freezing are commonly used in biomedical tissue preservation. It is experimentally verified that formalin fixing of the tissue can reduce the contrast in refractive index and absorption coefficient of different tissues (Sun et al. 2009), and freezing the sample will also reduce the absorption coefficient (Ashworth et al. 2007). Although the remaining contrast is still sufficient for us to distinguish tissue types, some techniques including dark field, phase imaging, and adding contrast agents could be used to improve the image contrast.

15.5.1 Dark-Field Technique

Dark-field imaging was first utilized to deal with image contrast enhancement in visible and near-infrared spectral range before being introduced into the THz range. This technique works by detecting the part of THz light that is scattered or deflected away from the main beam propagation direction. The image will be only formed by the light scattered from the sample, and the area around the sample will have no illumination at all. Therefore, the background of the image is completely dark. Considering the simplicity of the setup, the dark-field technique gives an impressive contrast performance (Löffler et al. 2001). This technique is especially useful to enhance the contrast of some samples of live or unstained biological tissues.

15.5.2 Phase Imaging

Phase imaging is a technique that uses an interferometer structure. Light propagating along the two arms of the interferometer is set to have a phase difference of π. When illuminating the areas of the sample plane that do not have any details, the phase difference will remain to be π, while in other places with details, the phase difference will change due to different refractive indexes of the detailed structures (Edward et al. 2008). Therefore, the background of the formed image will be black because of the destructive interference. Similarly to dark-field techniques, phase imaging is also well suited for biomedical samples.

15.5.3 Adding Contrast Agents

As reported in Huang et al. (2006), Lee et al. (2008), and Oh et al. (2009), gold nanorods (GNRs) can be used to enhance the contrast in cancer diagnosis. When injected into the tissue and illuminated with an infrared radiation source, plasma polaritons will be generated on the surface of the GNRs. This effect will cause the water temperature around the GNRs to rise, thus inducing a stronger THz signal. The GNRs are specially made that they are more likely to be absorbed by the cancer cells. Therefore, a more intense reflected signal will be achieved in cancer tissues.

15.6 Promising Studies and Directions for Future Work

The future of THz imaging for skin cancer assessment depends upon the successful realization of a number of technical and fundamental factors. To this end, several approaches are under investigation including a multimodal THz and optical approach, THz near-field imaging, and nanoparticle-enhanced THz contrast.

15.6.1 Multimodal THz and Optical Imaging

One of the disadvantages of THz imaging for biomedical applications is the wavelength-limited resolution: this prevents THz radiation from identifying tissue morphology. A research group at the University of Massachusetts Lowell (UML) has proposed combining THz imaging with an optical imaging modality that can provide the resolution to resolve tissue morphology. The proposed technology is wide-field PLI.

PLI is an optical technique that can be used to image superficial layers of thick tissue with a relatively large field of view at video rates with relatively high resolution (Yaroslavsky et al. 2003). The underlying principle is as follows: When linearly polarized light is incident on a sample, the reflection from the superficial layers is primarily copolarized with the incident beam. As light penetrates deeper into the tissue, the polarization is randomized by scattering. Thus, cross-polarized remitted signal is collected from deeper in the tissue volume and this is essentially equal to the copolarized signal from deep tissue. Thus, subtracting the cross-polarized remittance from the copolarized remittance is essentially an image of the signal from the superficial tissue layer. In this way, PLI can essentially be used to optically section thick tissue. Selecting the wavelength allows for variation in section thickness as tissue scattering increases with decreasing wavelength as can be seen in Figure 15.5.

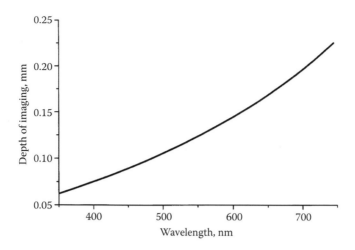

FIGURE 15.5   Imaging depth as a function of wavelength for PLI in skin.

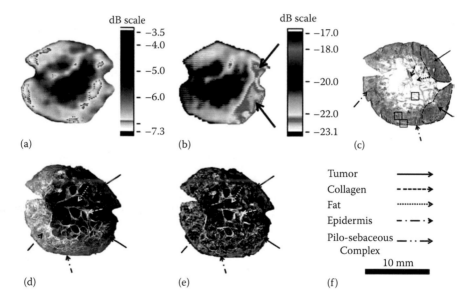

FIGURE 15.6   (a, b) Show co- and cross-polarized CW THz images, respectively, of a specimen with infiltrative BCC shown alongside optical cross-polarized and PLI images ([d] and [e], respectively) of the same sample and the H&E-stained 5 μm section of the same sample (c). (f) Shows the legends of the arrows in the figure. (From Joseph, C.S. et al., *J. Biophotonics*, 2012, DOI: 10.1002/jbio.201200111.)

The PLI system used by the researchers at UML had a field of view of 2.8 cm × 2.5 cm and a lateral resolution better than 15 μm. The wavelength selected for optical imaging was 440 nm, which corresponds to an imaging depth of 80 μm (Joseph et al. 2012).

PLI has been shown to offer high contrast between normal and cancerous tissue when used with an external contrast agent such as methylene blue. However, the intrinsic contrast offered by PLI is insufficient for reliable cancer detection (Yaroslavsky et al. 2003). Thus, combining PLI with THz imaging, which offers contrast but lacks resolution, is a possibility for intrinsic intraoperative margin delineation of skin cancers.

Figure 15.6 shows continuous-wave (CW) THz images of a sample of infiltrative BCC along with optical cross-polarized and PLI images of the specimen. The researchers showed that cross-polarized THz remittance can be used to identify tumor areas and optical images can be used to identify tissue morphology. The tumor area is characterized by low-THz cross-polarized reflectivity and a loss of structure in the optical image. Thus, after the THz is used to identify the cancerous region, the optical image can be used to outline the tumor margin.

The samples were kept hydrated during the measurement and the collected THz images were then compared to standard hematoxylin and eosin (H&E)-stained histology for evaluation. The images were collected using a CW THz imaging system operating at 584 GHz. The authors observed that cross-polarized THz imaging offers measurable contrast between cancerous and normal tissue. Moreover, in CW THz imaging (or frequency domain THz imaging), copolarized reflectance suffers from Fresnel artifacts at the air–window interface and at the surface of the tissue, thus obscuring contrast. The fact that repolarization of the backscattered beam requires several scattering events in the tissue implies that cross-polarized THz imaging penetrates deeper into tissue and the signal collected is representative of the tissue volume. The measured cross-polarized reflectance is low (<1%); however, it implies that scattering may not be negligible. Part of the contrast could be caused by clustering of cancerous cells into areas that are of the order of the wavelength, and as cancerous areas are more homogenous than normal skin (which contains sebaceous glands, hair follicles, and other structures), part of the observed THz contrast could be caused by scattering.

The observations of this work are consistent with the work of Loffler et al. where they found that while total loss in tumor areas is high, the scattering and deflected loss in tumor regions are low. Loffler et al. investigated THz dark-field imaging to enhance image contrast (Löffler et al. 2001). Dark-field imaging involves detecting the radiation that is deflected out of the sample due to scattering or diffraction. They imaged one sample of canine skin containing a mast cell tumor. The tissue was formalin fixed, alcohol dehydrated, and paraffin embedded and had a thickness of 3 mm. The sample was imaged in transmission modality using a time domain THz imaging system. The system was set up to collect deflected transmission THz radiation that did not include the ballistic transmitted component. They found that the total loss (including the ballistic component) was high in tumor areas and areas of skin with hairs and the connective tissue areas. However, they also found that deflection loss in tumor regions is low, especially compared to boundaries between different tissue structures and areas of skin with hairs. At 0.6 THz, scattering was found to be high in areas of skin with hairs and diffraction was the dominant effect at the tissue boundaries.

Combining THz imaging with other imaging modalities, such as OCT and confocal reflectance, can also offer complementary information to aid in cancer detection and demarcation.

15.6.2 THz Near-Field Imaging

A technique that has the potential to tackle the wavelength-limited resolution of a THz imager is THz near-field imaging. Near-field THz microscopes have been demonstrated using a variety of THz transceiver systems. These systems generally fall under one of two implementation designs: aperture-based near-field microscopy and apertureless near-field microscopy. Several research teams are currently pursuing both these designs (Chen et al. 2000, 2003; Chiu et al. 2009; Cho et al. 2005; Huber et al. 2008).

Aperture-based near-field techniques typically employ a subwavelength size aperture to image the sample. The resolution of the system is then determined by the size of the aperture. Near-field imaging using THz pulsed systems was first demonstrated in 1998 (Hunsche et al. 1998). The authors used an elliptical subwavelength aperture to achieve a resolution of 50 μm, which corresponds to $\lambda/4$. In general, the transmission through a subwavelength aperture scales as $(d/\lambda)^4$ where d is the aperture diameter. Selection of aperture shape and a series of concentric rings around the aperture (bull's-eye) can be used to enhance the transmission (Ishihara et al. 2006; Wang et al. 2011). A group in Taiwan has demonstrated a THz aperture-based near-field microscope and demonstrated its use by imaging thin sections of breast cancer tissue in transmission mode (Chiu et al. 2009). Figure 15.7 is an example image of the breast cancer specimen.

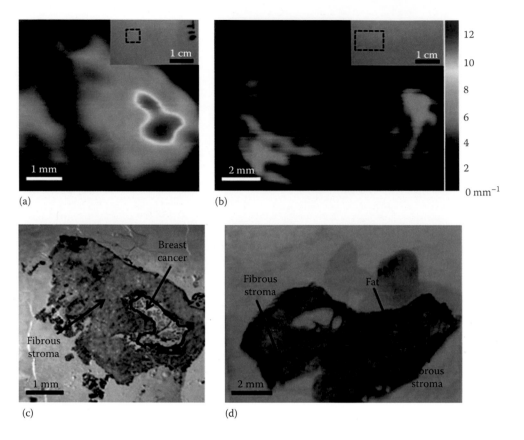

FIGURE 15.7   (a, b) show THz near-field transmission maps of thin (20 μm) sections of breast tissue, and (c, d) show the corresponding H&E-stained histology. (From Chiu, C.-M. et al. , *Opt. Lett.*, 34, 1084, 2009.)

The source used for this experiment was a CW Gunn diode oscillator with an output frequency of 312 GHz (962 μm). The detector was a room temperature Golay cell. Near-field imaging was implemented using a bull's-eye structure with an aperture of 250 μm. The achieved spatial resolution was 210 μm and the system field of view was 2 cm × 2 cm (Chiu et al. 2009).

The other implementation for near-field imaging is the apertureless or tip-based approach. According to Babinet's principle, a subwavelength size tip can be viewed as the inverse of a subwavelength size aperture. The tip is illuminated by the THz source and serves as a subwavelength scatterer. Thus, the resolution is determined by the diameter of the tip. As subwavelength tips can be manufactured fairly easily, this technique shows a lot of promise for near-field THz imaging. The first demonstration of tip-based THz near-field imaging was in 2002 (Van der Valk and Planken 2002). A resolution of 40 nm at 2.54 THz has been demonstrated using an atomic force microscopy (AFM) tip (Huber et al. 2008). Data interpretation for THz tip-based systems has shown the need to account for antenna properties of the tip probe (Li et al. 2009; Wang et al. 2004a,b).

15.6.3 Nanoparticle-Enhanced THz Imaging

One of the challenges facing THz imaging is the lack of contrast between malignant and benign lesions. A group in Seoul, South Korea, has been investigating the use of GNRs to enhance the sensitivity and contrast of THz imaging (terahertz molecular imaging [TMI]) (Oh et al. 2009, 2011). The basic principle is that near-infrared light illumination of GNRs induces a plasmon resonance that raises the temperature

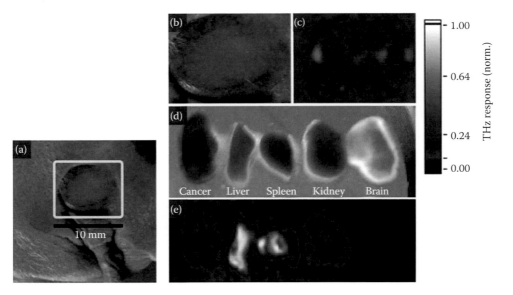

FIGURE 15.8 (a, b) show photographs of the tumor, while (c) shows the GNR-enhanced in vivo THz image. (d) shows excised samples and (e) shows the corresponding ex vivo THz images. (From Son, J.-H., Molecular imaging with terahertz waves, presented at *35th International Conference on Infrared Millimeter and Terahertz Waves*, 2010, p. 1.)

of water in cells and tissue and enhances their THz response (Oh et al. 2011; Tong et al. 2009). The sensitivity of the technique is based on the strong temperature dependence of the THz response of water.

Figure 15.8 shows an example TMI image of tumor in vivo. The THz image was collected using a reflectance time domain imaging system while simultaneously exciting the plasmon resonance in the GNRs using an 808 nm near-infrared diode laser. A GNR composite was conjugated with cetuximab for targeting an epidermal growth factor receptor (EGFR) specific to tumor cells. The mouse was injected with A431 epidermoid carcinoma tumor cells. Prior to imaging, the mouse was injected with 100 μL of 1 mM concentration of conjugated GNRs. The authors showed that the THz response increased linearly with GNR concentration and they were able to detect a minimum concentration of 15 μm in vivo (Oh et al. 2011).

GNR-enhanced THz imaging has the potential to facilitate early cancer diagnosis and monitor drug delivery processes. Moreover, localized cell-specific heating effects could eventually be used for cell-specific cancer treatment.

References

AACR and AIHW. 2008. *Cancer in Australia: An Overview*. Canberra, Australia: AIHW.

Åberg, P., P. Geladi, I. Nicander et al. 2005. Non-invasive and microinvasive electrical impedance spectra of skin cancer—A comparison between two techniques. *Skin Research and Technology* 11: 281–286.

Andersson-Engels, S., G. Canti, R. Cubeddu et al. 2000. Preliminary evaluation of two fluorescence imaging methods for the detection and the delineation of basal cell carcinomas of the skin. *Lasers in Surgery and Medicine* 26: 76–82.

Argenziano, G., S. Puig, I. Zalaudek et al. 2006. Dermoscopy improves accuracy of primary care physicians to triage lesions suggestive of skin cancer. *Journal of Clinical Oncology* 24: 1877–1882.

Argenziano, G., I. Zalaudek, R. Corona et al. 2004. Vascular structures in skin tumors: A dermoscopy study. *Archives of Dermatology* 140: 1485.

Ashworth, P. C., E. Pickwell-MacPherson, S. E. Pinder et al. 2007. Terahertz spectroscopy of breast tumors. In *Infrared and Millimeter Waves, 2007 and the 15th International Conference on Terahertz Electronics, 2007. IRMMW-THz. Joint 32nd International Conference on IEEE*, pp. 603–605.

Astner, S., S. Dietterle, N. Otberg et al. 2008. Clinical applicability of in vivo fluorescence confocal micros-copy for noninvasive diagnosis and therapeutic monitoring of nonmelanoma skin cancer. *Journal of Biomedical Optics* 13: 014003.

Brancaleon, L., A. J. Durkin, J. H. Tu et al. 2001. In vivo fluorescence spectroscopy of nonmelanoma skin cancer. *Photochemistry and Photobiology* 73: 178–183.

CDC. 2012. *Skin Cancer Rates by Race and Ethnicity.* Atlanta, GA: National Center for Chronic Disease Preven-ntion and Health Promotion. http://www.cdc.gov/cancer/dcpc/data/race.htm, accessed November 2012.

Chen, H.-T., R. Kersting, and G. C. Cho. 2003. Terahertz imaging with nanometer resolution. *Applied Physics Letters* 83: 3009–3011.

Chen, Q., Z. Jiang, G. Xu, and X.-C. Zhang. 2000. Near-field terahertz imaging with a dynamic aperture. *Optics Letters* 25: 1122–1124.

Chin, C. W. S., A. J. E. Foss, A. Stevens, and J. Lowe. 2003. Differences in the vascular patterns of basal and squamous cell skin carcinomas explain their differences in clinical behaviour. *The Journal of Pathology* 200: 308–313.

Chinn, S., E. Swanson, and J. Fujimoto. 1997. Optical coherence tomography using a frequency-tunable optical source. *Optics Letters* 22: 340–342.

Chiu, C.-M., H.-W. Chen, Y.-R. Huang et al. 2009. All-terahertz fiber-scanning near-field microscopy. *Optics Letters* 34: 1084–1086.

Cho, G. C., H.-T. Chen, S. Kraatz, N. Karpowicz, and R. Kersting. 2005. Apertureless terahertz near-field microscopy. *Semiconductor Science and Technology* 20: S286.

Choi, J., J. Choo, H. Chung et al. 2005. Direct observation of spectral differences between normal and basal cell carcinoma (BCC) tissues using confocal Raman microscopy. *Biopolymers* 77: 264–272.

Chu, A. C. and R. L. Edelson. 1999. *Malignant Tumours of the Skin.* London, U.K.: Arnold.

Cooper, S. M. and F. Wojnarowska. 2002. The accuracy of clinical diagnosis of suspected premalignant and malignant skin lesions in renal transplant recipients. *Clinical and Experimental Dermatology* 27: 436–438.

Davis, D. A., J. P. Donahue, J. E. Bost, and T. D. Horn. 2005. The diagnostic concordance of actinic kerato-sis and squamous cell carcinoma. *Journal of Cutaneous Pathology* 32: 546–551.

Desai, T. D., A. D. Desai, D. C. Horowitz, F. Kartono, and T. Wahl. 2007. The use of high-frequency ultra-sound in the evaluation of superficial and nodular basal cell carcinomas. *Dermatologic Surgery* 33: 1220–1227.

Edward, K., T. W. Mayes, B. Hocken, and F. Farahi. 2008. Trimodal imaging system capable of quantitative phase imaging without 2π ambiguities. *Optics Letters* 33: 216–218.

Ek, E. W., F. Giorlando, S. Y. Su, and T. Dieu. 2005. Clinical diagnosis of skin tumours: How good are we? *ANZ Journal of Surgery* 75: 415–420.

Ericson, M. B., J. Uhre, C. Strandeberg et al. 2005. Bispectral fluorescence imaging combined with texture analysis and linear discrimination for correlation with histopathologic extent of basal cell carci-noma. *Journal of Biomedical Optics* 10: 034009–0340098.

Fosko, S. W., W. Hu, T. F. Cook, and V. J. Lowe. 2003. Positron emission tomography for basal cell carci-noma of the head and neck. *Archives of Dermatology* 139: 1141.

Gambichler, T., A. Orlikov, R. Vasa et al. 2007a. In vivo optical coherence tomography of basal cell carci-noma. *Journal of Dermatological Science* 45: 167–173.

Gambichler, T., P. Regeniter, F. G. Bechara et al. 2007b. Characterization of benign and malignant mela-nocytic skin lesions using optical coherence tomography in vivo. *Journal of the American Academy of Dermatology* 57: 629–637.

Gerger, A., R. Hofmann-Wellenhof, U. Langsenlehner et al. 2008. In vivo confocal laser scanning micros-copy of melanocytic skin tumours: Diagnostic applicability using unselected tumour images. *British Journal of Dermatology* 158: 329–333.

Gerger, A., S. Koller, W. Weger et al. 2006. Sensitivity and specificity of confocal laser-scanning micros-copy for in vivo diagnosis of malignant skin tumors. *Cancer* 107: 193–200.

Gniadecka, M., H. Wulf, N. N. Mortensen, O. F. Nielsen, and D. H. Christensen. 1997. Diagnosis of basal cell carcinoma by Raman spectroscopy. *Journal of Raman Spectroscopy* 28: 125–129.

Gupta, A. K., D. H. Turnbull, F. S. Foster et al. 1996. High frequency 40-MHz ultrasound a possible noninvasive method for the assessment of the boundary of basal cell carcinomas. *Dermatologic Surgery* 22: 131–136.

Hallock, G. G. and D. A. Lutz. 1998. Prospective study of the accuracy of the surgeon's diagnosis in 2000 excised skin tumors. *Plastic and Reconstructive Surgery* 101: 1255–1261.

Har-Shai, Y., N. Hai, A. Taran et al. 2001. Sensitivity and positive predictive values of presurgical clinical diagnosis of excised benign and malignant skin tumors: A prospective study of 835 lesions in 778 patients. *Plastic and Reconstructive Surgery* 108: 1982–1989.

Harland, C., J. Bamber, B. Gusterson, and P. Mortimer. 1993. High frequency, high resolution B-scan ultrasound in the assessment of skin tumours. *British Journal of Dermatology* 128: 525–532.

Huang, X., I. H. El-Sayed, W. Qian, and M. A. El-Sayed. 2006. Cancer cell imaging and photothermal therapy in the near-infrared region by using gold nanorods. *Journal of the American Chemical Society* 128: 2115–2120.

Huber, A. J., F. Keilmann, J. Wittborn, J. Aizpurua, and R. Hillenbrand. 2008. Terahertz near-field nanoscopy of mobile carriers in single semiconductor nanodevices. *Nano Letters* 8: 3766–3770.

Hunsche, S., M. Koch, I. Brener, and M. Nuss. 1998. THz near-field imaging. *Optics Communications* 150: 22–26.

Ishihara, K., K. Ohashi, T. Ikari et al. 2006. Terahertz-wave near-field imaging with subwavelength resolution using surface-wave-assisted bow-tie aperture. *Applied Physics Letters* 89: 201120 (3 pages).

Joseph, C. S., R. Patel, V. A. Neel, R. H. Giles, and A. N. Yaroslavsky. 2012. Imaging of ex vivo nonmelanoma skin cancers in the optical and terahertz spectral regions optical and terahertz skin cancers imaging. *Journal of Biophotonics*. DOI: 10.1002/jbio.201200111.

Kreusch, J. F. 2002. Vascular patterns in skin tumors. *Clinics in Dermatology* 20: 248–254.

Löffler, T., T. Bauer, K. J. Siebert et al. 2001. Terahertz dark-field imaging of biomedical tissue. *Optics Express* 9: 616–621.

Lee, J. W., J. M. Yang, H. J. Ko et al. 2008. Multifunctional magnetic gold nanocomposites: Human epithelial cancer detection via magnetic resonance imaging and localized synchronous therapy. *Advanced Functional Materials* 18: 258–264.

Leffell, D. J., Y.-T. Chen, M. Berwick, and J. L. Bolognia. 1993. Interobserver agreement in a community skin cancer screening setting. *Journal of the American Academy of Dermatology* 28: 1003–1005.

Li, Y., S. Popov, A. T. Friberg, and S. Sergeyev. 2009. Rigorous modeling and physical interpretation of terahertz near-field imaging. *Journal of the European Optical Society-Rapid Publications* 4: 09007.

Lieber, C. A., S. K. Majumder, D. L. Ellis, D. D. Billheimer, and A. Mahadevan-Jansen. 2008. In vivo nonmelanoma skin cancer diagnosis using Raman microspectroscopy. *Lasers in Surgery and Medicine* 40: 461–467.

Menzies, S. W. 2002. Dermoscopy of pigmented basal cell carcinoma. *Clinics in Dermatology* 20: 268–269.

Miller, D. L. and M. A. Weinstock. 1994. Nonmelanoma skin cancer in the United States: Incidence. *Journal of the American Academy of Dermatology* 30: 774–778.

Mogensen, M. and G. B. Jemec. 2007. Diagnosis of nonmelanoma skin cancer/keratinocyte carcinoma: A review of diagnostic accuracy of nonmelanoma skin cancer diagnostic tests and technologies. *Dermatologic Surgery* 33: 1158–1174.

Mogensen, M., T. Joergensen, B. M. Nürnberg et al. 2009a. Assessment of optical coherence tomography imaging in the diagnosis of non-melanoma skin cancer and benign lesions versus normal skin: Observer-blinded evaluation by dermatologists and pathologists. *Dermatologic Surgery* 35: 965–972.

Mogensen, M., L. Thrane, T. M. Jørgensen, P. E. Andersen, and G. B. Jemec. 2009b. OCT imaging of skin cancer and other dermatological diseases. *Journal of Biophotonics* 2: 442–451.

Morrison, A., S. O'Loughlin, and F. C. Powell. 2001. Suspected skin malignancy: A comparison of diagnoses of family practitioners and dermatologists in 493 patients. *International Journal of Dermatology* 40: 104–107.

Na, R., I.-M. Stender, and H. C. Wulf. 2001. Can autofluorescence demarcate basal cell carcinoma from normal skin? A comparison with protoporphyrin IX fluorescence. *Acta Dermato-Venereologica-Stockholm* 81: 246–249.

Nehal, K. S., D. Gareau, and M. Rajadhyaksha 2008. Skin imaging with reflectance confocal microscopy. Presented at *Seminars in Cutaneous Medicine and Surgery* 27: 37–43.

Newell, B., A. Bedlow, S. Cliff et al. 2003. Comparison of the microvasculature of basal cell carcinoma and actinic keratosis using intravital microscopy and immunohistochemistry. *British Journal of Dermatology* 149: 105–110.

Nijssen, A., T. C. B. Schut, F. Heule et al. 2002. Discriminating basal cell carcinoma from its surrounding tissue by Raman spectroscopy. *Journal of Investigative Dermatology* 119: 64–69.

Nori, S., F. Rius-Díaz, J. Cuevas et al. 2004. Sensitivity and specificity of reflectance-mode confocal microscopy for in vivo diagnosis of basal cell carcinoma: A multicenter study. *Journal of the American Academy of Dermatology* 51: 923–930.

Oh, S. J., J. Kang, I. Maeng et al. 2009. Nanoparticle-enabled terahertz imaging for cancer diagnosis. *Optics Express* 17: 3469–3475.

Oh, S. J., J. Choi, I. Maeng et al. 2011. Molecular imaging with terahertz waves. *Optics Express* 19(5): 4009–4016.

Olmedo, J. M., K. E. Warschaw, J. M. Schmitt, and D. L. Swanson. 2006. Optical coherence tomography for the characterization of basal cell carcinoma in vivo: A pilot study. *Journal of the American Academy of Dermatology* 55: 408–412.

Onizawa, K., N. Okamura, H. Saginoya, and H. Yoshida. 2003. Characterization of autofluorescence in oral squamous cell carcinoma. *Oral Oncology* 39: 150–156.

Otis, L. L., D. Piao, C. W. Gibson, and Q. Zhu. 2004. Quantifying labial blood flow using optical Doppler tomography. *Oral Surgery, Oral Medicine, Oral Pathology, Oral Radiology, and Endodontology* 98: 189–194.

Panjehpour, M., C. E. Julius, M. N. Phan, T. Vo-Dinh, and S. Overholt. 2002. Laser-induced fluorescence spectroscopy for in vivo diagnosis of non-melanoma skin cancers. *Lasers in Surgery and Medicine* 31: 367–373.

Paoli, J., M. Smedh, A.-M. Wennberg, and M. B. Ericson. 2007. Multiphoton laser scanning microscopy on non-melanoma skin cancer: Morphologic features for future non-invasive diagnostics. *Journal of Investigative Dermatology* 128: 1248–1255.

Pickwell, E., B. E. Cole, A. J. Fitzgerald, M. Pepper, and V. P. Wallace. 2004. In vivo study of human skin using pulsed terahertz radiation. *Physics in Medicine and Biology* 49: 1595.

Pitris, C., A. Kartakoullis, and E. Bousi 2010. Optical coherence tomography theory and spectral time-frequency analysis. In *Handbook of Photonics for Biomedical Science. Series: Series in Medical Physics and Biomedical Engineering*, ed. Tučin, V. V., pp. 377–400. Boca Raton, FL: CRC Press.

Pye, R. J. 2000. The hidden epidemic of basal cell carcinoma. *Horizons in Medicine* 12: 339–346.

Querleux, B. 1995. Nuclear magnetic resonance (NMR) examination of the epidermis in vivo. In *Handbook of Non-Invasive Methods and the Skin*, eds. Serup, J. and Jemec, G. B. E., pp. 133–139. Boca Raton, FL: CRC Press.

Rajadhyaksha, M., S. González, J. M. Zavislan, R. R. Anderson, and R. H. Webb. 1999. In vivo confocal scanning laser microscopy of human skin II: Advances in instrumentation and comparison with histology. *Journal of Investigative Dermatology* 113: 293–303.

Rajadhyaksha, M., M. Grossman, D. Esterowitz, R. H. Webb, and R. R. Anderson. 1995. In vivo confocal scanning laser microscopy of human skin: Melanin provides strong contrast. *Journal of Investigative Dermatology* 104: 946–952.

Rapini, R. P. 1990. Comparison of methods for checking surgical margins. *Journal of the American Academy of Dermatology* 23: 288–294.

Rubin, A. I., E. H. Chen, and D. Ratner. 2005. Basal-cell carcinoma. *New England Journal of Medicine* 353: 2262–2269.

Saladi, R. N. and A. N. Persaud. 2005. The causes of skin cancer: A comprehensive review. *Drugs of Today* 41: 37–54.

Schmid-Wendtner, M.-H. and W. Burgdorf. 2005. Ultrasound scanning in dermatology. *Archives of Dermatology* 141: 217.

Schwartzberg, J. B., G. W. Elgart, P. Romanelli et al. 2005. Accuracy and predictors of basal cell carcinoma diagnosis. *Dermatologic Surgery* 31: 534–537.

Shriner, D. L., D. K. McCoy, D. J. Goldberg, and R. F. Wagner Jr. 1998. Mohs micrographic surgery. *Journal of the American Academy of Dermatology* 39: 79–97.

Sigurdsson, S., P. A. Philipsen, L. K. Hansen et al. 2004. Detection of skin cancer by classification of Raman spectra. *IEEE Transactions on Biomedical Engineering* 51: 1784–1793.

Stenquist, B., M. B. Ericson, C. Strandeberg et al. 2006. Bispectral fluorescence imaging of aggressive basal cell carcinoma combined with histopathological mapping: A preliminary study indicating a possible adjunct to Mohs micrographic surgery. *British Journal of Dermatology* 154: 305–309.

Strasswimmer, J., M. C. Pierce, B. H. Park, V. Neel, and J. F. de Boer. 2004. Polarization-sensitive optical coherence tomography of invasive basal cell carcinoma. *Journal of Biomedical Optics* 9: 292–298.

Sun, Y., B. M. Fischer, and E. Pickwell-MacPherson. 2009. Effects of formalin fixing on the terahertz properties of biological tissues. *Journal of Biomedical Optics* 14: 064017.

Tong, L., Q. Wei, A. Wei, and J. X. Cheng. 2009. Gold nanorods as contrast agents for biological imaging: Optical properties, surface conjugation and photothermal effects. *Photochemistry and Photobiology* 85: 21–32.

Tonouchi, M. 2007. Cutting-edge terahertz technology. *Nature Photonics* 1: 97–105.

Ulrich, M., E. Stockfleth, J. Roewert-Huber, and S. Astner. 2007. Noninvasive diagnostic tools for non-melanoma skin cancer. *British Journal of Dermatology* 157: 56–58.

Van der Valk, N. and P. Planken. 2002. Electro-optic detection of subwavelength terahertz spot sizes in the near field of a metal tip. *Applied Physics Letters* 81: 1558–1560.

Wallace, V. P., A. J. Fitzgerald, E. Pickwell et al. 2006. Terahertz pulsed spectroscopy of human basal cell carcinoma. *Applied Spectroscopy* 60: 1127–1133.

Wallace, V. P., A. J. Fitzgerald, S. Shankar et al. 2004. Terahertz pulsed imaging of basal cell carcinoma ex vivo and in vivo. *British Journal of Dermatology* 151: 424–432.

Wang, D., T. Yang, and K. B. Crozier. 2011. Optical antennas integrated with concentric ring gratings: Electric field enhancement and directional radiation. *Optics Express* 19: 2148–2157.

Wang, K., A. Barkan, and D. M. Mittleman. 2004a. Propagation effects in apertureless near-field optical antennas. *Applied Physics Letters* 84: 305–307.

Wang, K., D. M. Mittleman, N. C. van der Valk, and P. Planken. 2004b. Antenna effects in terahertz apertureless near-field optical microscopy. *Applied Physics Letters* 85: 2715–2717.

Welzel, J. 2001. Optical coherence tomography in dermatology: A review. *Skin Research and Technology* 7: 1–9.

Whited, J. D. and R. P. Hall. 1997. Diagnostic accuracy and precision in assessing dermatologic disease: Problem or promise? *Archives of Dermatology* 133: 1409.

Whited, J. D., R. D. Horner, R. P. Hall, and D. L. Simel. 1995. The influence of history on interobserver agreement for diagnosing actinic keratoses and malignant skin lesions. *Journal of the American Academy of Dermatology* 33: 603–607.

Williams, L. S., A. A. Mancuso, and W. M. Mendenhall. 2001. Perineural spread of cutaneous squamous and basal cell carcinoma: CT and MR detection and its impact on patient management and prognosis. *International Journal of Radiation Oncology, Biology, Physics* 49: 1061–1069.

Wolf, D. J. and J. A. Zitelli. 1987. Surgical margins for basal cell carcinoma. *Archives of Dermatology* 123: 340.

Woodward, R., V. Wallace, D. Arnone, E. Linfield, and M. Pepper. 2003a. Terahertz pulsed imaging of skin cancer in the time and frequency domain. *Journal of Biological Physics* 29: 257–259.

Woodward, R. M., V. P. Wallace, R. J. Pye et al. 2003b. Terahertz pulse imaging of ex vivo basal cell carcinoma. *Journal of Investigative Dermatology* 120: 72–78.

Yaroslavsky, A. N., V. Neel, and R. R. Anderson. 2003. Demarcation of nonmelanoma skin cancer margins in thick excisions using multispectral polarized light imaging. *Journal of Investigative Dermatology* 121: 259–266.

Yun, S.-H., G. J. Tearney, J. F. de Boer, N. Iftimia, and B. E. Bouma. 2003. High-speed optical frequency-domain imaging. *Optics Express* 11: 2953.

Zalaudek, I. 2005. Dermoscopy subpatterns of nonpigmented skin tumors. *Archives of Dermatology* 141: 532.

Zalaudek, I., G. Argenziano, B. Leinweber et al. 2004. Dermoscopy of Bowen's disease. *British Journal of Dermatology* 150: 1112–1116.

Zalaudek, I., G. Argenziano, H. Soyer et al. 2006. Three-point checklist of dermoscopy: an open internet study. *British Journal of Dermatology* 154: 431–437.

16

Application of Terahertz Technology to Breast Cancer

Maarten
R. Grootendorst
King's College London

Massimiliano
Cariati
King's College London

Philip C. Ashworth
University of Cambridge

Anthony
J. Fitzgerald
*University of Western
Australia*

Arnie Purushotham
King's College London

Vincent P. Wallace
*University of Western
Australia*

16.1 Introduction to Breast Cancer

16.1.1 Breast Cancer Key Facts

Breast cancer is by far the most common cancer among women, both in developed and in developing regions. Worldwide, an estimated 1.38 million women are diagnosed with the disease each year, making it the second most frequently diagnosed cancer after lung cancer (Ferlay et al. 2010). Among men, breast cancer is far less common, accounting for <1% of male cancer cases.

The incidence rate of breast cancer in women is higher in developed countries compared with other countries; Western Europe has the highest incidence rate (89.7 per 100,000) and Middle Africa the lowest (19.3 per 100,000). A range of factors contribute to this variation in incidence rates, particularly relating to lifestyle (Youlden et al. 2012). In the United Kingdom and other parts of Western Europe, the incidence rate of breast cancer has increased by almost 50% in the period 1980–2008. Currently, the estimated lifetime risk of developing breast cancer in women in the United Kingdom and the United States is one in eight (Cancer Research UK 2012a; Siegel et al. 2012).

Although the incidence rate is significantly higher in developed countries, the range in mortality rate is much less (6–19 per 100,000). This is due to a combination of factors, including earlier detection, favorable treatment modalities, and better access to these modalities. However, with approximately 189,000 deaths per annum, breast cancer is, together with lung cancer, the most frequent cause of cancer death in women in the world, both in developed and developing regions.

Breast cancer therefore remains a large health problem, despite the fact that early diagnosis and more effective treatment have lowered the mortality rate in several countries. Therefore, there is a need to improve and optimize breast cancer care globally.

16.1.2 Types of Breast Cancer

There are various types of breast cancer, each type presents with different symptoms and characteristics. Although it was previously believed that in some cases breast cancer arose in the ducts and in other cases in the lobules, it is now clear that this disease derives from the terminal ductal lobular unit (TDLU) (see Figure 16.1). Breast cancer can be roughly divided into noninvasive and invasive cancer.

Noninvasive breast cancers are cancers that have not yet broken through the myoepithelial layer and basement membrane and hence are confined within the ductolobular units and are termed carcinoma in situ. Ductal carcinoma in situ (DCIS) is the most common form of in situ breast cancer and starts within the duct system, involving almost always a single duct within the breast. DCIS generally has no signs or symptoms, although a small number of patients may present with a small palpable lump. If DCIS cells die and pile up, tiny specks of calcium form within the broken cells (called calcifications or microcalcifications). These calcifications are usually very small and therefore difficult to identify by imaging techniques.

Over time, cancer cells can penetrate through the wall of the ducts and lobules, thereby infiltrating the breast stroma. If this occurs, the cancer is called invasive. The most common form of invasive breast cancer,

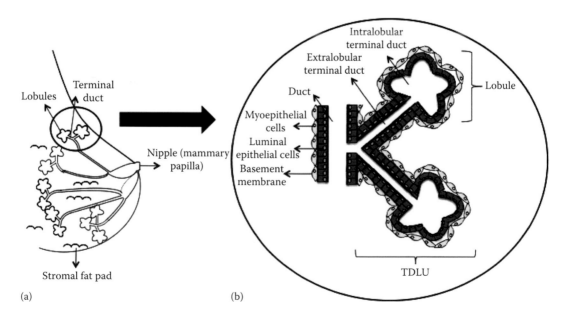

FIGURE 16.1 Schematic of the human mammary gland. (a) The human mammary gland is organized into lobules interconnected by a network of branched ducts. (b) The ducts consist of an inner layer of luminal epithelial cells and an outer layer of myoepithelial cells. The large duct separates into extralobular ducts, which then separate into intralobular ducts. The main functional unit of the breast is the TDLU that consists of the lobule, intralobular terminal ducts, and the extralobular terminal ducts.

and breast cancer in general, is invasive ductal carcinoma (IDC). IDC is diagnosed in more than 80% of all breast cancer patients (Cancer Research UK 2012b) and is more common as women grow older. Invasive lobular carcinoma is the second most common invasive cancer and accounts for 10% of all invasive cancers.

Invasive carcinomas tend to form hard, palpable lesions, while DCIS is mostly nonpalpable. This presents additional challenges for accurate preoperative and intraoperative identification.

16.1.3 Breast Cancer Surgery

Most women with breast cancer undergo some type of surgery to remove the primary tumor, which has moved from radical to more conservative over the last century. Breast cancer surgery can roughly be divided in breast-conserving surgery (BCS) and mastectomy (i.e., the complete removal of the breast). The choice between BCS or mastectomy depends on several factors, including the size of the tumor relative to the size of the breast, its location, the presence of diffuse microcalcifications (DCIS), and patient's preference. The purpose of adopting BCS is to minimize the psychological and physical morbidity associated with mastectomy.

16.1.3.1 Breast-Conserving Surgery

A combination of patient and physician awareness and increased use of screening mammography has significantly impacted on the stage at which cancers are diagnosed. Women with stage 0 (carcinoma in situ), stage 1 (invasive tumor <2 cm), or stage 2 (invasive tumor <5 cm) are ideal candidates for BCS, and due to the earlier stage of diagnoses, approximately two-thirds of newly diagnosed breast cancer patients in the United Kingdom and the United States undergo BCS as initial treatment (Jeevan et al. 2011; Katipamula et al. 2009).

The aim of BCS is to remove the primary tumor while conserving as much healthy breast tissue if possible to provide good cosmetic outcome and to reduce physical trauma. BCS is often followed by a course of postoperative radiotherapy, and the combination of BCS with radiotherapy provides similar survival rates to those achieved with mastectomy alone for women with invasive disease (Fisher et al. 2002). However, incomplete removal of the tumor results in involved tumor margins and involved margins are one of the main risk factors for local recurrence and affect disease-free survival (Singletary 2002; Veronesi et al. 2002).

There are various imaging techniques used to diagnose breast cancer, including x-ray mammography, ultrasound (US) imaging, and magnetic resonance imaging (MRI). These *preoperative* imaging techniques provide information on the location and size of the tumor, but these techniques have limited intraoperative capabilities. The correlation between tumor size estimation with preoperative imaging techniques and histopathological size remains suboptimal, thereby creating an element of uncertainty when deciding how much tissue is to be excised during surgery. Hence, surgeons can use this information only as a rough guide to define the margins of the tumor.

The techniques used to obtain intraoperative information on tumor margins depend on whether the tumor is palpable or nonpalpable. Palpable tumors are localized with the tip of the surgeon's thumb and index fingers, a technique known as palpation-guided surgery. By feeling for the tumor, the surgeon can determine the extent of the tumor, and after the boundaries are identified, the surgeon aims to excise the tumor together with a region of healthy tissue. One can appreciate that this is a very subjective method. Nonpalpable tumors cannot be identified by *touch*, so additional techniques are required to localize the lesion and to determine its boundaries. The various techniques that are available are outlined in Section 16.2 and make use of a wire, radioactive material, or US waves. The surgeon uses these techniques to guide the excision of the lesion, thereby aiming to obtain an adequate margin of healthy tissue.

The aforementioned surgical guidance techniques provide macroscopic information on tumor margins, and in order to determine if the tumor is excised completely, the tumor margins of the excised specimen need to be analyzed on a microscopic level. These so-called microscopic margins are determinative to whether a procedure is considered successful. There are currently two techniques available to assess the microscopic margins *intraoperatively*: frozen section analysis and touch print cytology.

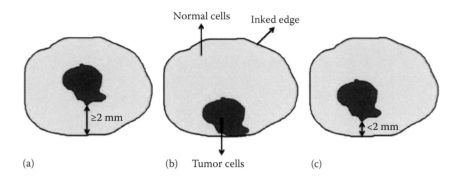

FIGURE 16.2   Negative margin (a), positive margin (b), and close margin (c).

However, as explained in more detail in Section 16.2, both techniques have disadvantages, and therefore, only a small number of hospitals utilize these techniques.

The golden standard for assessing microscopic margins is postoperative histological examination. During this procedure, the pathologist inks the surface of the excised specimen, and the closest distance of an inked surface to tumor cells is reported within several days after surgery. Most institutions consider a margin width of >2 mm as negative, 0–2 mm as close, and 0 mm as positive (Houssami et al. 2010; Morrow et al. 2012) (see Figure 16.2). Although there is no consensus on what constitutes an adequate microscopic margin, the majority of centers advise patients with positive or close margins to undergo a reoperation to remove residual disease.

During a reoperation, the surgeon reopens the site of the original surgery and removes a further slice of tissue of the margin reported to be involved. The reexcised tissue then undergoes a similar postoperative histological examination procedure as the BCS specimen to determine if all the cancer has been excised, and a second or third reexcision operation may be required if the reexcision did not result in clear margins. A mastectomy is performed in cases where an attempt to reexcise the residual disease fails multiple times. Approximately 10% of patients with BCS as primary surgery eventually undergo a mastectomy to obtain clear margins (McCahill et al. 2012; Waljee et al. 2008).

The need to perform a reoperation in patients with involved margins after initial BCS has various undesirable consequences. It most likely will cause a delay in adjuvant treatment, and there is evidence that patients with positive margins after initial BCS have a higher chance of local recurrence (Kouzminova et al. 2009). Reoperations will result in a significantly poorer cosmetic outcome, for both reexcisions and mastectomies (Munshi et al. 2009). The additional procedures also increase the health economic costs, thus representing a burden to the health-care system. The last and probably the most important consequence of reoperations is the emotional distress patients experience after being told their cancer is not removed completely. This may result in delays in recovery, causing adverse socioeconomic effects because patients cannot resume work or other activities.

Due to a lack of adequate intraoperative tools to identify and assess tumor margins in current clinical care, positive resection margins have been reported in up to 41% of patients and are more likely in patients with a carcinoma in situ component (Jeevan et al. 2012; Talsma et al. 2011; Waljee et al. 2008). Because of the high percentage of patients that currently need to undergo a reoperation as a consequence of positive margins, there is a clear need for new, intraoperative techniques to accurately assess resection margins in patients undergoing BCS.

16.1.3.2 Mastectomy

A mastectomy is a surgical procedure that involves removal of all breast tissue to ensure no disease is left behind. This procedure is more extensive than BCS and is associated with more postsurgery side effects and a longer recovery time. Reconstruction surgery can be carried out at the same time as the mastectomy (immediate reconstruction) or sometime after the initial mastectomy surgery (delayed reconstruction).

Mastectomies are generally performed in patients that are not suitable for BCS, including patients with a large tumor (particularly in women with small breasts), a tumor in the middle of the breast, multifocal disease or areas of DCIS in the rest of the breast, or in patients where the initial BCS along with one or more reexcisions has not completely removed the cancer.

Due to the earlier stage at which breast cancer is currently being diagnosed in most developed countries, mastectomies are performed much less often than BCS as initial treatment.

There are different types of mastectomies, varying from removal of only breast tissue to a more radical procedure where in addition to removal of the breast, the lymph nodes in the armpit (axilla) and the chest wall muscles under the breast are also removed. Each type of mastectomy encompasses the removal of all breast tissue, and therefore positive tumor margins resulting from incomplete removal of the tumor rarely occur.

16.1.4 Conclusion

A key problem that currently hampers the success of BCS is the lack of accurate, real-time techniques that provide intraoperative information on tumor margins. This leads to reexcision in up to 41% of patients. Reoperations have a significant physical and emotional impact on patients and cause a financial burden for the health-care sector. Therefore, there is a need for a technique that can accurately assess tumor margins intraoperatively, thereby reducing the number of reoperations.

THz radiation possesses a number of properties that make it a viable technique for tumor margin assessment. Section 16.3 will focus on the interaction of THz radiation with human breast tissue and cancer and the results that have been obtained with different THz imaging devices. The other techniques for intraoperative assessment of tumor margins are outlined in Section 16.2.

16.2 Other Techniques for Intraoperative Assessment of Tumor Margins

There are a number of techniques, either in current use, in clinical trials, or in the research phase, that provide the surgeon with intraoperative information on tumor margins. All these techniques aim to reduce the reexcision rate. However, in choosing or developing such techniques, there are important clinical criteria to consider, including time, detection depth, and sampling area.

This paragraph gives an overview of the performance, advantages, and limitations of techniques that are currently available as well as techniques that are in the development phase. The currently available techniques can be divided into intraoperative tumor localization techniques and intraoperative pathological techniques. The former provides information on tumor margins at a macroscopic level, while the latter assesses tumor margins microscopically.

16.2.1 Currently Available Intraoperative Tumor Localization Techniques

16.2.1.1 Wire-Guided Localization

Approximately 35% of breast cancers smaller than 5 cm are nonpalpable (Lovrics et al. 2009; Skinner et al. 2001), and the most commonly used technique to determine the location of these tumors is wire-guided localization (WGL).

The wire used for localizing the lesion is 20–25 cm long and bent at the tip to form a V-shaped hook. A needle is used to target the lesion guided by x-ray, US, or MRI. After introduction of the needle, a wire is threaded down through the needle (hook end first) to lodge at the target tissue. The needle is then removed, and the hook expands upon removal of the needle, thereby anchoring the wire in the lesion. Mammography is used to confirm the wire is positioned correctly, and after confirmation, the external part of the wire is taped to the body to prevent displacement.

During the operation, the surgeon views the mammograms to get an indication of the tumor localization, and the wire is used as a guide to find the nonpalpable lesion. Since the tip of the wire is in the lesion, the surgeon needs to deviate from the wire before he or she reaches the tip to obtain an adequate margin of healthy tissue.

Literature shows that 30%–37% of wire-guided procedures result in positive margins (Gajdos et al. 2002; Lovrics et al. 2009). From these results can be concluded that WGL is suboptimal in aiding complete excision of nonpalpable breast cancers.

One of the reasons for the high positive margin rate is that the guide wire does not provide a clear three-dimensional perspective on the various tumor edges, and the required extent of resection needed to achieve negative margins still remains an estimate. Furthermore, the guide wire is prone to move before or during surgery and may therefore provide inadequate information on tumor localization. Other disadvantages include the time required to place the wire and the associated patient discomfort, resulting in increased levels of stress and arousal (Kelly and Winslow 1996). Because of the high reexcision rates and patient discomfort, other techniques have been developed to assess tumor margins.

16.2.1.2 Radioisotope Labeled Localization

Radioisotope labeled localization is an alternative technique to WGL for intraoperative localization and simultaneous resection of nonpalpable breast tumors. There are two variations of this technique: radioguided occult lesion localization (ROLL) and radioguided seed localization (RSL). In ROLL, a radioactive tracer is injected into the tumor preoperatively, and a gamma probe is used to guide the resection during surgery. To facilitate correct positioning, the injection of the radioactive tracer is performed under stereotactic or ultrasonographic guidance. After the primary tumor is excised, the gamma probe can be used to search and identify residual disease in the breast cavity. RSL is almost identical to ROLL, but instead of a radiotracer, a radioactive seed is used.

Numerous studies have been performed to evaluate the oncologic safety of the ROLL and RSL procedures (Barentsz et al. 2013; Donker et al. 2013; Hughes et al. 2008; Lovrics et al. 2011; Medina-Franco et al. 2008; Monti et al. 2007; Sarlos et al. 2009; Thind et al. 2011). Positive margins varied from 4% to 27%, although Sarlos et al. reported a 35% positive margin rate in patients with DCIS. Hughes et al. compared RSL with WGL and found significantly lower positive margin rates with RSL as compared to wire localization, while Lovrics et al. reported similar positive margin rates. A recently published large multicenter clinical trial compared ROLL to WGL in patients with invasive breast carcinoma (Postma et al. 2012). They found similar positive margin rates for ROLL and WGL, and ROLL was considered comparable in terms of complete tumor excision and reexcision rates. However, they found a larger excision volume in the ROLL group and concluded that ROLL cannot replace WGL as standard of care.

Although the oncologic safety of ROLL and RSL might be similar to WGL, both techniques have several advantages over WGL. They are easy-to-perform radiological and surgical procedures, and the tumor can be identified in three dimensions allowing for more precise excision. More importantly, both techniques are more patient friendly; hence, the pain rankings reported by patients are significantly lower (Lovrics et al. 2011; Rampaul et al. 2004). However, these guidance tools are still invasive like WGL and therefore associated with patient discomfort. Moreover, the use of radioactive material exposes patients and health-care workers to radiation, is heavenly controlled by legislation, and is available only in hospitals with a nuclear medicine department.

16.2.1.3 Intraoperative Ultrasound–Guided Resection

Intraoperative US (IOUS) is a method of excising a tumor under direct visualization, thus providing the surgeon with real-time information of the tumor extent.

Immediately after excision, ex vivo US examination of the specimen is performed in the operating theater to check the completeness of the specimen. In case of positive or close margins, residual disease is removed by shaving the patient's cavity margins.

The COBALT trial compared IOUS-guided resection with palpation-guided resection for invasive *palpable* tumors and found tumor-involved margins in 11% of the IOUS-guided group and 28% in the palpation-guided group (p = 0.0031) (Krekel et al. 2012). Besides, IOUS-guided surgery resulted in significantly smaller excision volumes. Moore et al. also found a significant improvement in margin status in IOUS-guided surgery compared to palpation-guided surgery (Moore et al. 2001).

For *nonpalpable* invasive breast cancers, positive margins were found after IOUS-guided surgery in 4.3%–19% of patients (Bennett et al. 2005; Krekel et al. 2011; Ngô et al. 2007; Rahusen et al. 2002; Snider and Morrison 1999). Rahusen et al. reported significantly improved margin status using IOUS compared to WGL (11% vs. 45%, respectively), and this finding was supported by Krekel et al. Besides, Snider et al. found a smaller resection volume compared to WGL.

These results indicate that IOUS-guided surgery can lower the proportion of tumor-involved resection margins compared to palpation-guided and wire-guided localization, while decreasing resection volumes. US is widely available, does not require radiation, and minimizes patient trauma and discomfort as there is no need for additional interventions (i.e., a wire or material). However, IOUS also has some important disadvantages. The technique is not suitable for detecting microcalcifications as these are not sonographically visible, and DCIS can therefore not be identified. Besides, the lesion must be above a certain size to be imaged, and it is shown that 50% of nonpalpable tumors are missed (Klimberg 2003). Other possible restrictions are that the technique does not have the sensitivity and resolution to find residual disease after excision of the primary tumor and that a radiologist must be present in the operating theater during the procedure.

16.2.1.4 Intraoperative Specimen Radiography

Intraoperative specimen radiography is a technique often used to image the excised specimen in patients with nonpalpable breast tumors. In conventional specimen radiography, the surgical specimen is transported from the operating theater to the diagnostic imaging department, while intraoperative digital specimen mammography (IDSM) entails imaging in the operating theater. The former can entail significant time for transport (≈30 min). If specimen radiographs reveal involved margins, the surgeon can shave the associated cavity edges to remove any residual malignant disease.

In 2007, Kaufman et al. compared IDSM with CRF for assessing tumor margins. The sensitivity of IDSM and CRF for detecting positive margins was 36% and 31%; specificity was 71% and 74% (Kaufman et al. 2007). More recently, Bathla et al. evaluated the performance of an IDSM device for intraoperative margin assessment and found a sensitivity and specificity of 58.5% and 91.8%, respectively (Bathla et al. 2011). The positive predictive value was 82.7%; the negative predictive value was 76.7%.

A disadvantage of intraoperative specimen radiography is that in situ cancers will show up only if there is sufficient contrast or microcalcifications associated with it. Thus, even if the tumor seems to be adequately removed on the radiography image, reoperation may still be needed if histopathology identifies an in situ component near or at the margin. Besides, specimen radiography provides information on margin status at a macroscopic level and can therefore not be relied on solely.

16.2.2 Currently Available Intraoperative Pathological Techniques

A disadvantage related to all the aforementioned techniques is that no information is provided on the microscopic margin status. Therefore, some centers use additional intraoperative pathological techniques to provide the surgeons with information on the microscopic extent of the tumor. The most commonly used intraoperative pathological techniques are frozen section analysis (FSA) and touch imprint cytology (TIC).

16.2.2.1 Frozen Section Analysis

FSA is used to intraoperatively assess microscopic margin status in many oncologic procedures, including breast cancer. While the patient is still on the operating table, the excised specimen is inked, sliced, frozen, and analyzed microscopically by an experienced histopathologist (Weber et al. 2008).

This takes approximately 30 min. Surgery continues during the pathological analysis, and the surgeon closes the wound if the pathological result takes longer than the surgical procedure. In case involved margins are identified, the wound of the patient is reopened and additional cavity shaving is performed.

Reported sensitivity rates for the assessment of resection margin status range between 73% and 83%, whereas specificity rates are in between 87.5% and 99% (Esbona and Wilke 2012; Hunt et al. 2007; Olson et al. 2007; Weber et al. 2008). The lower sensitivity is mainly a result of unreliable detection of small tumors (<10 mm) and microcalcifications (i.e., DCIS). Two large studies evaluated the number of reoperations after FSA, and both studies demonstrated that frozen section leads to low reoperation rates (9% and 10%, respectively) (Esbona and Wilke 2012; Riedl et al. 2009). Apart from the unreliable detection of small tumors and microcalcifications, FSA does not provide the surgeon with information on tumor extent prior to excision, therefore not aiding in decreasing resection volumes. Furthermore, FSA is labor intensive and requires an experienced on-site pathologist and may therefore not be performed in hospitals where the pathology department is located outside of the hospital. Besides, FSA requires a relatively large part of the specimen, which may compromise postoperative evaluation by the pathologist, and freezing artifacts are common in fatty tissue, which may interfere with accurate identification of cell types.

16.2.2.2 Touch Imprint Cytology

Intraoperative TIC is a simple and rapid alternative to FSA. The excised specimen is oriented and pressed onto specifically coated slides, making an imprint of all six margins. Cells sticking to the glass surface are then fixed, stained, and microscopically analyzed (Klimberg et al. 1998). This technique is based on the difference in cellular surface characteristics between malignant cells and mammary fat; malignant cells will adhere to the slides and adipose cells will not.

The results of TIC are reported within 15 min, which is definitely quicker than FSA, and for most surgeries, the result is received before the wound is closed. Another advantage over FSA is that it saves tissue for permanent sectioning and histopathological examination, and that it is less expensive than FSA.

Esbona et al. performed a systematic review on reexcision rates, sensitivity, and specificity of TIC and compared the performance with FSA and postoperative histopathological evaluation (Esbona and Wilke 2012). The reexcision rate of TIC was 11%, against 10% and 35% for FSA and postoperative histopathology. The sensitivity and specificity were 72% and 92%, respectively, compared to 83% and 95% for FSA. In line with the results for FSA as described previously, most false-negative results with TIC were observed in tumors with in situ disease. However, there was a greater degree of variation present in the sensitivity of TIC. The difference in cytopathological proficiency between pathologists may account for the degree of variation, as the technique requires extensive cytology expertise.

Although the results of this technique seem very promising, so far it has not been as widely used as FSA. One of the explanations is that close margins are not taken into account, because only superficial tumor cells are detected with the technique. Therefore, no information is gathered on margin width, multifocality, and quantity of cancerous cells approaching the cut edge (Pleijhuis et al. 2009). There is also the potential of artifacts caused by draught and surface cautery, and TIC seems less effective in identifying lobular carcinoma (Valdes et al. 2007).

16.2.3 Techniques under Development

The previously described techniques all have certain disadvantages and result in positive resection margins in a significant amount of patients. The main disadvantage of all these techniques is the inaccuracy in detecting DCIS, and the need to detect DCIS is significant as it is considered a more challenging intraoperative assessment target. Currently, several techniques are being developed for assessing tumor margins intraoperatively, with a particular focus on techniques that are able to identify DCIS. A nonexhaustive list of these new, innovative techniques is provided later.

16.2.3.1 Radiofrequency Spectroscopy: MarginProbe®

The MarginProbe (Dune Medical Devices, Framingham, USA) uses radiofrequency spectroscopy to detect minute differences in electromagnetic properties of tissue with and without cancer, and provides this key information as a *positive* or *negative* readout to breast surgeons intraoperatively. The device consists of a console and a disposable handheld probe with an effective measurement area of 7 mm and a detection depth of about 1 mm.

A prospective, randomized, multicenter trial was conducted to evaluate the performance of the device in assessing surgical margins (Allweis et al. 2008). After excision of the main lumpectomy specimen, patients were randomized to a device and a control arm. In the device arm, the surgeon applied the device to the six margins (medial, lateral, superior, inferior, deep, and anterior), and reexcised tissue if the device indicated positive margins. In the device arm, 60% of the involved margins were correctly identified and reexcised, compared with 40% in the control arm ($p = 0.044$). The percentage of patients with correctly identified margins in a nonpalpable subgroup was also higher (69% vs. 39%, respectively). The need for reoperation in case of positive margins was not defined or dictated by the study, so the actual decrease in reexcision rate from using the device cannot be determined from this study.

One of the advantages of the technique is the potential to detect DCIS (Pappo et al. 2010; Thill et al. 2011), so the MarginProbe can be of special interest for this group of patients. Other advantages include a short measurement time (1–2 s per measurement) and controlled, user-independent tissue measurements using a vacuum-based mechanism. The main disadvantage of this technique is that the performance of the device decreases for tissue with a more heterogeneous composition, that is, the probe is less sensitive to measurement sites with a small cancer feature size (Pappo et al. 2010). Consequently, patients with small tumors might still need to undergo a second surgery to obtain clear margins. Besides, the technique samples only an area of 7 mm, and positive margins can be missed if the surgeon does not accurately cover each entire margin.

16.2.3.2 Diffuse Reflectance Spectroscopy

Diffuse reflectance spectroscopy (DRS) can identify tissue characteristics by measuring their intrinsic light absorption and scattering properties at different wavelengths in the UV–visible range. Diffuse reflectance spectra can be obtained by illuminating tissue with a selected light spectrum, and these spectra reflect the absorption and scattering properties of the tissue. The absorption coefficient is directly related to the concentration of physiologically relevant absorbers in the tissue, which include oxygenated and deoxygenated hemoglobin. The scattering coefficient reflects the size and density of scattering centers in tissue, such as cells and nuclei. Since changes in human tissue associated with malignant transformation include alterations in cellular composition, metabolic rate, and tissue morphologic characters, the reflectance spectra can be used to differentiate tumor from normal tissue.

Bigio et al. performed one of the earliest studies on the applicability of DRS for assessing tumor margins (Bigio et al. 2000). They used elastic scattering spectroscopy (a variant of DRS) implemented in a fiber-optic probe assembly and measured diffuse reflectance spectra from the tumor cavity in vivo after resection. The probe had a sensing depth of 300 μm, which is sufficient to detect disease at the surface. The measured tissue was biopsied for pathologic correlation, and the spectra were fed into classification algorithms to provide estimations for the presence of residual disease. They measured 72 breast tissue sites, and the sensitivity and specificity were 69% and 85%, respectively.

More recently, a biomedical group from the Duke University has developed a diffuse reflectance imaging device and studied 55 resection margins in an ex vivo setting in 48 patients. They were able to detect positive margins, regardless of pathology or depth from the margin, with 79% sensitivity and 67% specificity (Ramanujam et al. 2009; Wilke et al. 2009) (see Figure 16.3). Interestingly, positive margins for DCIS were correctly identified in eight of nine margins, corresponding with a sensitivity of 89%. However, of the eight patients who received neoadjuvant therapy, only four patients had their margins assessed correctly, indicating the inability of the current device to identify positive margins in this patient group.

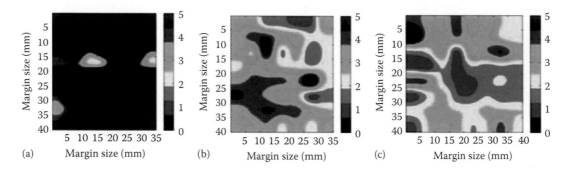

FIGURE 16.3 **(See color insert.)** Diffuse reflectance spectroscopy for tumor margin assessment. Parameter maps per margin were obtained from the ratio of β-carotene (absorption parameter) and the wavelength-averaged reduced scattering coefficient (scattering parameter). Blue areas generally represent healthy tissue, whereas red areas represent tumor. (a) Pathologically confirmed negative margin. (b) Pathologically confirmed margin positive for DCIS. (c) Pathologically confirmed margin positive for IDC. (Reproduced from Wilke, L. G., J. Brown, T. M. Bydlon et al., *Am. J. Surg.*, 198, 566, 2009, with permission from Elsevier.)

Another group combined diffuse reflectance and autofluorescence spectroscopy (Keller et al. 2010). Excised specimens were measured by a fiber-optic probe, and diffuse reflectance and autofluorescence spectra were then classified as benign or malign using a two-part classification method. Tissues from 32 patients were measured, and the sensitivity and specificity were 85% and 96%, respectively.

These preliminary results suggest that DRS (in combination with autofluorescence) could provide a useful tool to assess tumor margins intraoperatively. The potential to detect DCIS is of particular interest, as all currently available techniques lack the ability to accurately detect DCIS. However, due to the small scan area of the current probes, which is restricted by the output of the optical fibers, only a small tissue volume can be scanned. Hence, in order to cover the entire margin, multiple sites must be measured, which is time consuming and subject to user errors. Therefore, an important challenge for this technique is to increase the scan volume to make the technique suitable for intraoperative use.

16.2.3.3 Spatially Offset Raman Spectroscopy

Another optical technique that can potentially be used for breast tumor margin evaluation is Raman spectroscopy. Raman spectroscopy is a form of vibrational spectroscopy. Laser light (mostly in the visible, near-infrared (NIR), or near-ultraviolet range) is nonelastically scattered from molecules that are excited by the incident light, and this inelastic scattering can be used to identify organic molecules, such as fat, collagen, cell cytoplasm, and nucleus. These molecules have been shown to provide a signature associated with abnormality (Frank et al. 1995), thus providing the possibility for margin assessment.

Keller et al. used spatially offset Raman spectroscopy (SORS), which is a variant of Raman spectroscopy, for breast tumor surgical margin evaluation (Keller et al. 2011). They developed a SORS probe with one source fiber and multiple detection fibers, allowing the detection of spectral signatures from tumor within the first 2 mm in depth from the tissue surface. A total of 35 excised specimens were scanned: 15 samples with negative margins (>2 mm) and 20 samples with positive invasive tumor margins (<2 mm). The spectra were then classified to predict if a margin was positive or negative. They found a sensitivity, specificity, negative predictive value, and positive predictive value of 95%, 100%, 94%, and 100%. Although their findings are based on a small data set and did not include actual reexcision rates, their results seem very promising. However, the diameter of their probe is approximately 5 mm, thus covering only a small tissue area. The biggest challenge for this technique is therefore similar to DRS: adapt the probe to interrogate larger areas of tissue in order to meet the clinical criteria of sufficient area sampling within practical time constraints.

16.2.3.4 Optical Coherence Tomography

In the last decade, optical coherence tomography (OCT) has emerged as a high-resolution optical diagnostic imaging modality that is currently being used in ophthalmology and dermatology. More recently, OCT has been investigated as a technique to assess breast tumor margins. OCT is an optical equivalent of US, but instead of acoustic waves, it is based on the reflection of light from a low-coherence broadband light source (typically employing NIR light). By measuring backscattered or back-reflected light, cross-sectional tomographic images of the tissue microstructure can be obtained with a similar resolution as histopathology (μm).

Nguyen et al. at the University of Illinois used OCT images to intraoperatively assess exposed tumor margins (Nguyen et al. 2009). Their system operates in the spectral region around 1300 nm, providing a lateral and axial resolution of 35 and 5.9 μm in tissue, respectively. They scanned excised specimens (including DCIS) from 37 patients. The OCT images from the first 17 patients were used to establish imaging protocols and OCT criteria for identifying positive margins, and the images from the remaining patients were used for the study set. Eleven margins were identified as positive and nine as negative on OCT imaging, giving a sensitivity, specificity, and overall accuracy of 100%, 82%, and 90%, respectively. The increased scattering, due to strong reflections from tightly packed cells, was the main feature for classifying margins as positive.

The main challenge for this technique is to decrease acquisition and processing time, allowing to scan larger areas in a shorter time interval. Recent developments in OCT technology may make it possible to decrease scanning times, but these advances will result in significant increases in data volume. This requires automated classification algorithms, as individual analysis and interpretation of the OCT images would be too time consuming. However, once these challenges are overcome, the technique may provide useful intraoperative information to reduce the number of reoperations.

16.2.3.5 Near-Infrared Fluorescence Imaging

In near-infrared fluorescence (NIRF) imaging, an NIR fluorophore is administered, after which the breast tissue is irradiated by an external laser in the NIR spectral range (650–900 nm). Upon excitation, the fluorophore will release photons of a higher NIR wavelength, which are captured by an NIRF camera system and digitally converted into a visible image. In the 650–900 nm range, the absorption coefficient in tissue is minimum, resulting in decreased light scattering and autofluorescence, and increased penetration depth compared to visible and UV light.

Fluorophores can be conjugated to a specific targeting ligand or monoclonal antibody to image tumor-targeted molecules that have been identified with breast cancer, including Her2/Neu receptor, vascular endothelial growth factor (EGF) receptor, and EGF. The tumor-specific binding properties of these so-called tumor-targeted fluorophores seem perfectly suitable for image-guided surgery, as it provides the surgeon with real-time, tumor-specific information on the location and extent of the lesion.

However, the need to administer exogenous fluorophores limits the clinical application of the technique due to the requirement to be approved for human use. Indocyanine green (ICG) is currently the most often used NIR fluorophore, but ICG does not offer the possibility of tumor-specific antigen coupling. Another reason why the development of this technique has been hampered is the lack of dedicated intraoperative imaging systems.

To date, a minimal number of studies have been published that use NIRF imaging to guide surgical excision in breast cancer. Mieog et al. used NIRF in a breast cancer rat model. Seventeen rats were operated on, resulting in a complete resection of seventeen out of seventeen tumors (Mieog et al. 2011). Moreover, the technique enabled the identification of remnant tumor in the surgical cavity, which is an important advantage over WGL and US. Aydogan et al. performed a feasibility study in two patients. They injected nontargeted ICG in the breast lesion under US guidance. Surgery was performed 1 h after injection, and in both patients, clear margins were obtained (Aydogan et al. 2012).

Overall, NIRF imaging seems a suitable candidate for early intraoperative introduction, as it is a fast and simple-to-operate technique, with a sufficient penetration depth to assess tumor margins. However, progress needs to be made with clinically available fluorophores and dedicated imaging systems. Furthermore, the technique does not provide information on microscopic margin status. Thus, it is likely that it will be used together with a technique that provides microscopic information on tumor margins.

16.2.4 Conclusion

The intraoperative techniques that are currently being used in BCS all have their limitations. The overall disadvantage of all these techniques is the inaccurate detection of DCIS, and this limitation will become of greater importance in the near future, as patients will be diagnosed at an earlier stage. The techniques that are currently in the clinical trial or research phase have the potential to assess tumor margins with a similar or even better accuracy than current techniques. Moreover, most of these techniques seem to be able to detect DCIS, thus providing additional value in this specific patient population. However, technological developments are required for these techniques to meet the clinical criteria of sufficient area sampling within practical time constraints, and further studies must be performed to elucidate their value in decreasing reoperation rates in BCS.

16.3 THz Technology in Breast Cancer

16.3.1 Introduction

In recent years, there has been growing interest in the application of terahertz (THz) technology to the field of biomedical research (Berry et al. 2004; Fitzgerald et al. 2006; Knab et al. 2007; Pickwell et al. 2004; Reid et al. 2007; Woodward et al. 2003). THz has been used to successfully characterize DNA (Nagel et al. 2006) and proteins (Knab et al. 2007) in this range, demonstrating the ability of the radiation to probe intermolecular interactions. Typical THz imaging systems produce radiation of wavelengths in the range of 3 mm to 80 μm; this is longer than that of the visible spectrum or infra-red, so THz radiation is less susceptible to scattering within biological tissue (Han et al. 2000), and in general, scattering is assumed to be negligible. Due to the unique sensitivity of THz to water and the safe nature of the wavelengths used, THz techniques have been investigated for imaging of tissues (Arnone et al. 1999; Ashworth et al. 2006, 2008; Berry et al. 2004; Brucherseifer et al. 2000; Fitzgerald et al. 2006; Han et al. 2000; Markelz et al. 2000; Nagel et al. 2006; Nakajima et al. 2007; Pickwell et al. 2004). Many tissues in the human body are made up of about 70% water, and the adult human body is about 57% water by mass (Hall 2010). It is known that many cancers have a higher concentration of water than normal tissue. THz radiation is uniquely sensitive to water, which, together with the advantages listed earlier, makes it a viable tool for medical imaging and in particular of cancers.

Wallace et al. working at TeraView Ltd (Cambridge, UK) demonstrated the application of THz reflection images to differentiate between normal skin and cancerous skin on both ex vivo and in vivo samples (Fitzgerald et al. 2004; Pickwell et al. 2004; Wallace et al. 2006) and additionally performed THz spectroscopy to characterize the properties of different tissue types. Further work has been done to identify contrast between healthy colon tissue, dysplasia, and cancer by Reid et al. at University College, London (Reid 2009).

16.3.2 Application of THz Technology to Breast Cancer

The application of THz technology to breast cancer was first proposed by Fitzgerald et al. in 2004 (Fitzgerald et al. 2004). Using a portable THz pulsed imaging (TPI) system developed for use in a hospital environment, they measured samples of excised breast tissue. The system used photoconduction to generate and detect THz pulses with a frequency content from 0.1 to 4 THz. Several freshly excised breast samples were imaged. Two parameters from the time domain impulse functions were used to produce images: the minimum of the THz impulse function, E_{min}, and the ratio of the E_{min} to the maximum of the impulse function, E_{max}. The images showed contrast between healthy and cancerous tissues with good agreement to histology. Interestingly, the images also highlighted DCIS. This preliminary study demonstrated the potential of TPI to image breast tumors and encouraged further studies to determine the ability of the technique to discriminate between different types of breast tissue (Fitzgerald et al. 2004).

A more comprehensive study performed in 2006 by Fitzgerald et al. (2006) confirmed that TPI could be used to identify contrast between healthy breast tissue and breast cancer by using E_{min} and E_{min}/E_{max}. The shape and size of the tumor on TPI correlated well with histopathology (see Figure 16.4). This work showed that THz can be used to distinguish invasive lobular and ductal breast cancer from normal and adipose breast tissue, and that it is likely that DCIS can be imaged as well (Fitzgerald et al. 2006).

Given it quasi-3D nature, THz imaging produces large volumes of data due to the two spatial and temporal components both being recorded. Fitzgerald et al. (2012) investigated data reduction methods prior to the classification of THz data from freshly excised breast cancer tissues. THz images have typically been formed using a range of parameters (or features), derived from the pulse or spectral profiles (Fitzgerald et al. 2002; Woodward et al. 2002). This heuristic approach to data reduction was compared to an unsupervised method that has traditionally been used in other areas, called principal component analysis (PCA) (Hutchings et al. 2009). PCA provides a theoretically optimal linear reduction, which requires no underlying assumptions about the statistical nature of the data. The PCA method was applied to the THz pulses and compared with the heuristic parameters. Classification for this study was performed using the support vector machine algorithm, which is well suited to finding complicated decision boundaries and has been used with good effect on other THz data sets (Yin et al. 2007). The classification results from the THz signals were then compared with histopathology. It was shown that using appropriate data reduction methods, based on parametric features and/or principal components, THz signals reflected from freshly excised breast cancer tissue can be classified with accuracies up to 92%.

This chapter provides further evidence on the efficacy of the technique and points to methods to improve classification of signals obtained when using THz in breast cancer. With these encouraging results, it was suggested that this technique of imaging could potentially be used to intraoperatively assess tumor margins in patients undergoing BCS, eventually aiming to reduce the number of reoperations (Wallace et al. 2005).

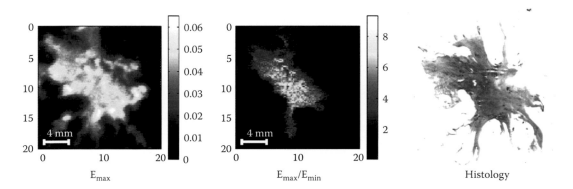

FIGURE 16.4 **(See color insert.)** This figure shows two terahertz images generated using the maximum of the reflected pulse (E_{max}) and the ratio between maximum and minimum of the reflected pulse (E_{max}/E_{min}). In the E_{max} image, all the tissues are shown, tumor with surrounding adipose tissue. In E_{max}/E_{min}, only the tumor is visible and correlated well with the tumor shown in the histology image. (Reproduced from Ashworth, P. C., E. Pickwell-MacPherson, E. Provenzano et al., *Opt. Exp.*, 17, 12444, 2009. With permission of Optical Society of America.)

16.3.3 Handheld THz Probe for Intraoperative Use

In order to use THz technology for intraoperative assessment of breast tumor margins, a handheld THz imaging probe has been developed (see Figure 16.5) (Wallace et al. 2005). The system uses a photoconductive emitter and receiver to produce and detect THz pulses of bandwidth 0.1–2.0 THz. The THz pulse beam is focused to the probe tip, from which they are reflected back, through a silicon lens and a system of Risley beam steering prisms. The THz pulse beam is scanned back and forth across the active area by controlled rotation of the Risley prisms, resulting in an imaging window with a length of 8 mm.

This probe has been used to study tissue samples, obtained from freshly excised lumpectomy and mastectomy specimens, from 37 patients at Guy's Hospital in London (Ashworth et al. 2008). Figure 16.6 shows typical impulse responses observed upon the reflection of THz pulses from the three main breast tissue types—adipose, fibrous, and cancerous. A number of distinguishing parameters (features) were identified, and Figure 16.7 shows the average values for four parameters for each of the three key tissue types. It is clear from these two figures that it is easy to identify differences between adipose tissue and cancer with the probe. However, as expected, the discrimination between fibrous tissue and cancer is more challenging. By examining the pulse integral and full width half maximum in Figure 16.7, we can see subtle differences in the average values of these parameters between fibrous tissue and cancer. PCA was applied to the full-time domain THz pulse data, and then linear discriminant analysis was used to predict the tissue types of individual specimens as listed in Table 16.1. This shows that a promising value of 90% was found for the sensitivity and 81% for specificity (Ashworth 2010).

It is clear that these studies with a handheld THz probe are promising; THz technology is now becoming sufficiently versatile that it is possible to build such a probe and achieve a good level of performance in a clinical environment. One of the limitations of the handheld probe is that due to the small imaging window, only a small tissue volume can be scanned. If in the near future the THz probe is used to scan complete lumpectomy specimens, multiple measurements per margin must be performed, which is time-consuming and sensitive to user errors. A potential limitation of the use of THz for assessing breast margins in general is the limited penetration depth as a result of high attenuation of THz in water. Tumor cells located at a depth >1 mm may therefore not be detected, but further studies are needed to determine the actual penetration depth of THz in breast tissue. There are still improvements that can be made in order to increase the SNR, reduce errors in data registration, and remove contaminant data from the training data set for discriminate analysis. These improvements may well bring the performance of the technique closer to routine use during surgery.

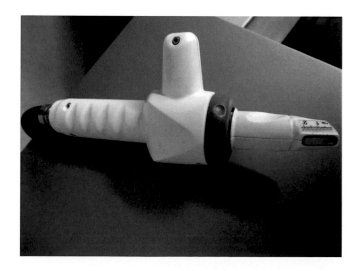

FIGURE 16.5   THz handheld intraoperative probe; courtesy of TeraView Ltd. (patented).

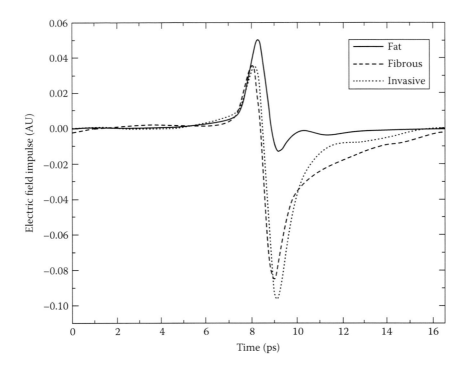

FIGURE 16.6 Graph showing typical pulses obtained from three key tissue types found in tissue excised from a breast cancer patient.

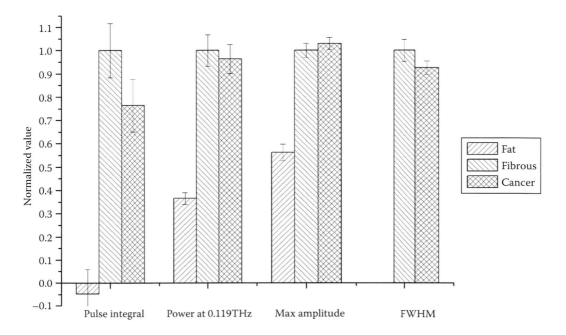

FIGURE 16.7 Bar graph showing the average values of four parameters for each of the three tissue types. Error bars represent 95% confidence intervals.

TABLE 16.1 Table Listing the Results of Linear Discriminant Analysis on Data Collected Using Handheld Probe

	Histology Result	
THz Prediction	Cancer	Healthy
Cancer	27	34
Healthy	3	142
Total N	30	176
Total correct	27	142
	Sensitivity	Specificity
	90%	81%

16.3.4 Understanding THz Signals from Breast Tissue

There is a need to understand the mechanisms that cause contrast in THz images of breast cancer and relate reflected pulse data to changes in tissue pathology. As a first step, Ashworth et al. measured the absorption and refractive indices of both healthy and diseased breast tissues in the THz region via time-domain THz pulsed spectroscopy (TPS) (Ashworth et al. 2009). THz transmission spectroscopy or TPS has previously been used to obtain the THz optical characteristics of skin tissue and basal cell carcinomas (BCCs) (Pickwell et al. 2004; Wallace et al. 2006), including the refractive index and absorption coefficient spectra. In the Ashworth study, 74 fresh breast tissue samples were measured, of these 33 were classified as *cancer*, 22 as *healthy fibrous*, and 19 as *healthy fat*. Figures 16.8 and 16.9 show the measured average absorption coefficient and refractive index for each of these groups, respectively. The error bars displayed

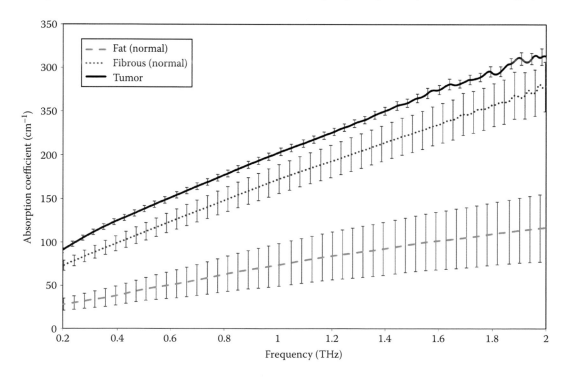

FIGURE 16.8 Graph showing average absorption coefficient spectra of groups of tissue types; error bars represent 95% confidence intervals.

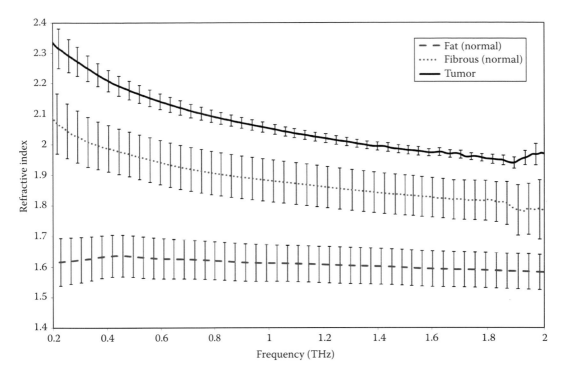

FIGURE 16.9 Graph showing average refractive index spectra of groups of tissue types; error bars represent 95% confidence intervals.

represent 95% confidence intervals derived from the standard error of the mean. The error bars for the refractive index are constant for fibrous tissue and fat but are slightly larger at high and low frequencies for the cancer samples. This is due to low signal in these ranges caused by attenuation through the sample.

Due to the nature of fibrous tissue within the breast, it was a challenging task isolating sufficient fibrous tissue from fatty tissue intact such that it would be suitable for our spectroscopy needs. As a result, the fibrous tissue group had a higher percentage fat than the tumor group, as tumors formed solid lumps that proved easier to separate.

There is a clear difference in the refractive indices between all pure tissue types and also a difference between the absorption coefficient of fat and the other two tissue types. In Figure 16.8, the shape of the spectrum of the cancer is very similar to fibrous tissue, but the transmitted signal is lower due to the increased attenuation in the cancer. The attenuation is a combination of absorption and scattering, and it would be impossible with this type of measurement to determine the contribution from each mechanism. As discussed previously, increased water content in the cancer tissue would lead to increased absorption, and in cancer tissue samples, we therefore assume absorption to be dominant and scattering negligible. But if one looks at studies on dehydrated, wax-embedded samples, attenuation also increases for regions of cancer, which can only result from structural changes, which, in turn, change the scattering properties (Berry et al. 2004; Nakajima et al. 2007). It is seen that breast tumor tissue has a higher refractive index than both healthy fatty and healthy fibrous tissues with the greatest difference being at 0.32 THz. Furthermore, around 0.32 THz, the absorption coefficient of breast tumor tissue increases slightly compared with fibrous breast tissue. Simulations were carried out to predict the impulse responses from performing THz pulsed reflection imaging on these tissues. A large difference was seen, as expected, between the impulse responses of healthy fat and those of healthy fibrous and breast tumor tissues. A difference in peak height of about 60% was seen on the impulse responses between cancer and healthy fibrous tissues.

In conclusion, the contrast seen when imaging breast cancer in reflection is mainly caused by an increase in the refractive index between cancer and healthy tissues and also in part due to an increase in the absorption coefficient. The root cause of the increases in these fundamental properties is yet to be determined.

Acquiring spectroscopic THz data allows us to simulate the reflected time domain response functions of healthy fibrous breast tissue and breast cancer tissue. Pickwell et al. have looked at the refractive index and absorption coefficient separately to characterize tissues. They have used the same model to investigate correlations between THz data and pathology and identify parameters of the simulated reflected impulse response function and corresponding spectroscopic properties with a view to improving our ability to distinguish between the fibrous and cancer tissues in the breast (Pickwell-Macpherson et al. 2012).

16.4 Conclusion

The field of medical imaging using THz technology is very much in its infancy, and there is a great deal of work yet to be done to show its true potential before the modality would be widely adopted by the medical community. At present, only skin, colon, and breast cancers have been examined in an ex vivo setting where the water content of the tissues is close to that of in vivo tissue. It would be interesting to look at the potential use of THz in identifying other cancers that can either be accessed topically or in cases where a conservative excision is required. These may, for example, include head and neck cancers such as mouth, throat, or brain cancers, cervical cancer, or rectal cancer. However, it is important not to lose focus and to ensure that the development of THz technology with a view to intraoperatively diagnosing those cancers for which contrasts have already been identified continues, and current techniques are improved.

There are several challenges to using this technique during surgery, such as the presence of blood and other fluids in the region, maintaining good contact of the probe with the breast tissue, and inter-patient variability. Although the presence of blood and other fluids can be managed during surgery, for example, through cauterization, this may also affect the THz response of the tissue to a varying degree and thus will require further study.

References

Allweis, T. M., Z. Kaufman, S. Lelcuk et al. 2008. A prospective, randomized, controlled, multicenter study of a real-time, intraoperative probe for positive margin detection in breast-conserving surgery. *The American Journal of Surgery* 196: 483–489.

Arnone, D. D., C. M. Ciesla, A. Corchia et al. 1999. Applications of terahertz (THz) technology to medical imaging. Presented at *Proceedings of SPIE* 3828: 209–219.

Ashworth, P. C. 2010. Biomedical applications of terahertz technology. Cambridge, U.K.: University of Cambridge. PhD.

Ashworth, P. C., P. O'Kelly, A. D. Purushotham et al. 2008. An intra-operative THz probe for use during the surgical removal of breast tumors. Presented at *33rd International Conference on Infrared, Millimeter and Terahertz Waves*, Pasadena, CA, pp. 1–3.

Ashworth, P. C., E. Pickwell-MacPherson, E. Provenzano et al. 2009. Terahertz pulsed spectroscopy of freshly excised human breast cancer. *Optics Express* 17: 12444–12454.

Ashworth, P. C., J. A. Zeitler, M. Pepper, and V. P. Wallace 2006. Terahertz spectroscopy of biologically relevant liquids at low temperatures. Presented at *31st International Conference on Infrared Millimeter Waves and 14th International Conference on Terahertz Electronics*, Shanghai, China, p. 184.

Aydogan, F., V. Ozben, E. Aytac et al. 2012. Excision of nonpalpable breast cancer with indocyanine green fluorescence-guided occult lesion localization (IFOLL). *Breast Care* 7: 48–51.

Barentsz, M. W., M. A. A. J. van den Bosch, W. B. Veldhuis et al. 2013. Radioactive seed localization for non-palpable breast cancer. *British Journal of Surgery* 100: 582–588.

Bathla, L., A. Harris, M. Davey, P. Sharma, and E. Silva. 2011. High resolution intra-operative two-dimensional specimen mammography and its impact on second operation for re-excision of positive margins at final pathology after breast conservation surgery. *The American Journal of Surgery* 202: 387–394.

Bennett, I. C., J. Greenslade, and H. Chiam. 2005. Intraoperative ultrasound-guided excision of nonpalpable breast lesions. *World Journal of Surgery* 29: 369–374.

Berry, E., J. W. Handley, A. J. Fitzgerald et al. 2004. Multispectral classification techniques for terahertz pulsed imaging: An example in histopathology. *Medical Engineering & Physics* 26: 423–430.

Bigio, I. J., S. G. Bown, G. Briggs et al. 2000. Diagnosis of breast cancer using elastic-scattering spectroscopy: Preliminary clinical results. *Journal of Biomedical Optics* 5: 221–228.

Brucherseifer, M., M. Nagel, P. Haring Bolivar et al. 2000. Label-free probing of the binding state of DNA by time-domain terahertz sensing. *Applied Physics Letters* 77: 4049–4051.

Cancer Research UK. 2012a. Breast cancer incidence statistics—lifetime risks. http://www.cancerresearchuk.org/cancer-info/cancerstats/types/breast/incidence/#risk, accessed June 25, 2013.

Cancer Research UK. 2012b. Invasive ductal breast cancer. http://www.cancerresearchuk.org/cancer-help/type/breast-cancer/%20about/types/invasive-ductal-breast-cancer, accessed June 25, 2013.

Donker, M., C. A. Drukker, R. A. V. Olmos et al. 2013. Guiding breast-conserving surgery in patients after neoadjuvant systemic therapy for breast cancer: A comparison of radioactive seed localization with the ROLL technique. *Annals of Surgical Oncology* 20: 1–7.

Esbona, K. and L. G. Wilke. 2012. Intraoperative imprint cytology and frozen section pathology for margin assessment in breast conservation surgery: A systematic review. *Annals of Surgical Oncology* 19: 3236–3245.

Ferlay, J. S., H. R. Shin, F. Bray et al. 2010. GLOBOCAN 2008, cancer incidence and mortality worldwide: IARC CancerBase No. 10. Presented at *International Agency for Research on Cancer 2010*, Lyon, France, p. 29.

Fisher, B., S. Anderson, J. Bryant et al. 2002. Twenty-year follow-up of a randomized trial comparing total mastectomy, lumpectomy, and lumpectomy plus irradiation for the treatment of invasive breast cancer. *New England Journal of Medicine* 347: 1233–1241.

Fitzgerald, A. J., E. Berry, N. N. Zinovev et al. 2002. An introduction to medical imaging with coherent terahertz frequency radiation. *Physics in Medicine and Biology* 47: R67.

Fitzgerald, A. J., S. Pinder, A. D. Purushotham et al. 2012. Classification of terahertz-pulsed imaging data from excised breast tissue. *Journal of Biomedical Optics* 17: 0160051 01600510.

Fitzgerald, A. J., V. P. Wallace, M. Jimenez-Linan et al. 2006. Terahertz pulsed imaging of human breast tumors. *Radiology* 239: 533–540.

Fitzgerald, A. J., V. P. Wallace, R. Pye et al. 2004. *Terahertz Imaging of Breast Cancer, a Feasibility Study*. New York: IEEE.

Frank, C. J., R. L. McCreery, and D. C. B. Redd. 1995. Raman spectroscopy of normal and diseased human breast tissues. *Analytical Chemistry* 67: 777–783.

Gajdos, C., P. I. Tartter, I. J. Bleiweiss et al. 2002. Mammographic appearance of nonpalpable breast cancer reflects pathologic characteristics. *Annals of Surgery* 235: 246.

Hall, J. E. 2010. *Guyton and Hall Textbook of Medical Physiology: Enhanced E-Book*. Philadelphia, PA: Elsevier Health Sciences.

Han, P. Y., G. C. Cho, and X.-C. Zhang. 2000. Time-domain transillumination of biological tissues with terahertz pulses. *Optics Letters* 25: 242–244.

Houssami, N., P. Macaskill, M. L. Marinovich et al. 2010. Meta-analysis of the impact of surgical margins on local recurrence in women with early-stage invasive breast cancer treated with breast-conserving therapy. *European Journal of Cancer* 46: 3219–3232.

Hughes, J. H., M. C. Mason, R. J. Gray et al. 2008. A multi-site validation trial of radioactive seed localization as an alternative to wire localization. *The Breast Journal* 14: 153–157.

Hunt, K. K., A. A. Sahin, H. M. Kuerer et al. 2007. Role for intraoperative margin assessment in patients undergoing breast-conserving surgery. *Annals of Surgical Oncology* 14: 1458–1471.

Hutchings, J., C. Kendall, B. Smith et al. 2009. The potential for histological screening using a combination of rapid Raman mapping and principal component analysis. *Journal of Biophotonics* 2: 91–103.

Jeevan, R., D. Cromwell, J. Browne, and J. Van Der Meulen. 2011. *National Mastectomy and Breast Reconstruction Audit 2011*. Leeds, U.K.: The NHS Information Centre.

Jeevan, R., D. A. Cromwell, M. Trivella et al. 2012. Reoperation rates after breast conserving surgery for breast cancer among women in England: Retrospective study of hospital episode statistics. *British Medical Journal* 345: e4505.

Katipamula, R., A. C. Degnim, T. Hoskin et al. 2009. Trends in mastectomy rates at the Mayo Clinic Rochester: Effect of surgical year and preoperative magnetic resonance imaging. *Journal of Clinical Oncology* 27: 4082–4088.

Kaufman, C. S., L. Jacobson, B. A. Bachman et al. 2007. Intraoperative digital specimen mammography: Rapid, accurate results expedite surgery. *Annals of Surgical Oncology* 14: 1478–1485.

Keller, M. D., S. K. Majumder, M. C. Kelley et al. 2010. Autofluorescence and diffuse reflectance spectroscopy and spectral imaging for breast surgical margin analysis. *Lasers in Surgery and Medicine* 42: 15–23.

Keller, M. D., E. Vargis, N. de Matos Granja et al. 2011. Development of a spatially offset Raman spectroscopy probe for breast tumor surgical margin evaluation. *Journal of Biomedical Optics* 16: 077006.

Kelly, P. and E. H. Winslow 1996. Needle wire localization for nonpalpable breast lesions: Sensations, anxiety levels, and informational needs. Presented at *Oncology Nursing Forum* 23: 639–645.

Klimberg, V. S. 2003. Advances in the diagnosis and excision of breast cancer. *The American Surgeon* 69: 11.

Klimberg, V. S., K. C. Westbrook, and S. Korourian. 1998. Use of touch preps for diagnosis and evaluation of surgical margins in breast cancer. *Annals of Surgical Oncology* 5: 220–226.

Knab, J. R., J.-Y. Chen, Y. He, and A. G. Markelz. 2007. Terahertz measurements of protein relaxational dynamics. *Proceedings of the IEEE* 95: 1605–1610.

Kouzminova, N. B., S. Aggarwal, A. Aggarwal, M. D. Allo, and A. Y. Lin. 2009. Impact of initial surgical margins and residual cancer upon re-excision on outcome of patients with localized breast cancer. *The American Journal of Surgery* 198: 771–780.

Krekel, N., M. H. Haloua, A. M. Lopes Cardozo et al. 2012. Intraoperative ultrasound guidance for palpable breast cancer excision (COBALT trial): A multicentre, randomised controlled trial. *The Lancet Oncology* 14: 48–54.

Krekel, N. M. A., B. M. Zonderhuis, H. B. A. C. Stockmann et al. 2011. A comparison of three methods for nonpalpable breast cancer excision. *European Journal of Surgical Oncology* 37: 109–115.

Lovrics, P. J., S. D. Cornacchi, F. Farrokhyar et al. 2009. The relationship between surgical factors and margin status after breast-conservation surgery for early stage breast cancer. *The American Journal of Surgery* 197: 740–746.

Lovrics, P. J., D. McCready, G. Gohla, C. Boylan, and M. Reedijk. 2011. A multicentered, randomized, controlled trial comparing radioguided seed localization to standard wire localization for nonpalpable, invasive and in situ breast carcinomas. *Annals of Surgical Oncology* 18: 3407–3414.

Markelz, A. G., A. Roitberg, and E. J. Heilweil. 2000. Pulsed terahertz spectroscopy of DNA, bovine serum albumin and collagen between 0.1 and 2.0 THz. *Chemical Physics Letters* 320: 42–48.

McCahill, L. E., R. M. Single, E. J. A. Bowles et al. 2012. Variability in reexcision following breast conservation surgery. *The Journal of the American Medical Association* 307: 467–475.

Medina-Franco, H., L. Abarca-Pérez, M. N. García-Alvarez et al. 2008. Radioguided occult lesion localization (ROLL) versus wire-guided lumpectomy for non-palpable breast lesions: A randomized prospective evaluation. *Journal of Surgical Oncology* 97: 108–111.

Mieog, J. S. D., M. Hutteman, J. R. van der Vorst et al. 2011. Image-guided tumor resection using real-time near-infrared fluorescence in a syngeneic rat model of primary breast cancer. *Breast Cancer Research and Treatment* 128: 679–689.

Monti, S., V. Galimberti, G. Trifiro et al. 2007. Occult breast lesion localization plus sentinel node biopsy (SNOLL): Experience with 959 patients at the European Institute of Oncology. *Annals of Surgical Oncology* 14: 2928–2931.

Moore, M. M., L. A. Whitney, L. Cerilli et al. 2001. Intraoperative ultrasound is associated with clear lumpectomy margins for palpable infiltrating ductal breast cancer. *Annals of Surgery* 233: 761.

Morrow, M., J. R. Harris, and S. J. Schnitt. 2012. Surgical margins in lumpectomy for breast cancer—Bigger is not better. *The New England Journal of Medicine* 367: 79–82.

Munshi, A., S. Kakkar, R. Bhutani et al. 2009. Factors influencing cosmetic outcome in breast conservation. *Clinical Oncology* 21: 285–293.

Nagel, M., M. Först, and H. Kurz. 2006. THz biosensing devices: Fundamentals and technology. *Journal of Physics: Condensed Matter* 18: S601.

Nakajima, S., H. Hoshina, M. Yamashita, C. Otani, and N. Miyoshi. 2007. Terahertz imaging diagnostics of cancer tissues with a chemometrics technique. *Applied Physics Letters* 90: 041102.

Ngô, C., A. G. Pollet, J. Laperrelle et al. 2007. Intraoperative ultrasound localization of nonpalpable breast cancers. *Annals of Surgical Oncology* 14: 2485–2489.

Nguyen, F. T., A. M. Zysk, E. J. Chaney et al. 2009. Intraoperative evaluation of breast tumor margins with optical coherence tomography. *Cancer Research* 69: 8790–8796.

Olson, T. P., J. Harter, A. Munoz, D. M. Mahvi, and T. M. Breslin. 2007. Frozen section analysis for intraoperative margin assessment during breast-conserving surgery results in low rates of re-excision and local recurrence. *Annals of Surgical Oncology* 14: 2953–2960.

Pappo, I., R. Spector, A. Schindel et al. 2010. Diagnostic performance of a novel device for real-time margin assessment in lumpectomy specimens. *Journal of Surgical Research* 160: 277–281.

Pickwell, E., B. E. Cole, A. J. Fitzgerald, M. Pepper, and V. P. Wallace. 2004. In vivo study of human skin using pulsed terahertz radiation. *Physics in Medicine and Biology* 49: 1595.

Pickwell-Macpherson, E., A. J. Fitzgerald, and V. P. Wallace 2012. Breast cancer tissue diagnosis at terahertz frequencies. In *Optical Interactions with Tissue and Cells XXIII*, eds. Jansen, E. D. and R. J. Thomas, Bellingham, WA: SPIE.

Pleijhuis, R. G., M. Graafland, J. De Vries, J. Bart, J. S. De Jong, and G. M. Van Dam. 2009. Obtaining adequate surgical margins in breast-conserving therapy for patients with early-stage breast cancer: current modalities and future directions. *Annals of Surgical Oncology* 16: 2717–2730.

Postma, E. L., H. M. Verkooijen, S. van Esser et al. 2012. Efficacy of "radioguided occult lesion localisation" (ROLL) versus "wire-guided localisation" (WGL) in breast conserving surgery for non-palpable breast cancer: A randomised controlled multicentre trial. *Breast Cancer Research and Treatment* 136: 469–478.

Rahusen, F. D., A. J. A. Bremers, H. F. J. Fabry, and R. P. A. Boom. 2002. Ultrasound-guided lumpectomy of nonpalpable breast cancer versus wire-guided resection: A randomized clinical trial. *Annals of Surgical Oncology* 9: 994–998.

Ramanujam, N., J. Q. Brown, T. M. Bydlon et al. 2009. Quantitative spectral reflectance imaging device for intraoperative breast tumor margin assessment. Presented at *Annual International Conference of the IEEE Engineering in Medicine and Biology Society*: 6554–6556.

Rampaul, R. S., M. Bagnall, H. Burrell et al. 2004. Randomized clinical trial comparing radioisotope occult lesion localization and wire-guided excision for biopsy of occult breast lesions. *British Journal of Surgery* 91: 1575–1577.

Reid, C. 2009. Spectroscopic methods for medical diagnosis at terahertz wavelengths. Doctoral Thesis, University College London, London, U.K.

Reid, C., A. P. Gibson, J. C. Hebden, and V. P. Wallace 2007. The use of tissue mimicking phantoms in analysing contrast in THz pulsed imaging of biological tissue. Presented at *Joint 32nd International Conference on Infrared and Millimeter Waves and the 15th International Conference on Terahertz Electronics*, Cardiff, U.K., pp. 567–568.

Riedl, O., F. Fitzal, N. Mader et al. 2009. Intraoperative frozen section analysis for breast-conserving therapy in 1016 patients with breast cancer. *European Journal of Surgical Oncology* 35: 264–270.

Sarlos, D., L. D. Frey, H. Haueisen et al. 2009. Radioguided occult lesion localization (ROLL) for treatment and diagnosis of malignant and premalignant breast lesions combined with sentinel node biopsy: A prospective clinical trial with 100 patients. *European Journal of Surgical Oncology* 35: 403–408.

Siegel, R., D. Naishadham, and A. Jemal. 2012. Cancer statistics, 2012. *CA: A Cancer Journal for Clinicians* 62: 10–29.

Singletary, S. E. 2002. Surgical margins in patients with early-stage breast cancer treated with breast conservation therapy. *The American Journal of Surgery* 184: 383–393.

Skinner, K. A., H. Silberman, and M. J. Silverstein. 2001. Palpable breast cancers are inherently different from nonpalpable breast cancers. *Annals of Surgical Oncology* 8: 705–710.

Snider Jr, H. C. and D. G. Morrison. 1999. Intraoperative ultrasound localization of nonpalpable breast lesions. *Annals of Surgical Oncology* 6: 308–314.

Talsma, A. K., A. M. J. Reedijk, R. A. M. Damhuis, P. J. Westenend, and W. J. Vles. 2011. Re-resection rates after breast-conserving surgery as a performance indicator: Introduction of a case-mix model to allow comparison between Dutch hospitals. *European Journal of Surgical Oncology* 37: 357–363.

Thill, M., K. Röder, K. Diedrich, and C. Dittmer. 2011. Intraoperative assessment of surgical margins during breast conserving surgery of ductal carcinoma in situ by use of radiofrequency spectroscopy. *Breast* 20: 579–580.

Thind, C. R., S. Tan, S. Desmond et al. 2011. SNOLL. Sentinel node and occult (impalpable) lesion localization in breast cancer. *Clinical Radiology* 66: 833–839.

Valdes, E. K., S. K. Boolbol, I. Ali, S. M. Feldman, and J.-M. Cohen. 2007. Intraoperative touch preparation cytology for margin assessment in breast-conservation surgery: Does it work for lobular carcinoma? *Annals of Surgical Oncology* 14: 2940–2945.

Veronesi, U., N. Cascinelli, L. Mariani et al. 2002. Twenty-year follow-up of a randomized study comparing breast-conserving surgery with radical mastectomy for early breast cancer. *New England Journal of Medicine* 347: 1227–1232.

Waljee, J. F., E. S. Hu, L. A. Newman, and A. K. Alderman. 2008. Predictors of re-excision among women undergoing breast-conserving surgery for cancer. *Annals of Surgical Oncology* 15: 1297–1303.

Wallace, V. P., A. J. Fitzgerald, E. Pickwell et al. 2006. Terahertz pulsed spectroscopy of human basal cell carcinoma. *Applied Spectroscopy* 60: 1127–1133.

Wallace, V. P., A. J. Fitzgerald, B. Robertson, E. Pickwell, and B. Cole 2005. Development of a hand-held TPI system for medical applications. In *2005 IEEE MTT-S International Microwave Symposium*, Vols. 1–4, Long Beach, CA, pp. 637–639. IEEE.

Weber, W. P., S. Engelberger, C. T. Viehl et al. 2008. Accuracy of frozen section analysis versus specimen radiography during breast-conserving surgery for nonpalpable lesions. *World Journal of Surgery* 32: 2599–2606.

Wilke, L. G., J. Brown, T. M. Bydlon et al. 2009. Rapid noninvasive optical imaging of tissue composition in breast tumor margins. *The American Journal of Surgery* 198: 566–574.

Woodward, R. M., B. E. Cole, V. P. Wallace et al. 2002. Terahertz pulse imaging in reflection geometry of human skin cancer and skin tissue. *Physics in Medicine and Biology* 47: 3853.

Woodward, R. M., V. P. Wallace, R. J. Pye et al. 2003. Terahertz pulse imaging of ex vivo basal cell carcinoma. *Journal of Investigative Dermatology* 120: 72–78.

Yin, X., B.-H. Ng, B. M. Fischer, B. Ferguson, and D. Abbott. 2007. Support vector machine applications in terahertz pulsed signals feature sets. *IEEE Sensors Journal* 7: 1597–1608.

Youlden, D. R., S. M. Cramb, N. A. Dunn et al. 2012. The descriptive epidemiology of female breast cancer: An international comparison of screening, incidence, survival and mortality. *Cancer Epidemiology* 36: 237–248.

17

Frozen Terahertz Imaging of Oral Cancer

Jae Yeon Park
University of Seoul

Yookyeong
Carolyn Sim
Princeton University

Joo-Hiuk Son
University of Seoul

17.1 Introduction

Terahertz (THz) imaging has been evaluated for the application of cancer diagnosis, especially skin and breast cancers (Fitzgerald et al. 2006; Wallace et al. 2006), as thoroughly reviewed in Chapters 15 and 16. The reason to focus on such cancers using THz technology is that THz radiation cannot penetrate deeply into biological tissues, most of which have abundant water. The oral area is also easily accessible with a THz endoscope, as described in detail in Chapter 6, although internal organs such as the stomach and colon can now be reached following the development of more compact THz endoscopes (Ji et al. 2009). Most oral cancers are in the epidermis of the oral cavity and are produced from melanocytes or carcinoma cells; in their early stages, they seem to be potentially less dangerous to life. However, some papers on oral cancer metastasis have reported a higher death rate than that of skin cancer, so the early diagnosis of oral cancer is important to prevent the proliferation of cancer cells (Lozano et al. 2013). Therefore, the detection of oral tumors using THz radiation is an interesting subject and could have wide-ranging clinical applications.

Another important issue in cancer imaging with THz radiation is to understand the contrast mechanism and to enhance the contrast for an accurate diagnosis. It is known that the contrast between benign and malignant tissues originates from both cell structure deformation and changes in water content. Water not only contributes to the distinction but also limits the interaction between the THz radiation and the tissue. Therefore, it is imperative to compare the imaging results of wet tissues with dry samples in which the liquid water has been eliminated. There are some methods available to remove liquid water. One is a paraffin-embedding method, but it requires many processing steps and takes a long time (Miura et al. 2011; Park et al. 2011). A simple way of drying biological tissues is lyophilization at a frozen temperature (Booth and Kenny 1974).

In this chapter, we review the THz imaging results of human oral melanoma and carcinoma and assess the effect of lyophilization by comparison with images of wet samples.

17.2 Oral Cancer

Oral cancer is a term used to define malignant neoplasms or tumors that originate in the mucous membrane lining of the oral cavity. The word "cancer" is a generic term that encompasses all malignant tumors. There are two general classes of malignant tumors based on the tissue of origin: epithelial and mesodermal. Carcinoma is a malignant tumor of epithelial origin, and oral cancer belongs to this category. As a matter of fact, carcinomas comprise over 90% of all malignancies known to humans. Sarcoma affects mesodermal tissues such as muscle, cartilage, and bone.

Oral cancer invades the various structures of the head and neck and is a significant portion of the entire cancer incidence. Malignancy of the oropharynx alone comprises about 5% of all human cancer and results in nearly 70% of all malignancies occurring in the upper respiratory tract (Posner 2011). Carcinoma of the oropharynx is the most frequently encountered malignancy of the head and neck, with the exception of skin cancer. Therefore, oropharyngeal and cutaneous cancers are important cancers of the head and neck in frequency. However, carcinoma of the oral cavity is a much more lethal disease than skin cancer, and its management is much more difficult (Wein et al. 2010). Because of its rapid course, early recognition of oral cancer is essential if treatment is to be successful. The stage to which the disease has advanced at the outset of management is frequently a more important factor in the prognosis than the skill of the surgeon or the treatment modalities used.

Oral cancer accounts for 2%–5% of all cancers diagnosed annually in the United States. According to the National Cancer Institute, the estimated new cases and deaths from oral cancer in the United States in 2013 are 41,380 and 7,890, respectively. Oral cancer can develop in any region of the oral cavity or oropharynx. Although no age group is immune to oral cancer, it is essentially a disease of those in middle and advanced age groups. The majority of patients are over the age of 40 at the time of discovery, but it is now occurring more frequently in younger patients. In general, cancer of the oral cavity is more prevalent in men. The male population is affected twice as frequently as the female population, and the age of the patients is known to be positively correlated with the frequency and severity of oral cancer. The survival rate of oral malignancy depends on the site of the oral cancer, and overall, the 5-year survival rate is reported to be less than 50% for stage III or IV disease (Felix et al. 2012).

It appears that the vast majority of oral cancers are associated with chronic stimulation. There are some predisposing causes that can be recognized. It is well documented that irritation from a long-existing abnormal condition may eventuate in carcinoma. In fact, several forms of chronic irritation contribute to the production of precancerous or cancerous lesions of the lip and oral mucous membrane. An important example of chronic irritation is sunburn. Excessive exposure to sunshine accounts for the relatively high incidence of skin and lip cancer among sailors and outdoor workers. It is also believed by some researchers that tissue complexion plays a part, with fair and ruddy individuals being more susceptible. There is convincing evidence that tobacco smoke directly stimulates the development of malignant oral lesions. Chewing tobacco held in the gingival sulcus is a well-known cause of oral cancer. The risk is even higher for tobacco users who drink alcohol heavily. Over 75% of oral cancers are found in those who use tobacco and/or alcohol. Studies report that certain viral infections are related to oral cancer, such as the human papilloma virus (Hashibe and Sturgis 2013).

Superficial soft tissue lesions of the oral mucosa are usually benign and, in most instances, lend themselves to simple surgical removal using biopsy techniques. All benign lesions are overgrowths of normally present histological elements in the oral mucosa and submucosa. However, oral cancer of the floor of the mouth is likely to be an invasive malignancy. When oral cancer metastasizes, it usually travels through the lymphatic system. When this happens, a new neoplasm occurs with the same type of abnormal cells as in the primary tumor location. If oral cancer spreads to the lung, the cancer cells in the lung are actually oral cancer cells (Dequanter et al. 2013). No anatomical barrier exists to retard its progress. Cancers of the floor of the mouth, the middle and the posterior tongue, and the soft palate all tend to be invasive. Lesions in a more posterior position have a higher rate and a less orderly method of metastasis. Not all oral cancers grow at the same speed.

The treatment of oral cancer is often a multidisciplinary approach. Oral cancers are treated with surgery, radiation, chemotherapy, or a combination of these methods. The treatment for any cancer patient depends on several factors, including the histopathological diagnosis, the location of the tumor, the presence of metastasis, the radiosensitivity or chemical sensitivity of the tumor, and the physical condition of patient (Kumar and Manjunatha 2013). The surgical procedures for the excision of oral cancer vary with the type and extent of the lesion. Small carcinomas that are in accessible locations and are not associated with lymph nodes can be removed with a simple excision. Malignancies of the oral cavity that have diagnosed lymph node involvement need extensive surgery for complete resection, along with the surrounding tissues. This procedure may produce large defects of the jaws and extensive loss of soft tissues, making functional and esthetic rehabilitation a long and harsh process (Jones 2012).

A squamous cell layer covers most surfaces of the mouth, tongue, and lips. Malignancies of oral cancer may arise from a variety of tissues such as the salivary gland, muscle, and blood vessels. It may even present as a metastasized form from a distant site. The most common forms are squamous cell carcinomas of the oral mucosa (Figure 17.1). Squamous cell carcinomas of the lip and tongue are the lesions most frequently encountered. These two lesions together make up about 50% of all oral malignancies (Laudenbach 2013; Terada 2012). In the advanced stages, the cancer cells often invade nearby lymph nodes and spread to other parts of the body such as the lung, liver, or brain (Figure 17.2). Thus, it is critical to diagnose and detect cancerous regions early to increase the survival outcome for patients. However, extensive ablative surgeries

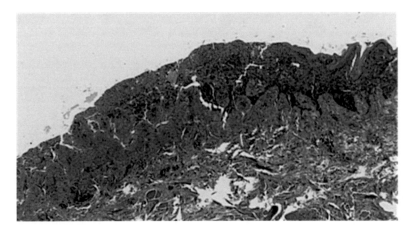

FIGURE 17.1 Squamous cell carcinoma of the tongue. Note the elevated and destructive surface changes.

FIGURE 17.2 Mucoepidermoid carcinoma of the tongue. Note the invasive nodules of cancerous cells.

are required to provide sufficient margins for most oral cancer patients, resulting from late diagnosis. It is important to diagnose early because the growth rate of the tumor influences the prognosis.

The extent of the cancer resection is determined by preoperative magnetic resonance imaging (MRI), ultrasonography, and visual examination. A safety margin of at least 1.5 cm should be obtained to prevent recurrence. However, it is difficult for surgeons to know whether all tumors have been completely removed until the excised tissue samples are sent for a final pathologic examination. A frozen section of the specimen is randomly selected during the operation, so it cannot cover the entire lesion (Shah and Gil 2009). There is a strong clinical need to define the margins of oral cancer accurately during surgery.

17.3 THz Imaging of Human Oral Melanoma

Human oral melanoma is developed from melanocytes and mostly found on the squamous tissues in the oral cavity. The incidence rate of oral cancer is just 1.4% of the total number of melanomas, but its 5-year survival rate is approximately 20%, and it is difficult to define the surgical margin at the stage when the oral cancer cells infiltrate into the muscle layer (Hicks and Flaitz 2000; Manganaro et al. 1995; Mihajlovic et al. 2012). THz imaging was conducted with a freeze-and-thaw process for the detection of melanoma. A piece of oral tissue about 15 mm² in size with malignant melanoma was excised from the oral cavity of a 75-year-old woman and was immediately frozen using dry ice. The THz imaging system is equipped with a temperature controller and measured the oral tissue at −20°C and 20°C with a raster scan in 250 μm steps.

Figure 17.3a and b shows THz images of the frozen and thawed tissues, respectively (Sim et al. 2013a). The scale bar represents the peak amplitudes of the measured temporal THz waveforms. The THz image

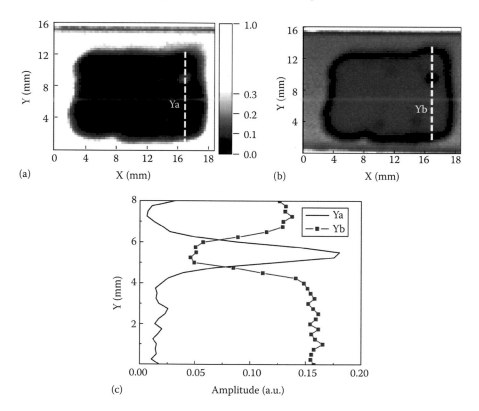

FIGURE 17.3 (See color insert.) THz images of an oral melanoma specimen at (a) −20°C and (b) 20°C, and (c) their cross-sectional amplitudes for the lines Ya and Yb. (Reprinted with permission from Sim, Y. C., K.-M. Ahn, J. Y. Park, C. Park, and J.-H. Son, *IEEE J. Biomed. Health Inform.*, 17, 779, 2013a. Copyright © 2013 IEEE.)

of the frozen tissue has a lower amplitude than that of the thawed tissue. This implies that the THz radiation is reflected by the frozen tissue less than by the thawed tissue, which absorbs and reflects THz radiation well because of the water molecules present. In the THz images, noticeable spots are displayed in the top right corner of the specimen. Comparing the spots in Figure 17.3a and b, the image contrast of Ya is higher than that of Yb, as shown in Figure 17.3c. The full-width at half-maximum (FWHM) of a peak in Ya is about 1 mm, which was compared with its real size as identified in the histological analysis.

Figure 17.4 shows a histological image of the spot in the THz image of Figure 17.3 (Sim et al. 2013a). The oral melanoma cell group is marked by a dashed circle and surrounded by normal oral cells. The melanoma group has a size of about 300 μm along its long axis, which is smaller than the FWHM of Ya in Figure 17.3c. The size difference comes from the THz measurement method using an obliquely focused THz beam. A focused beam with a frequency range of 0.2–1.2 THz has a beam diameter of 800 μm resulting from its diagonal incidence with respect to the specimen. The size of the melanoma group was measured to be 600 μm by a deconvolution process using a focused THz beam diameter, which can provide information about surgical margin close to the histological image within an error of several hundred micrometers.

The temporal THz waveforms of the normal mucosa and malignant melanoma extracted from the pixels of the THz images in Figure 17.3a and b are shown in Figure 17.5. The specimen was large enough to obtain independent waveforms from 20 pixels of the normal mucosa area and 5 pixels from the region of the oral melanoma. The error bars represent 95% confidence interval. In the measurements taken at 20°C, as shown in Figure 17.5a, the THz waveforms reflected by the normal mucosa have greater amplitudes than those reflected by the oral melanoma. The reflection amplitudes were opposite in the specimens measured at −20°C, with a greater reflection from the oral melanoma region, as shown in Figure 17.5b, because the liquid water in the tissue froze into ice, which has a smaller index of refraction and a lower absorption than those of water (Sim et al. 2013a).

B-scan images are reconstructed as shown in Figure 17.6 by using the temporal waveforms in Figure 17.5, as the time interval of the waveforms represents the distance in the specimens (Sim et al. 2013a). Figure 17.6a shows the boundary between the substrate and the tissue at a frozen temperature of −20°C across the dashed line shown with the two-dimensional image in the inset. The boundary was unvarying where the line crossed along the surface of the normal mucosa; however, the B-scan image

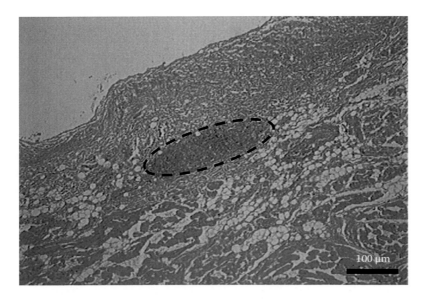

FIGURE 17.4   A pathological image of an oral melanoma cell group in the dashed circle. (Reprinted with permission from Sim, Y. C., K.-M. Ahn, J. Y. Park, C. Park, and J.-H. Son, *IEEE J. Biomed. Health Inform.*, 17, 779, 2013a. Copyright © 2013 IEEE.)

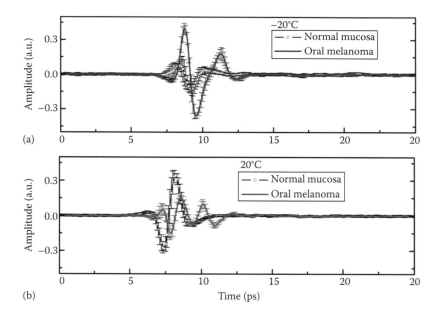

FIGURE 17.5 Temporal THz waveforms reflected from areas of normal mucosa and oral melanoma acquired at (a) −20°C and (b) 20°C. (Reprinted with permission from Sim, Y. C., K.-M. Ahn, J. Y. Park, C. Park, and J.-H. Son, *IEEE J. Biomed. Health Inform.*, 17, 779, 2013a. Copyright © 2013 IEEE.)

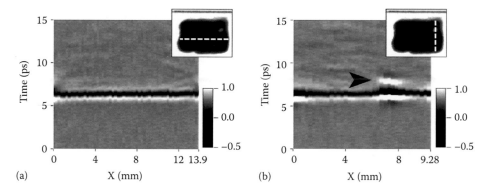

FIGURE 17.6 THz B-scan images of the areas of (a) normal mucosa and (b) oral melanoma at frozen temperatures. (Reprinted with permission from Sim, Y. C., K.-M. Ahn, J. Y. Park, C. Park, and J.-H. Son, *IEEE J. Biomed. Health Inform.*, 17, 779, 2013a. Copyright © 2013 IEEE.)

crossing the line through the melanoma nodule revealed a spot inside the tissue surface, designated by an arrow in Figure 17.6b. This spot exactly corresponded to a melanoma nodule that was 150 μm beneath the tissue surface, as seen in the histological image in Figure 17.4.

17.4 THz Imaging of Human Oral Carcinomas

Mucoepidermoid carcinoma and squamous cell carcinoma have represented more than 90% of all oral cancer cases for the last five decades. Their death rates are also higher than those of skin or breast cancers. Oral tissues with carcinomas of about 10 mm² in size, listed in Table 17.1, were imaged using THz radiation at frozen and room temperatures, and the results from the two temperatures were compared with histology.

TABLE 17.1 List of Human Oral Tissues with Carcinomas

Identification No.	Cancer	Cancer Type
1	Y	MEC
2	Y	MEC
3	Y	SCC
4	Y	SCC
5	Y	SCC
6	Y	SCC
7	Y	SCC

Source: Reprinted from Sim, Y. C., J. Y. Park, K.-M. Ahn, C. Park, and J.-H. Son, *Biomed. Opt. Exp.*, 4, 1413, 2013b. Copyright © 2013 by Optical Society of America. With permission of Optical Society of America.

MEC, mucoepidermoid carcinoma; SCC, squamous cell carcinoma.

Figure 17.7a through d shows visible images, THz images of 0.5 THz at −20°C and 20°C, and histological images, respectively, of the oral tissues with carcinoma from no. 1 to no. 6 in Table 17.1 (Sim et al. 2013b). Specimen no. 7 had a malignant tumor inside the tissue, as verified by histological sectioning, and will be dealt with separately. The visible images of the excised tissues in row (a) show the red color of blood, but no noticeable sign of cancer. THz images at frozen and room temperatures in rows (b) and (c),

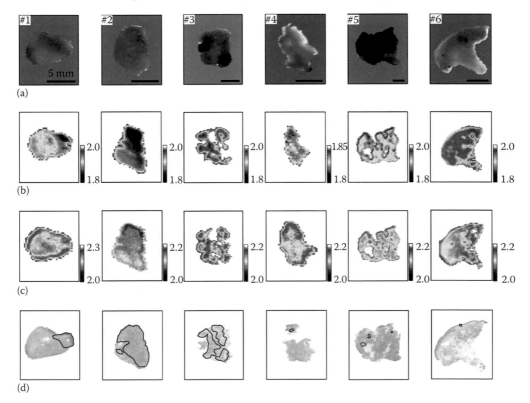

FIGURE 17.7 **(See color insert.)** (a) Visible images, THz images at (b) −20°C and (c) 20°C, and (d) pathological images of oral carcinoma specimens. (Reprinted from Sim, Y. C., J. Y. Park, K.-M. Ahn, C. Park, and J.-H. Son, *Biomed. Opt. Exp.*, 4, 1413, 2013b. Copyright © 2013 by Optical Society of America. With permission of Optical Society of America.)

respectively, are composed of image pixels mapped to the indices of refraction at 0.5 THz with a rainbow scale bar. The purple pixels, representing the points on the indices from 1.80 to 1.85 in the THz images in row (b), are well correlated with the cancerous regions marked by the blue loops on the histological images in row (d), although little correlation was found between the images in rows (b) and (d).

The pathological examination shows the microscopic appearance of the morphological characteristics of the disease's origin. Figure 17.8 shows the magnified pathological images of the marked cancer area in Figure 17.7d. Figure 17.8a is the image magnifying the center of the oral carcinoma area. There are enlarged nuclei composed of pleomorphic and hyperchromatic cells connected by intercellular bridges. This deformation of the cellular structure emerged from hypermitosis, which increases the density of cells with abnormal forms and hyperchromatic nuclei. Around the cancerous cell clusters, significant inflammatory spots were found, as shown in Figure 17.8b and c. The tissue damage came from the rapid increase in the population of macrophages, spindle cells, and lymph cells, as well as from the attack of eosinophilic and polyangular cells into the epithelium. This substantiates that the purple area in the THz images at a frozen temperature was introduced from the interaction between the deformed cellular structure and the THz radiation (Sim et al. 2013b).

Figure 17.9 shows the average refractive indices and absorption coefficients of oral cancer and normal mucosa at two temperatures with the error bars representing the 95% confidence interval. At a frozen temperature of −20°C, the THz index and absorbance from 0.5 to 1 THz have distinct gaps between the normal mucosa and oral cancer, as shown in Figure 17.9a. This implies that the THz radiation senses the

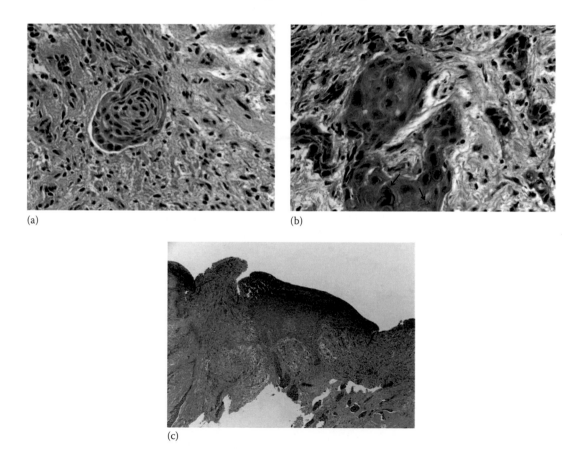

(a)   (b)

(c)

FIGURE 17.8 Magnified pathological images in (a) center of, (b) adjacent, and (c) facet of cancer areas from Figure 17.7d. (Reprinted from Sim, Y. C., J. Y. Park, K.-M. Ahn, C. Park, and J.-H. Son, *Biomed. Opt. Exp.*, 4, 1413, 2013b. Copyright © 2013 by Optical Society of America. With permission of Optical Society of America.)

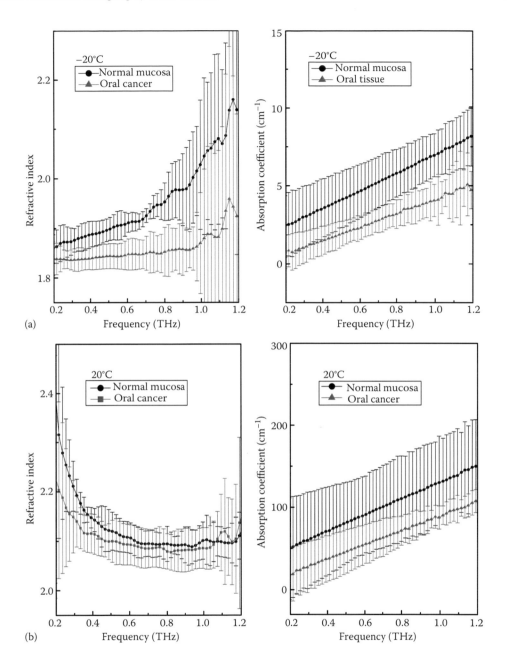

FIGURE 17.9 Refractive indices and absorption coefficients in 0.2–1.2 THz of normal mucosa and oral cancer at (a) –20°C and (b) 20°C. (Reprinted from Sim, Y. C., J. Y. Park, K.-M. Ahn, C. Park, and J.-H. Son, *Biomed. Opt. Exp.*, 4, 1413, 2013b. Copyright © 2013 by Optical Society of America. With permission of Optical Society of America.)

deformation of the cellular structure, as mentioned in the previous paragraph, without being absorbed as strongly by ice as by liquid water. However, at a room temperature of 20°C, the average values of the THz spectra for the cancerous region are smaller than those for the normal mucosa region, and their error bars overlap each other, as shown in Figure 17.9b. The behavior of their refractive indices and absorption coefficients is also similar to that of liquid water. This suggests that THz radiation on thawed tissues mostly reacts with the liquid water, which results in poor sensitivity distinguishing the cancerous region from the normal area (Sim et al. 2013b).

The cancerous tumor below the tissue surface of specimen no. 7 listed in Table 17.1 was also detected at a frozen temperature by monitoring temporal THz pulses, as the cell structures of oral cancer and normal mucosa are different. Figure 17.10a–e shows the visible image, THz images acquired at two temperatures, a histological image, and the temporal THz waveforms of the seventh sample, respectively. The surface images, as shown in Figure 17.10a–c did not disclose any sign of tumor. The tumor was found during the horizontal sectioning in histology, as shown in Figure 17.10d, which was located 1.3 mm below the tissue surface. The depth was also estimated by the path length of the THz radiation as it entered the mucosa tissue at a frozen temperature. Figure 17.10e shows the temporal THz waveforms taken at the areas designated by the red arrows in Figure 17.10b and c. The first pulses reflected from the tissue surface were measured at both temperatures; but the second pulse arrived 17 ps later and was detected only at the frozen temperature. This delay corresponds to 1.2 mm in depth, considering the approximate index of refraction of the frozen tissue and the angle of incidence of the focused THz beam, which was in good agreement

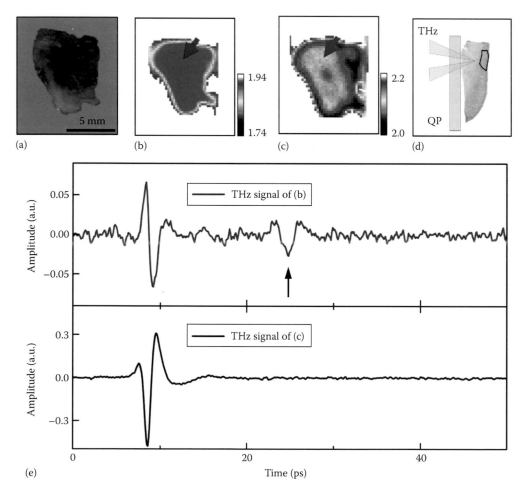

FIGURE 17.10 **(See color insert.)** THz detection of an infiltrative cancerous tumor hidden within a tissue sample. (a) A visible image, and THz images at (b) –20°C and (c) 20°C of the surface of the specimen and (d) a pathological image from a vertical section. (e) Temporal THz waveforms extracted from the areas designated by the red arrows in (b) and (c). (THz: THz radiation, QP: quartz plate). (Reprinted from Sim, Y. C., J. Y. Park, K.-M. Ahn, C. Park, and J.-H. Son, *Biomed. Opt. Exp.*, 4, 1413, 2013b. Copyright © 2013 by Optical Society of America. With permission of Optical Society of America.)

with the histology measurement. This result demonstrates that the frozen THz imaging technique can identify an area of cancerous tumor 1 mm below the tissue surface by distinguishing the differences in cell structure (Sim et al. 2013b).

Some papers have reported that the freeze–thaw process can cause damage to acellular lung or vary the viscoelastic properties of the cells (Chan and Titze 2003; Nonaka et al. 2013). However, our pathological results did not show any evidence of damage to the organelles or cellular structures of the specimens after the THz measurements, including the freeze–thaw process.

17.5 Discussion: Frozen Sectioning vs. Frozen THz Imaging

Frozen sectioning is a histological procedure performed in a laboratory to analyze a tissue specimen using a rapid microscopic examination. This method is only used for a gross evaluation of the specimen resulting from the poor quality of the tissue preparation compared to a conventional fixed-tissue processing method. Although frozen section diagnosis has high accuracy rates, certain relative limitations must be understood, especially when a speedy diagnosis is of the essence.

The surgeon in an operating room submits a specimen to the pathology laboratory. The excised tissue sample is rapidly frozen with liquid nitrogen or isopentane after appropriate measurements are obtained. In order to eliminate ice crystal formations resulting in artifacts, it is essential to freeze the specimen as fast as possible (Wilson 1905). The recommended size of the specimen is 20 mm × 20 mm × 20 mm. The tissue specimen is placed on a metal disk and secured in a chuck at an operating temperature between −20°C and −30°C (Gal and Cagle 2005). The specimen is embedded in a cryocompound consisting of polyethylene glycol and polyvinyl alcohol. The tissue is cut while frozen with a cryostat and placed on a glass slide and stained with hematoxylin and eosin. The preparation of the tissue is much quicker than that in conventional histology, 15–20 min compared with 15–16 h.

Pathological diagnoses using a frozen section are one of the most important but difficult procedures in the practice of surgical pathology. The most common indications for frozen sections are to establish and determine the nature of a lesion during surgery and for the assessment of resection margins (Ganly et al. 2012; Shah et al. 2003). An intraoperative consultation from a pathologist using a frozen section is limited to a benign or malignant status of the excised tissue or the clearance of a surgical margin from a cancerous change. Nevertheless, the frozen section diagnosis will influence and change the surgical procedures. The limited tissue preparation of a frozen section requires an experienced pathologist to confirm the malignancy of the excised tissue and the marginal clearance.

The process of frozen sectioning is performed with the resection and staining of a specimen and the expert decision of pathologists by microscopic examination in approximately 10 min of time. The THz imaging technique at frozen temperatures might help remove some of the steps such as the resection and staining of specimens. The decision could also be made by operating surgeons on the spot without pathological knowledge, as demonstrated in Sections 17.3 and 17.4. The rapid development of THz technology, such as the acquisition of THz images, has decreased to real time (Behnken et al. 2008; Blanchard et al. 2013; Lee et al. 2006), and a miniaturized THz endoscope can be inserted into human organs (Ji et al. 2009), which makes this technique more attractive.

17.6 Conclusion

In this chapter, we reviewed the results of THz imaging of oral cancer to assess the feasibility of diagnostic clinical applications. THz images at frozen temperatures have a good correlation with histological images. This implies that THz radiation can sense that the deformation of the structure of cancer cells is different from that of normal cells without monitoring the water content change caused by the proliferation of blood vessels around the cancerous tumor. The lyophilization technique also enables the detection of a cancerous tumor deep below the tissue surface by observing the pulse reflected by the tumor. The freeze-dried specimens also provided better contrast between the normal mucosa and oral cancer than the wet samples, which make the THz radiation interact mostly with water molecules. The authors expect that the frozen THz imaging technique might be able to substitute for frozen sectioning, allowing surgeons to identify cancerous tumors in the operating room without transferring tissues for histological confirmation.

References

Behnken, B. N., G. Karunasiri, D. R. Chamberlin, P. R. Robrish, and J. Faist. 2008. Real-time imaging using a 2.8 THz quantum cascade laser and uncooled infrared microbolometer camera. *Optics Letters* 33: 440–442.

Blanchard, F., A. Doi, T. Tanaka, and K. Tanaka. 2013. Real-time, subwavelength terahertz imaging. *Annual Review of Materials Research* 43: 237–259.

Booth, A. G. and A. J. Kenny. 1974. A rapid method for the preparation of microvilli from rabbit kidney. *Biochemical Journal* 142: 575–581.

Chan, R. W. and I. R. Titze. 2003. Effect of postmortem changes and freezing on the viscoelastic properties of vocal fold tissues. *Annals of Biomedical Engineering* 31: 482–491.

Dequanter, D., M. Shahla, P. Paulus, and P. Lothaire. 2013. Long term results of sentinel lymph node biopsy in early oral squamous cell carcinoma. *OncoTargets and Therapy* 6: 799.

Felix, D. H., J. Luker, and C. Scully. 2012. Oral medicine: 3. Ulcers: Cancer. *Dental Update* 39: 664.

Fitzgerald, A. J., V. P. Wallace, M. Jimenez-Linan et al. 2006. Terahertz pulsed imaging of human breast tumors1. *Radiology* 239: 533–540.

Gal, A. A. and P. T. Cagle. 2005. The 100-year anniversary of the description of the frozen section procedure. *JAMA: The Journal of the American Medical Association* 294: 3135–3137.

Ganly, I., S. Patel, and J. Shah. 2012. Early stage squamous cell cancer of the oral tongue—Clinicopathologic features affecting outcome. *Cancer* 118: 101–111.

Hashibe, M. and E. M. Sturgis. 2013. Epidemiology of oral-cavity and oropharyngeal carcinomas: Controlling a tobacco epidemic while a human papillomavirus epidemic emerges. *Otolaryngologic Clinics of North America* 46: 507–520.

Hicks, M. J. and C. M. Flaitz. 2000. Oral mucosal melanoma: Epidemiology and pathobiology. *Oral Oncology* 36: 152–169.

Ji, Y. B., E. S. Lee, S.-H. Kim, J.-H. Son, and T.-I. Jeon. 2009. A miniaturized fiber-coupled terahertz endoscope system. *Optics Express* 17: 17082–17087.

Jones, D. L. 2012. Oral cancer: Diagnosis, treatment, and management of sequela. *Texas Dental Journal* 129: 459.

Kumar, G. and B. Manjunatha. 2013. Metastatic tumors to the jaws and oral cavity. *Journal of Oral and Maxillofacial Pathology* 17: 71.

Laudenbach, J. M. 2013. Oral medicine update: Oral cancer—Screening, lesions and related infections. *Journal of the California Dental Association* 41: 326–328.

Lee, A. W., B. S. Williams, S. Kumar, Q. Hu, and J. L. Reno. 2006. Real-time imaging using a 4.3-THz quantum cascade laser and a 320/spl times/240 microbolometer focal-plane array. *IEEE Photonics Technology Letters* 18: 1415–1417.

Lozano, R., M. Naghavi, K. Foreman et al. 2013. Global and regional mortality from 235 causes of death for 20 age groups in 1990 and 2010: A systematic analysis for the Global Burden of Disease Study 2010. *The Lancet* 380: 2095–2128.

Manganaro, A. M., H. L. Hammond, M. J. Dalton, and T. P. Williams. 1995. Oral melanoma: Case reports and review of the literature. *Oral Surgery, Oral Medicine, Oral Pathology, Oral Radiology, and Endodontology* 80: 670–676.

Mihajlovic, M., S. Vlajkovic, P. Jovanovic, and V. Stefanovic. 2012. Primary mucosal melanomas: A comprehensive review. *International Journal of Clinical and Experimental Pathology* 5: 739.

Miura, Y., A. Kamataki, M. Uzuki et al. 2011. Terahertz-wave spectroscopy for precise histopathological imaging of tumor and non-tumor lesions in paraffin sections. *The Tohoku Journal of Experimental Medicine* 223: 291–296.

Nonaka, P. N., N. Campillo, J. J. Uriarte et al. 2013. Effects of freezing/thawing on the mechanical properties of decellularized lungs. *Journal of Biomedical Materials Research Part A* 102: 413–419.

Park, J. Y., H. J. Choi, K.-S. Cho, K.-R. Kim, and J.-H. Son. 2011. Terahertz spectroscopic imaging of a rabbit VX2 hepatoma model. *Journal of Applied Physics* 109: 064704.

Posner, M. L. 2011. Head and neck cancer. In *Cecil Medicine*, eds. Goldman, L. D. and A. I. Schafer, pp. 1257–1263. Philadelphia, PA: Saunders Elsevier.

Shah, J. P. and Z. Gil. 2009. Current concepts in management of oral cancer-surgery. *Oral Oncology* 45: 394–401.

Shah, J. P., N. W. Johnson, and J. G. Batsakis. 2003. *Oral Cancer*. London, U.K.: Martin Dunitz.

Sim, Y. C., K.-M. Ahn, J. Y. Park, C. Park, and J.-H. Son. 2013a. Temperature-dependent terahertz imaging of excised oral malignant melanoma. *IEEE Journal of Biomedical and Health Informatics* 17: 779–784.

Sim, Y. C., J. Y. Park, K.-M. Ahn, C. Park, and J.-H. Son. 2013b. Terahertz imaging of excised oral cancer at frozen temperature. *Biomedical Optics Express* 4: 1413.

Terada, T. 2012. Adenoid squamous cell carcinoma of the oral cavity. *International Journal of Clinical and Experimental Pathology* 5: 442.

Wallace, V. P., A. J. Fitzgerald, E. Pickwell et al. 2006. Terahertz pulsed spectroscopy of human basal cell carcinoma. *Applied Spectroscopy* 60: 1127–1133.

Wein, R. O., Malone, J. P., and Weber, R. S. 2010. Malignant neoplasms of the oral cavity. In *Cummings Otolaryngology—Head and Neck Surgery: Head and Neck Surgery*, eds. P. W. Flint, B. H. Haughey, J. K. Niparko, M. A. Richardson, V. J. Lund, K. T. Robbins, M. M. Lesperance, and J. R. Thomas, pp. 1293–1318. St. Louis, MI: Elsevier Health Sciences.

Wilson, L. B. 1905. A method for the rapid preparation of fresh tissues for the microscope. *Journal of the American Medical Association* 45: 1737.

18

Terahertz Molecular Imaging and Its Clinical Applications

Kwang Sung Kim
University of Seoul

Hyuk Jae Choi
Asan Medical Center

Jae Yeon Park
University of Seoul

Joo-Hiuk Son
University of Seoul

18.1 Introduction

In many diseases, structural or anatomical changes that can be detected by conventional imaging modalities, such as ultrasound, computed tomography (CT), or magnetic resonance imaging (MRI), appear later than molecular change. Recent advances in molecular biology and imaging techniques have allowed the visualization of individual molecules in living animals (Kherlopian et al. 2008). Using these techniques, researchers can trace molecules within cells or tissues under physiologic conditions; accordingly, clinicians can perform early detection and management of diseases, which improves the clinical outcome.

Molecular imaging is also necessary for monitoring the response of tumors to therapy. The currently used Response Evaluation Criteria in Solid Tumors utilizes structural changes in tumors such as changes in the size and shape after treatment (Costelloe et al. 2010). Molecular alteration after treatment occurs before structural change; therefore, clinicians can improve patient management by using molecular imaging.

The currently available molecular imaging techniques include MRI, single-photon emission CT, positron emission tomography, and optical imaging using fluorescence or bioluminescence techniques (Choi et al. 2011; Massoud and Gambhir 2003; Weissleder and Pittet 2008). However, each technique has advantages and disadvantages in terms of spatial resolution, sensitivity, quantification, and other features, as summarized in Table 18.1.

We have developed a novel molecular imaging technique using terahertz (THz) radiation by adopting nanoparticles. THz imaging has recently been evaluated for application as a medical imaging

TABLE 18.1   Characteristics of Various Molecular Imaging Techniques

Technique	Spatial Resolution	Sensitivity	Quantification	Advantages	Disadvantages
MRI	50–500 μm	10^{-3}–10^{-5} mol/L	★★	Combine morphological and functional imaging	Long scan and postprocessing time
CT	50 μm	Not well characterized	Not applicable	Image lungs and bones	Limited to soft tissue, radiation risk
Ultrasound	50–500 μm	Not well characterized	★	Real time, low cost	Limited spatial resolution
PET	1–2 mm	10^{-11}–10^{-12} mol/L	★★★	High sensitivity, quantitative translation	Low resolution, use of radioactive isotopes
Bioluminescence/ fluorescence imaging	2–10 mm	Not well characterized, possibly 10^{-9}–10^{-16} mol/L	★	High sensitivity, low cost	Low resolution, limited depth
NIR absorption imaging	0.5–3 mm	10^{-4} mol/L	Not applicable	Fast thermal imaging	Low resolution, limited depth
TMI	100–500 μm	10^{-5} mol/L	★★★	Nonionizing, spectroscopic capability	Limited depth

★, fair; ★★, good; ★★★, excellent.

modality, in particular for the diagnosis of various types of cancers, as described in Chapters 15 through 17 (Fitzgerald et al. 2006; Sim et al. 2013b; Woodward et al. 2003), because it has numerous advantages such as safety (unlike ionizing x-rays), diffraction-limited spatial resolution in tissues, and high sensitivity to water molecules. However, the contrast between malignant and benign tissues is rather small when only THz radiation is used for imaging, even for highly developed cancers. This poor contrast or sensitivity has been dramatically improved by using antibody-conjugated nanoparticle contrast agent probes (Oh et al. 2009, 2011), which have been applied to enhance the MRI sensitivity (Lee et al. 2006).

In this chapter, we will review the principle of THz molecular imaging (TMI) with nanoparticle probes (NPPs) and discuss its biomedical applications including cancer diagnosis, drug delivery monitoring, and stem cell (SC) tracking.

18.2 Operating Principle

18.2.1 Terahertz Molecular Imaging Using Nanoparticle Probes

The diagnostic performance of molecular imaging has been enhanced by the use of NPPs (McCarthy and Weissleder 2008). The most widely researched nanoparticles are related to a group of iron oxides designed for MRI. Magnetism-engineered, iron oxide NPPs can boost the image contrast by reducing the relaxation time of water protons; these NPPs have been employed for both humans and animals for more than a decade (Frank et al. 2002; Wang 2011). Oh et al. also took a similar approach to achieve high sensitivity and contrast in THz imaging by adopting gold nanorods (GNRs) as NPPs (Oh et al. 2009). The GNRs first employed for TMI were a few tens of nanometers in length and had an aspect ratio of approximately 4 as shown in Figure 18.1a–c. Their absorption peak was approximately 800 nm and could be modified by changing the aspect ratio as shown in Figure 18.1d.

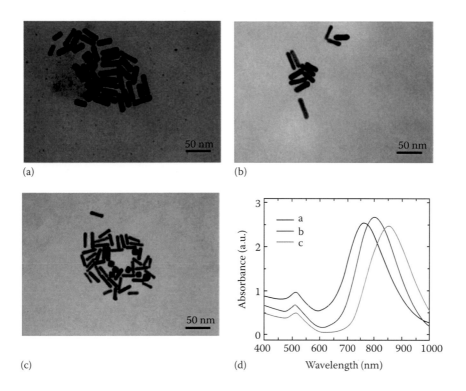

FIGURE 18.1 Transmission emission microscope (TEM) images of GNRs with aspect ratios of (a) 3.2, (b) 4.0, and (c) 4.2; (d) UV-visible absorption spectra of the GNRs in (a), (b), and (c). (Reprinted from Oh, S. J., J. Kang, I. Maeng et al., *Opt. Exp.*, 17, 3469, 2009. Copyright © 2009 by Optical Society of America. With permission of Optical Society of America.)

To simulate the THz response of cells after the uptake of NPPs, the GNRs were placed in water, which is a major component of biological cells, and the response was measured by a typical reflection-mode THz time domain spectroscopy system employing an InAs generator and a low-temperature-grown GaAs photoconductive detector described in Chapter 2. The propagating THz radiation was reflected by specimen containers holding water with GNRs and water only, the second of which was used as a reference. The THz waveforms reflected by the samples with and without GNRs were almost identical, because the GNRs caused insignificant reflection or absorption of THz radiation, as their dimensions are three to four orders of magnitude smaller than THz wavelengths. A continuous-wave infrared (IR) Ti:sapphire laser was employed to induce surface plasmons on the GNRs, which could increase the ambient temperature of the water around them, as demonstrated by a previous study on hyperthermia treatment of cancer cells. A temperature increase of more than 10 K was observed with the generation of surface plasmons around magnetic gold nanocomposites, as shown in Figure 18.2 (Lee et al. 2008). The temperature variation reveals that the power absorption of THz waves by water changes is greatly owing to the low-frequency vibrations and hydrogen bond stretching of water molecules occurring in the THz region (Rønne et al. 1997). Therefore, the surface plasmon effect induced by IR irradiation increased the reflectance of the THz pulses, as shown in Figure 18.3. The GNR concentration in water was 3.2 mg/mL and the IR intensity was 10 W/cm^2 (Oh et al. 2009).

To demonstrate the principle with cancer cells, epidermoid carcinoma A431 cells were endocytosed with the GNRs, and the THz reflectance of the cells was measured under IR irradiation. The reflectance from the A431 cells without GNRs was also measured for comparison (Oh et al. 2009). The result showed a change in the reflectance of the cells with GNRs under IR intensities of 10 and 20 W/cm^2, whereas that of the cells without GNRs was changed, as shown in Figure 18.4. A higher reflectance at a higher

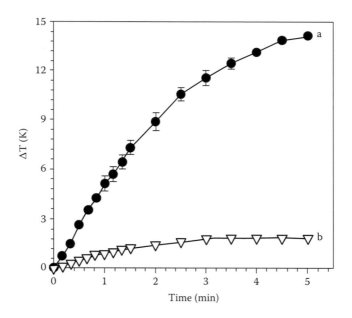

FIGURE 18.2 Temperature change in (a) multifunctional magnetic gold nanocomposite solution (5 mL, 4 mg/mL) and (b) pure water (5 mL) induced by NIR laser irradiation. (From Lee, J. W., J. M. Yang, H. J. Ko et al., *Adv. Funct. Mater.*, 18, 258, 2008. Copyright © 2008 by Advanced Functional Materials. Reprinted by permission of Advanced Functional Materials.)

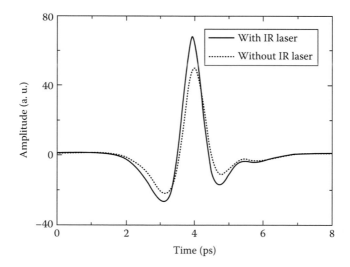

FIGURE 18.3 Temporal THz waveforms reflected by water with GNRs before (dotted line) and 5 min after illumination with an IR laser (line). (Reprinted from Oh, S. J., J. Kang, I. Maeng et al., *Opt. Exp.*, 17, 3469, 2009. Copyright © 2009 by Optical Society of America. With permission of Optical Society of America.)

IR beam power was also observed. This indicates that the operating principle of TMI is proven with biological cells as well as with the GNR-filled water. For a feasibility test of cancer imaging, the A431 cells with and without GNRs were imaged using a reflection-mode THz imaging system as shown in Figure 18.5. Figure 18.5a shows the cancer cells without GNRs and endocytosed with GNRs in containers; they are clearly distinguishable in the photo because of the GNR uptake. However, THz images of both types of cells without IR illumination appeared indistinguishable despite the GNR uptake, as

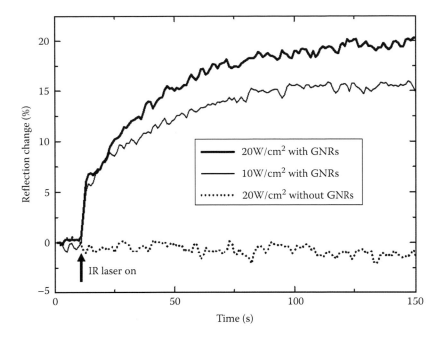

FIGURE 18.4 Changes in peak THz amplitudes reflected from live cancer cells with and without GNRs at IR intensities of 10 and 20 W/cm². (Reprinted from Oh, S. J., J. Kang, I. Maeng et al., *Opt. Exp.*, 17, 3469, 2009. Copyright © 2009 by Optical Society of America. With permission of Optical Society of America.)

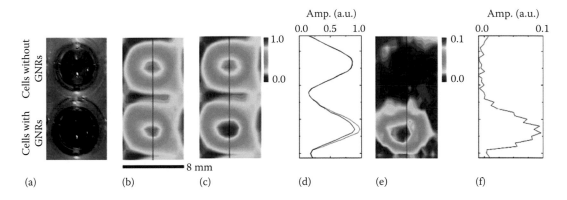

FIGURE 18.5 **(See color insert.)** Images of cancer cells with and without GNRs. (a) Visible image; (b) THz image without IR illumination; (c) THz image with IR illumination; (d) amplitudes along the lines in (b) (black) and (c) (red); (e) differential image between (b) and (c); and (f) amplitude along the line in (e). (Reprinted from Oh, S. J., J. Kang, I. Maeng et al., *Opt. Exp.*, 17, 3469, 2009. Copyright © 2009 by Optical Society of America. With permission of Optical Society of America.)

shown in Figure 18.5b. Under IR illumination, the THz image of the cells with GNRs became brighter, whereas no change was observed in the cells without GNRs as shown in Figure 18.5c. Figure 18.5d shows the amplitudes along the solid lines in Figure 18.5b and c. The reflectance from the cells with GNRs is boosted by 10% under IR illumination. This enhancement is too small to support the application of nanoparticle contrast agents that might cause adverse side effects. However, the excellence of this TMI technique was obviously demonstrated when the image in Figure 18.5b was subtracted from that in Figure 18.5c, as shown in Figure 18.5e. The difference was almost zero for the cells without GNRs, although the image of the cells with GNRs was clear. Figure 18.5f shows the amplitude along the line in Figure 18.5e; the ratio of the reflectance from the cells with GNRs to that from cells without GNRs was approximately 30, which indicates that the TMI technique can realize high-contrast, target-specific sensing of cancer cells targeted with NPPs (Oh et al. 2009).

18.2.2 Differential Detection via Direct Modulation

TMI is an extremely sensitive technique for imaging molecules and cells targeted with NPPs. However, the principle described in the previous subsection requires two sets of THz images with and without IR irradiation, which may diminish the significance of the TMI technique for application to biomedical objects as they are not usually fixed during imaging and it requires much more time to measure a sample twice. To overcome this disadvantage, the IR beam can be modulated using a mechanical chopper that is added to the IR laser beam path, in order to perform the mathematical subtraction seen in Figure 18.6 (Oh et al. 2011, 2012). Figure 18.6 shows the amplitude modulation of the IR laser and the results. The THz reflectance from water with GNRs under IR irradiation exhibited a large variation in the signals corresponding to the IR beam modulation (bold line), whereas the amplitudes from the water without GNRs (line) and from water with GNRs without IR illumination (gray line) were almost zero (Oh et al. 2011). The direct IR modulation approach was compared with the numerical difference method described in the previous subsection; THz imaging by direct modulation exhibited superior contrast to numerical subtraction, in addition to having the advantage of requiring only a single measurement. The areal correlation between the direct-modulated THz image and the sample, which is critical in the demarcation of cancer imaging, was also excellent (Oh et al. 2012).

A practical example of TMI with direct modulation is shown with animal organ tissues in Figure 18.7. GNRs were injected into a mouse through the tail vein, as will be described in the next section, while another mouse was not injected to serve as a control. The GNRs were phase conjugated to reach cancerous tumors, but most of them were captured by the liver, spleen, and other organs on the way to the

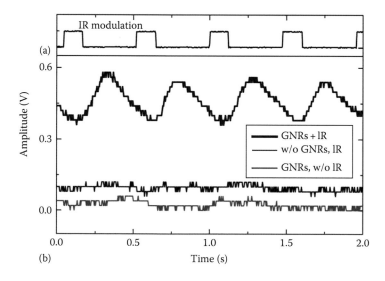

FIGURE 18.6   THz response under NIR modulation. (a) NIR signal modulated by a mechanical chopper. (b) THz responses from water with GNRs at 100 μM under NIR illumination (bold line), with the GNRs without NIR illumination (gray line), and without GNRs under NIR illumination (line). The intensity of the NIR beam was 15 W/cm². The THz amplitudes were acquired by an oscilloscope triggered by the modulation signal of the NIR beam. (Reprinted from Oh, S. J., J. Choi, I. Maeng et al., *Opt. Exp.*, 19, 4009, 2011. Copyright © 2011 by Optical Society of America. With permission of Optical Society of America.)

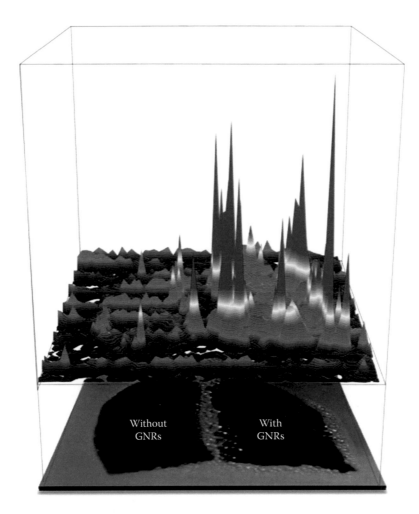

FIGURE 18.7   TMI signals of spleens taken from a set of mice without and with GNR injection. (Reprinted from Son, J.-H., Nanotechnology, 24, 214001, 2013. Copyright © 2013 by Institute of Physics Publishing. With permission of Institute of Physics Publishing.)

tumor (Oh et al. 2011; Son et al. 2011). The spleens extracted from a set of mice were imaged with THz radiation under modulated IR illumination as shown in Figure 18.7. The figure shows a dramatic THz response from the spleen that absorbed the GNRs.

18.2.3 Characteristics of Terahertz Molecular Imaging

Figure 18.8 demonstrates that TMI yields a sensitive response. To characterize the minimum detection sensitivity, GNRs were diluted with water at various concentrations. Although there was almost no reflection from the water alone, the solution with a concentration of 10 μM could be imaged as shown in Figure 18.8a. Figure 18.8b shows the linear relationship between the reflectivity and the GNR concentration with a correlation coefficient of 0.99. The minimum detection sensitivity and quantification linearity were also examined using an artificial tumor composed of Matrigel in a mouse. The characteristics of TMI with the artificial tumor also demonstrated an excellent linear relation with a minimum detection concentration of 15 μM (Oh et al. 2011).

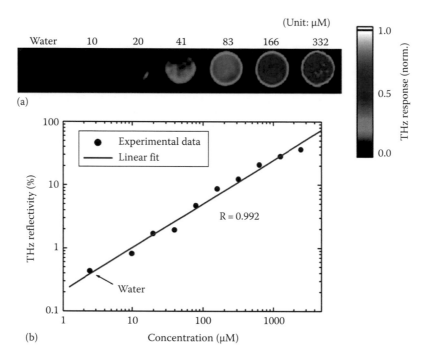

FIGURE 18.8 Quantification of TMI in solution. (a) THz molecular images of solutions with various concentrations of GNRs from 332 to 10 μM (n/2 for each concentration). (b) THz reflectivity of solutions with various concentrations of GNRs (solid circles). Experimental data were fitted by a linear line. (Reprinted from Oh, S. J., J. Choi, I. Maeng et al., *Opt. Exp.*, 19, 4009, 2011. Copyright © 2011 by Optical Society of America. With permission of Optical Society of America.)

18.3 Applications

18.3.1 Cancer Diagnosis

Cancer is a major cause of death worldwide. In the management of many cancer patients, early cancer detection can decrease the mortality rate. Conventional clinical imaging techniques including x-ray, ultrasound, CT, and MRI, however, offer images only at the macroscopic level. For high-resolution surface imaging, many optical imaging modalities are being investigated, including THz imaging.

THz imaging is a suitable technique for cancer imaging because THz radiation is sensitive to polar molecules such as water, and the blood supply is usually increased in cancer tissue. The change in water content provides one of the contrast mechanisms between a cancer tumor and adjacent normal tissue in THz imaging. The contrast also originates in the structural changes in tumor cells, as confirmed by THz imaging of paraffin-embedded or frozen cancer tissues (Park et al. 2011; Sim et al. 2013a,b). The combined contrast in all cases, however, is rather small, which limits the practicality of THz imaging for cancer diagnosis.

TMI can be a good candidate for an accurate diagnostic tool if NPPs are effectively delivered to cancerous tumors. Many studies have been conducted to transport magnetic NPPs to tumors for the application of MRI to prostate cancers (McCarthy and Weissleder 2008), epithelial cancers (Lee et al. 2008), and other types of cancers. As a type of NPPs, GNRs have also been used for THz imaging and hyperthermia treatment by delivering them to cancerous tumors (Choi et al. 2012; Oh et al. 2011). GNRs were injected through the tail vein of mice for targeting to tumors using antibody phase conjugation. To improve the specificity of targeting to cancerous cells, the exterior of the polyethylene glycolylated GNRs (PGNRs) was reformed by cetuximab (CET) (CET-PGNRs), an antiepidermal growth factor receptor and chimeric monoclonal antibody (Choi et al. 2012).

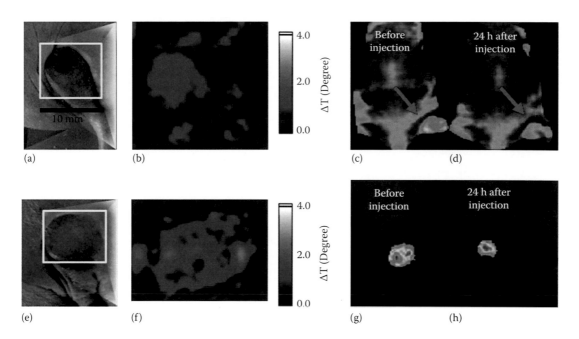

FIGURE 18.9 **(See color insert.)** In vivo THz molecular images with NIR absorption images for comparison. (a) Photograph of a mouse with an A431 cancer tumor; (b) THz molecular image of the mouse shown in (a); (c) and (d) NIR absorption images of the mouse shown in (a) before and 24 h after NPP injection, respectively; (e) photograph of a mouse with a smaller tumor (size = 2.1 cm³); (f) THz molecular image of the mouse shown in (e); (g) and (h) NIR absorption images of the mouse shown in (e) before and 24 h after NPP injection, respectively. ((a,b) With kind permission from Springer Science+Business Media: Oh, S. J., Y.-M. Huh, J.-S. Suh et al., *J. Infrared Millim. Terahertz Waves*, 33, 74, 2012. Copyright © 2012; (c,d) Reprinted with permission from Son, J.-H., Nanotechnology, 24, 214001, 2013. Copyright © 2013 by Institute of Physics Publishing; (e–h) Reprinted from Oh, S. J., J. Choi, I. Maeng et al., *Opt. Exp.*, 19, 4009, 2011. Copyright © 2011. With permission of Optical Society of America.)

The phase-conjugated GNRs were injected into male nude mice with an A431 epidermoid carcinoma cancer tumor in the proximal thigh region. Figure 18.9 shows two from tens of xenograft mouse models at various concentrations of CET-PGNRs and the results of imaging with THz radiation and near-infrared (NIR) beams. A mouse, shown in Figure 18.9a, was injected with 54 μM of CET-PGNRs at a concentration of 1 mM (Oh et al. 2012). Figure 18.9b shows a THz image taken 24 h after the injection. Because of the surface plasmon effect under IR irradiation, the temperature was raised by 2.5°. For comparison with a conventional technique, the tumor was also imaged by near-infrared absorption imaging (NAI), an in vivo optical imaging technique used to measure NPPs. Unlike TMI via differential modulation, NAI requires acquisition of images before and after NPP injection and measurement of the relative qualitative change in the distribution of the delivered NPPs. The NAI results display a qualitative change before and 24 h after the injection, as shown in Figure 18.9c and d, respectively. Figure 18.9e shows a smaller tumor 2.1 cm³ in size. Because the smaller tumor endocytoses a smaller amount of NPPs, the injection dose was raised to 100 μM at a concentration of 1 mM. Regardless of the tumor size, a distinct THz molecular image was acquired with a quantitative information, as shown in Figure 18.9f, whereas NAI could not discriminate the images acquired before and after NPP injection, as seen in Figure 18.9g and h.

18.3.2 Imaging of Nanoparticle Drug Delivery

Various types of nanoparticle drugs are currently being developed, and studies for tracking them are also being (Prokop 2011; Son et al. 2011). TMI can be a good solution for imaging the delivery and distribution of such drugs. To measure the concentrations of nanoparticles reaching specific locations, the

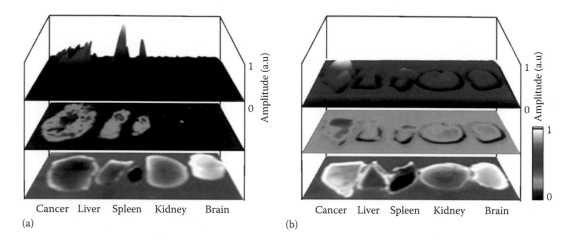

Cancer Liver Spleen Kidney Brain
(a)

Cancer Liver Spleen Kidney Brain
(b)

FIGURE 18.10 THz images of cancerous tumors and organs. (a) From a mouse after NPP injection and (b) from a mouse without injection. (With kind from permission from Springer Science+Business Media: Son, J.-H., S. J. Oh, J. Choi et al., *Intracellular Delivery: Fundamentals and Applications*, pp. 701–711, Springer, New York, 2011. Copyright © 2011.)

organs of a mouse injected with nanoparticles through the tail vein, as described in Section 18.3.1, were surgically removed (Son et al. 2011). Nanoparticles, like to ordinary drugs, were expected to be taken up by the liver and spleen on the way to the cancerous tumor by antibody conjugation targeting. The distributions of nanoparticles in several organs are displayed in Figure 18.10a, which shows that the liver captured most of the nanoparticles, and the spleen captured some of them. Figure 18.10b shows reference THz images of organs from a mouse without nanoparticle injection. The result indicates that quantitative imaging of nanoparticle drug delivery can be realized by TMI (Son et al. 2011).

18.3.3 Stem Cell Tracking

SCs divide through mitosis; differentiate into various, specialized cell types; and self-renew to produce more SCs. It is important to track the migration of SCs because they modulate multiple processes, such as the healing of injured tissue, and have the potential to address all of the tissues at risk from tumors or metastasis (Modo et al. 2002; Weissman 2000). TMI was used to trace the migration of mesenchymal stem cells (MSCs) moving toward a burn spot to heal it. As shown in Figure 18.11, MSCs endocytosed

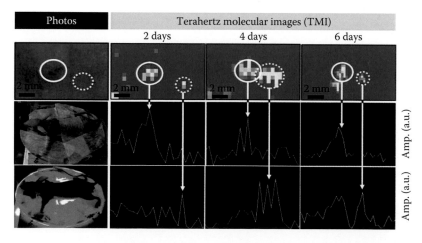

FIGURE 18.11 **(See color insert.)** Tracking of MSCs transplanted into a mouse by TMI of tagged SPIOs. See text for details.

with superparamagnetic iron oxide nanoparticles (SPIOs) were injected below the skin of a mouse (solid circle) a few millimeters away from a burn spot (dashed circle). In vivo images were acquired during 1 week; they showed the migration of the MSCs tagged by SPIOs. Until the second day, most of the MSCs stayed in the original injection location, although they moved to the burn spot on the fourth day. On the sixth day, the signal was diminished because the MSCs were dispersed after healing. The use of TMI made it possible to track MSCs below the skin, whereas conventional imaging modalities, such as MRI and CT, have difficulties in imaging the skin surface.

18.4 Comparison with Conventional Medical Imaging Techniques

Although THz imaging is a promising medical imaging modality, it should exhibit superiority over conventional medical imaging techniques in some regards. Conventional techniques, such as ultrasound, CT, MRI, and positron emission tomography, can provide tomography with better penetration by their individual contrast mechanisms. However, THz imaging can better differentiate malignant from benign lesions located on the surface.

TMI is a novel technique that should be evaluated and compared with conventional molecular imaging techniques, whereas MRI is the most widely used molecular imaging technique using SPIOs (Lee et al. 2008; McCarthy and Weissleder 2008). To compare TMI with MRI, a dual-mode nanoparticle contrast agent, a commercially available SPIO called Feridex®, was used, which has been approved for human use by the US Food and Drug Administration. The SPIOs were endocytosed into ovarian SKOV3

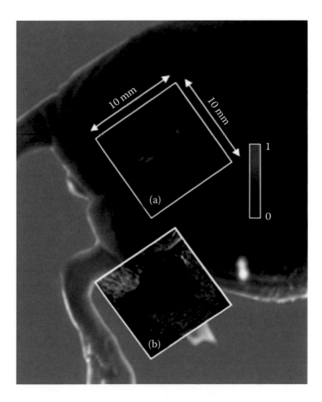

FIGURE 18.12 **(See color insert.)** In vivo (a) TMI and (b) MRI images of a mouse 24 h after transfection of SPIOs into SKOV3 ovarian cancer cells. (From Park, J. Y., H. J. Choi, K.-S. Cho, K.-R. Kim, and J.-H. Son, *J. Appl. Phys.*, 109, 064704, 2011. Copyright © 2011 by IEEE. Reprinted by permission of IEEE.)

cancer cells, and the cells were later injected into the right thighs of mice. Figure 18.12 shows in vivo TMI and MRI images of an injected mouse. Both show a similar, crescent-shaped tumor, which displays an intensity increase in TMI but image darkening in MRI. This result indicates that the TMI technique can be applied to monitor surgical processes in an operating room, whereas MRI has been used for pre- and postoperative imaging (Park et al. 2012).

18.5 Conclusion and Prospect

The principle of molecular imaging with THz radiation is reviewed and some of its biomedical applications are also presented in this chapter. Molecular imaging is a promising new field in clinical medicine because of huge demands for early disease detection, treatment monitoring, and personalized management of patients. Imaging techniques using THz radiation have recently undergone substantial development, and the results of investigations have demonstrated its potential for future clinical use, especially for cancer detection and characterization. TMI using NPPs has been developed even further to exploit the merits of conventional THz imaging and the advantages of NPPs analogous to the contrast agents used in conventional medical imaging techniques, which offer very high measurement sensitivity and enable target-specific sensing of biomedical materials. However, the limited number of clinical trials of THz imaging is a weak point for clinical use of this technique. TMI should also be advanced further and validated by comparison with current conventional imaging modalities.

In conclusion, we have begun to test the feasibility of THz imaging as a novel medical imaging modality. More studies should be performed to improve and, even more important, realize breakthroughs in detection, characterization, diagnosis, and aftertreatment monitoring. With these developments, the clinical application of THz imaging can be encouraged.

References

Choi, J., J. Yang, D. Bang et al. 2012. Targetable gold nanorods for epithelial cancer therapy guided by near-IR absorption imaging. *Small* 8: 746–753.

Choi, J. H., J. M. Yang, J. S. Park et al. 2011. Specific near-IR absorption imaging of glioblastomas using integrin-targeting gold nanorods. *Advanced Functional Materials* 21: 1082–1088.

Costelloe, C. M., H. H. Chuang, J. E. Madewell, and N. T. Ueno. 2010. Cancer response criteria and bone metastases: RECIST 1.1, MDA and PERCIST. *Journal of Cancer* 1: 80.

Fitzgerald, A. J., V. P. Wallace, M. Jimenez-Linan et al. 2006. Terahertz pulsed imaging of human breast tumors. *Radiology* 239: 533–540.

Frank, J. A., H. Zywicke, E. K. Jordan et al. 2002. Magnetic intracellular labeling of mammalian cells by combining (FDA-approved) superparamagnetic iron oxide MR contrast agents and commonly used transfection agents. *Academic Radiology* 9: S484–S487.

Kherlopian, A. R., T. Song, Q. Duan et al. 2008. A review of imaging techniques for systems biology. *BMC Systems Biology* 2: 74.

Lee, J.-H., Y.-M. Huh, Y.-W. Jun et al. 2006. Artificially engineered magnetic nanoparticles for ultra-sensitive molecular imaging. *Nature Medicine* 13: 95–99.

Lee, J. W., J. M. Yang, H. J. Ko et al. 2008. Multifunctional magnetic gold nanocomposites: Human epithelial cancer detection via magnetic resonance imaging and localized synchronous therapy. *Advanced Functional Materials* 18: 258–264.

Massoud, T. F. and S. S. Gambhir. 2003. Molecular imaging in living subjects: Seeing fundamental biological processes in a new light. *Genes and Development* 17: 545–580.

McCarthy, J. R. and R. Weissleder. 2008. Multifunctional magnetic nanoparticles for targeted imaging and therapy. *Advanced Drug Delivery Reviews* 60: 1241–1251.

Modo, M., D. Cash, K. Mellodew et al. 2002. Tracking transplanted stem cell migration using bifunctional, contrast agent-enhanced, magnetic resonance imaging. *Neuroimage* 17: 803–811.

Oh, S. J., J. Choi, I. Maeng et al. 2011. Molecular imaging with terahertz waves. *Optics Express* 19: 4009–4016.

Oh, S. J., Y.-M. Huh, J.-S. Suh et al. 2012. Cancer diagnosis by terahertz molecular imaging technique. *Journal of Infrared, Millimeter, and Terahertz Waves* 33: 74–81.

Oh, S. J., J. Kang, I. Maeng et al. 2009. Nanoparticle-enabled terahertz imaging for cancer diagnosis. *Optics Express* 17: 3469–3475.

Park, J. Y., H. J. Choi, K.-S. Cho, K.-R. Kim, and J.-H. Son. 2011. Terahertz spectroscopic imaging of a rabbit VX2 hepatoma model. *Journal of Applied Physics* 109: 064704.

Park, J. Y., H. J. Choi, G.-E. Nam, K.-S. Cho, and J.-H. Son. 2012. In vivo dual-modality terahertz/magnetic resonance imaging using superparamagnetic iron oxide nanoparticles as a dual contrast agent. *IEEE Transactions on Terahertz Science and Technology* 2: 93–98.

Prokop, A. 2011. *Intracellular Delivery: Fundamentals and Applications*. New York: Springer.

Rønne, C., L. Thrane, P.-O. Åstrand et al. 1997. Investigation of the temperature dependence of dielectric relaxation in liquid water by THz reflection spectroscopy and molecular dynamics simulation. *The Journal of Chemical Physics* 107: 5319.

Sim, Y. C., K.-M. Ahn, J. Y. Park, C. Park, and J.-H. Son. 2013a. Temperature-dependent terahertz imaging of excised oral malignant melanoma. *IEEE Journal of Biomedical and Health Informatics* 17: 779–784.

Sim, Y. C., J. Y. Park, K.-M. Ahn, C. Park, and J.-H. Son. 2013b. Terahertz imaging of excised oral cancer at frozen temperature. *Biomedical Optics Express* 4: 1413.

Son, J.-H. 2013. Principle and applications of terahertz molecular imaging. *Nanotechnology* 24: 214001.

Son, J.-H., S. J. Oh, J. Choi et al. 2011. Imaging of nanoparticle delivery using terahertz waves. In *Intracellular Delivery: Fundamentals and Applications*, ed. A. Prokop, pp. 701–711. New York: Springer.

Wang, Y.-X. J. 2011. Superparamagnetic iron oxide based MRI contrast agents: Current status of clinical application. *Quantitative Imaging in Medicine and Surgery* 1: 35.

Weissleder, R. and M. J. Pittet. 2008. Imaging in the era of molecular oncology. *Nature* 452: 580–589.

Weissman, I. L. 2000. Stem cells: Units of development, units of regeneration, and units in evolution. *Cell* 100: 157–168.

Woodward, R. M., V. P. Wallace, R. J. Pye et al. 2003. Terahertz pulse imaging of ex vivo basal cell carcinoma. *Journal of Investigative Dermatology* 120: 72–78.

19

Prospects in Medical Applications of Terahertz Waves and Conclusions

Joo-Hiuk Son
University of Seoul

As discussed previously in this book, terahertz (THz) electromagnetic waves, with their unique advantageous features, have been utilized to study biological phenomena. They have also been applied to the field of medicine, especially for testing THz technology as a future imaging modality. However, there are some drawbacks of THz waves with respect to their application to medical imaging. One of them is that it is difficult to obtain specific characteristic signals from biological tissues, cells, and macromolecules because the resonant characteristics are smeared out because of the inhomogeneous broadening of macromolecules and the large absorption by water molecules, which are abundant in living cells (Son 2009). The effect of water can be removed by lyophilizing cells and tissues (Png et al. 2008, Globus et al. 2012, Sim et al. 2013). The freezing technique, as described in Chapter 17, is also beneficial to the enhancement of THz wave penetration into biological samples. Oh et al. developed another measurement depth enhancement technique in the imaging of biological samples with water by utilizing chemical gels, which have a lower absorption of THz waves than water and a refractive index that is flat in the THz frequency range (Oh et al. 2013).

To demonstrate the enhancement of imaging depth, glycerol was used as a THz penetration-enhancing agent (THz PEA). Figure 19.1a and b shows a mouse's abdominal skin placed on a metal blade target with (*1) and without (*2) glycerol application and the measurement scheme, respectively. Glycerol was only applied on the right half of the tissue, which had a thickness of 220 μm. THz waveforms reflected by the tissue were acquired 30 min after applying glycerol, and they are shown in Figure 19.1c. The THz images shown in Figure 19.1d and e were reconstructed using the first and second peaks of temporal domain waveforms, respectively, in Figure 19.1c. The surface image, Figure 19.1d, does not adequately reveal the effect of glycerol on both the left and right sides. However, the image out of the second peak, Figure 19.1e, clearly showed the distinction between the left and right sides of the tissues. The left image of the metal target below the tissue without glycerol application was blurred, whereas the contrast of the right image of glycerol-applied tissue was considerably better. The amplitude of the second peak passing through the tissue with glycerol was almost twice that without glycerol. The time interval between the first and second peaks also decreased. These results suggest that the THz optical characteristics of tissues, such as absorbance and refractivity, were modified by replacing some of the interstitial water molecules with glycerol (Oh et al. 2013). By using PEAs, THz waves can reach deeper into the wet samples under diagnosis and relieve the limitation of THz waves in the measurement of biological tissues. I also believe that better PEAs, which should be biocompatible materials having a lower absorption of THz radiation, will be developed if THz imaging of biomedical samples is widely practiced.

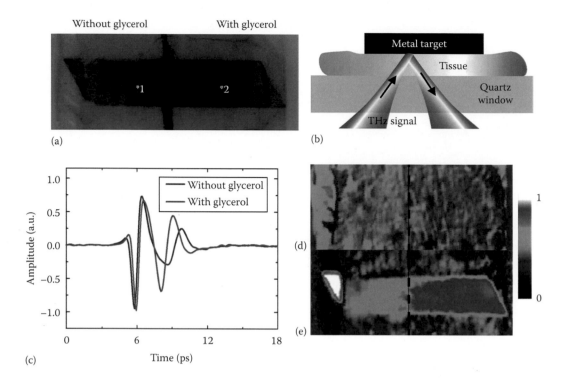

FIGURE 19.1   (See color insert.) Enhancement of measurement depth of mouse tissue through the application of glycerol as a THz PEA. (a) Photo of a mouse's abdominal skin placed on a metal target with (*1) and without (*2) glycerol application, (b) measurement scheme, (c) temporal THz waveforms reflected by the tissue, and THz images reconstructed using (d) the first and (e) second peaks of waveforms in (c). The tissue size is 5 × 3 cm². (From Oh, S. J., Kim, S.-H., Jeong, K. et al., Measurement depth enhancement in terahertz imaging of biological tissues, *Opt. Exp.*, 21, 21299, 2013. Copyright © 2013 by Optical Society of America. Reprinted by permission of Optical Society of America.)

One of the most important advantages in the employment of THz waves is the spectroscopic capability that's emphasized in this book, and this property should be exploited to find specific signals to establish the superiority of THz imaging over other medical imaging techniques such as magnetic resonance imaging or computed tomography. Until the submission of the manuscript of this book, no prominent spectral features have been found in medical imaging, especially in the measurement of cancers. However, it is well known that many diseases accompany the chemical changes in molecules (Cooper and Youssoufian 1988, Ehrlich 2002, Pineda et al. 2010), and this might be characterized by THz waves that give particular signatures.

One significant aspect of medical applications is the advancement of components with which affordable systems can be realized. A variety of sources, detectors, modulators, filters, and other devices based on electronic and photonic technologies are being developed to fulfill potential applications. In particular, continuous-wave solid-state devices are promising in terms of cost, although laser-based pulsed systems have a better spectroscopic capability. With maturing THz device technology, several specific areas of applications, other than cancer imaging, can emerge in real-world clinics. Some such examples are the measurements of blood components (Jeong et al. 2011, Reid et al. 2013), the characterization of burn wounds (Dougherty et al. 2007, Taylor et al. 2011, Baughman et al. 2013), the sense of cornea hydration level (Singh et al. 2010, Taylor et al. 2011), the diagnosis of dental pulp vitality (Hirmer et al. 2012), and the study of insulin amyloid fibrillation (Liu et al. 2010).

To realize the utilization of THz technology in medical clinics, it is essential to have close collaborations with researchers and clinicians in the field of medicine to the extent of focusing efforts toward device and system development. As we continue to strive to push the limits, the editor hopes to see THz medical equipment operated in hospitals in the near future.

References

Baughman, W. E., Yokus, H., Balci, S., Wilbert, D. S., Kung, P., and Kim, S. 2013. Observation of hydrofluoric acid burns on osseous tissues by means of terahertz spectroscopic imaging. *IEEE Journal of Biomedical and Health Informatics* 17: 798–805.

Cooper, D. N. and Youssoufian, H. 1988. The CpG dinucleotide and human genetic disease. *Human Genetics* 78: 151–155.

Dougherty, J. P., Jubic, G. D., and Kiser, W. L. 2007. Terahertz imaging of burned tissue. Paper presented at the *International Conference on Terahertz and Gigahertz Electronics and Photonics VI*, San Jose, CA.

Ehrlich, M. 2002. DNA methylation in cancer: Too much, but also too little. *Oncogene* 21: 5400–5413.

Globus, T., Dorofeeva, T., Sizov, I. et al. 2012. Sub-THz vibrational spectroscopy of bacterial cells and molecular components. *American Journal of Biomedical Engineering* 2: 143–154.

Hirmer, M., Danilov, S. N., Giglberger, S. et al. 2012. Spectroscopic study of human teeth and blood from visible to terahertz frequencies for clinical diagnosis of dental pulp vitality. *Journal of Infrared, Millimeter, and Terahertz Waves* 33: 366–375.

Jeong, K., Huh, Y.-M., Kim, S.-H. et al. 2011. Characterization of blood cells by using terahertz waves. Paper presented at the *International Conference on Infrared, Millimeter and Terahertz Waves*, Houston, TX.

Liu, R., He, M., Su, R. et al. 2010. Insulin amyloid fibrillation studied by terahertz spectroscopy and other biophysical methods. *Biochemical and biophysical research communications* 391: 862–867.

Oh, S. J., Kim, S.-H., Jeong, K. et al. 2013. Measurement depth enhancement in terahertz imaging of biological tissues. *Optics Express* 21: 21299–21305.

Pineda, M., González, S., Lázaro, C., Blanco, I., and Capellá, G. 2010. Detection of genetic alterations in hereditary colorectal cancer screening. *Mutation Research* 693: 19–31.

Png, G. M., Choi, J. W., Ng, B. W. et al. 2008. The impact of hydration changes in fresh bio-tissue on THz spectroscopic measurements. *Physics in Medicine and Biology* 53: 3501–3517.

Reid, C. B., Reese, G., Gibson, A. P., and Wallace, V. P. 2013. Terahertz time-domain spectroscopy of human blood. *IEEE Journal of Biomedical and Health Informatics* 17: 774–778.

Sim, Y. C., Park, J. Y., Ahn, K.-M., Park, C., and Son, J.-H. 2013. Terahertz imaging of excised oral cancer at frozen temperature. *Biomedical Optics Express* 4: 1413–1421.

Singh, R. S., Tewari, P., Bourges, J. L. et al. 2010. Terahertz sensing of corneal hydration. Paper presented at the *Annual International Conference of the IEEE Engineering in Medicine and Biology Society*, Buenos Aires, Brazil.

Son, J. H. 2009. Terahertz electromagnetic interactions with biological matter and their applications. *Journal of Applied Physics* 105: 102033.

Taylor, Z. D., Singh, R. S. Bennett, D. B. et al. 2011. THz medical imaging: In vivo hydration sensing. *IEEE Transactions on Terahertz Science and Technology* 1: 201–219.

Index

Printed and bound by CPI Group (UK) Ltd, Croydon, CR0 4YY

24/10/2024

01778309-0005